ROCK MECHANICS PRINCIPLES

D.F. Coates
   Mining Research Laboratories
Fuels and Mining Practice Division

MINES BRANCH
Department of Mines and Technical Surveys, Ottawa

Mines Br. Monograph 874

1965

© Crown Copyrights reserved

Available by mail from the Queen's Printer, Ottawa,
and at the following Canadian Government bookshops:

OTTAWA
Daly Building, Corner Mackenzie and Rideau

TORONTO
Mackenzie Building, 36 Adelaide St. East

MONTREAL
Æterna-Vie Building, 1182 St. Catherine St. West

WINNIPEG
Mall Center Bldg., 499 Portage Avenue

VANCOUVER
657 Granville Avenue

or through your bookseller

A deposit copy of this publication is also available
for reference in public libraries across Canada

Price $5.00   Catalogue No. M32—874

Price subject to change without notice

ROGER DUHAMEL, F.R.S.C.
Queen's Printer and Controller of Stationery
Ottawa, Canada
1965

"It is the mark of an educated man to look for precision in each class of things just so far as the nature of the subject admits; it is evidently equally foolish to accept probable reasoning from a mathematician and to demand from a rhetorician scientific proofs"

Aristotle.

FOREWORD

It is Mines Branch policy to promote research and to disseminate information on subjects of importance to the mineral industry. In the winning of mineral resources from the crust of the earth, the problems distinguishing mining from other industrial pursuits arise from the need to break and remove rock while maintaining control over the surrounding rock, i.e., rock breakage and ground control. For this reason, the basic science for mining activity is an integration of "materials" science and structural engineering into the subject of rock mechanics.

Considerable effort has been made over the past few years to build up research facilities in this area at the Mines Branch and, in conformity with our policy, to make these facilities available for cooperative work with both universities and industry. In this way graduate student projects oriented towards field research can make use of existing facilities and also industry can have access to the activities and guidance of an active research group concerned with ground problems.

Rock mechanics is a difficult scientific area for quick results. At this stage, it requires encouragement, much research, and the training not only of mining engineers but also of professional personnel being recruited from time to time by the Mines Branch from the fields of geology, physics, mechanical, and civil engineering. It is considered important to obtain the combination of the ability to analyse stress mathematically with the knowledge of the peculiar physical properties of rocks. The purpose of this monograph is to establish the principles along which rock mechanics is likely to develop; consequently, it can be used as an introduction to the subject for persons associated with research in this area.

The task of editing and preparing the manuscript for publication was successfully achieved through the good services of Mr. F.T. Rabbitts of the Mines Branch and Mr. D.A. Shenstone of the Editorial and Information Division of the department.

John Convey,
Director.

PREFACE

The purpose of this monograph is to provide some guidance for the young engineer or scientist entering the field of rock mechanics. The emphasis is on the application of engineering mechanics to problems arising from the needs either to prevent or to cause rock failure with particular reference to those problems encountered in mining. As the individuals entering this field are from a variety of disciplines - geology, physics, engineering - the treatment starts at a rather elementary level.

The various theories in mechanics and the rock properties that are important in this subject are reviewed for those requiring such background information. The main groups of rock problems are then examined to determine the mechanics pertinent to the various geometries and loadings of these cases. Some indication of the degree of practical applicability is also provided, although the monograph is not intended as an engineer's manual for use by planners or operators.

The objective of rock mechanics is to endeavour to obtain the predictability enjoyed in other engineering disciplines. In this early period of the science, some areas can be treated quantitatively while others can only be appraised qualitatively, which consequently provides a challenge to reduce as many of these areas as possible to a quantitative basis. For example, in the past in structural engineering the dispersion of actual strength values of various test samples of a material from the assumed average value was either ignored or taken into account by the use of judgment. Now it is becoming accepted that the dispersion of material properties can be expressed quantitatively by the use of statistical and probability mathematics so that explicit decisions can be made on the degree of risk to be taken. At the same time, whereas little can be predicted with certainty in rock mechanics, comparative analyses can still be very valuable; for example, although it is not possible to predict the critical slope angle for an open pit in hard rock, it is possible to analyse which of two situations would be the more stable - a 40 degree slope with the ground water level at the crest elevation or a 45 degree slope with the ground water level behind the slope drawn down to the elevation of the toe. Such comparative studies can provide rational bases for economic decisions.

Those working in rock mechanics should continue to appreciate that the scepticism of the practising engineer towards new, unproven theory is not a bad thing. The rationale of theory is to be able to generalise experience; for example, if rock failure occurs and can be satisfactorily explained by analysing the pertinent stresses, then this experience provides a good basis for using the same analysis to predict other cases of rock failure. In other words, the only reason for analysing stress is to be able to predict failure, otherwise it serves no useful purpose. On the other hand, some other factor such as deformation might be more easily related to failure of rock masses. Our assumption that the analysis of stress is the key to predicting rock behaviour is, at this stage, merely inherited from other related subjects, but it has not yet been proven to be the best procedure.

Finally, it is well to recognize that even in science good judgment is an extremely valuable faculty. Decisions must be made in research similar to the decisions that we make in everyday life; this judgment is a matter of experienced guessing on what is likely to be feasible, recognizing the nature of things as they are as opposed to the idealizations that are used in calculations. When this important ingredient of any practical decision is recognized, then the corollary can be accepted that no matter how brilliant the scientist nor how rigorous the analysis the resultant answer can be wrong if good judgment is absent.

To my colleagues in Mines Branch, besides expressing my pleasure in sharing their scientific community, I sincerely congratulate Dr. J. Convey and Mr. A. Ignatieff for creating such a favourable atmosphere for research in this field. To my colleagues at McGill University, particularly Professor R.G.K. Morrison a leader in the development of studies in rock mechanics, I am indebted for stimulation and assistance.

Ottawa, Canada
January 1965

D.F. Coates

CONTENTS

CHAPTER | PAGE

1 - MECHANICAL NATURE OF ROCKS
1-1 to 1-51

Introduction. Some definitions. Stress at a point. Mohr's Circle. Behaviour of rocks. Mohr's Strength Theory (Cohesionless ground, Cohesive ground, Effective stresses). Ultimate bearing stresses. Griffith's Strength Theory. Other strength theories. Testing (Uniaxial compression test. Triaxial compression test. Shear tests. Tension test. In situ shear test. In situ seismic velocity test. Laboratory seismic velocity test. Plate-load testing). Classification of rocks. References. (Figures 1-1 to 1-24).

2 - ELASTIC PROTOTYPES
2-1 to 2-32

Introduction. Equilibrium. Plane stress and plane strain. Thick-walled cylinders. Hole in an infinite elastic solid. Semi-infinite elastic solid. Beams. Plates. Arches. References. (Figures 2-1 to 2-17).

3 - SHAFTS, DRIFTS AND TUNNELS
3-1 to 3-39

Introduction. Stresses around shafts (Circular shafts in elastic ground subjected to gravity stresses. Circular shafts in non-ideal ground. Circular shafts in residual stress fields). Ground support around shafts (Excavation and lining to prevent failure. Uniform pressure from granular or broken ground. Concentrated pressures from granular or broken ground. Pressure from broken ground on rectangular shafts. Pressure from viscous ground. Effects of location with respect to stopes). Stresses around drifts and tunnels (Elastic ground. Non-ideal ground). Ground support around drifts and tunnels (Pressure from broken ground. Deformation from viscous rocks. Rock bolts. Gunite. Ground water effect on weak or soft ground. Pressure tunnels). References. (Figures 3-1 to 3-17).

4 - PILLARS
4-1 to 4-22

Introduction. Stress distributions. Pillar loading in deep, long mining zones. Loads due to gravity acting on yielding overlying ground. Pillar strength. Roof reactions. References. (Figures 4-1 to 4-15).

5 - STOPES, CAVING AND SUBSIDENCE
5-1 to 5-18

Introduction. Caving (Gravitational stress field. Residual stress field. Induced caving). Drawing. Subsidence. References. (Figures 5-1 to 5-11).

CHAPTER	PAGE
6 - ROCK SLOPES	6-1 to 6-23

    Introduction. Infinite slopes. Elastic stress distribution. Slopes in rocks that yield. Types of slope failure (Rock falls. Rotational shear. Plane shear. Block flow). Prevention and control. References. (Figures 6-1 to 6-18).

7 - FOUNDATIONS	7-1 to 7-26

    Introduction. Settlement. Bearing capacity. Rock anchors (Anchorage of piles). References. (Figures 7-1 to 7-15).

8 - ROCK DYNAMICS	8-1 to 8-69

    Introduction. Simple harmonic motion. Wave transmission. Explosions. Cratering. Blasting. Excavating and blasting with nuclear explosives. Underground openings. Rock bursts. References. (Figures 8-1 to 8-37).

APPENDIX

A - SYMBOLS AND ABBREVIATIONS	A-1 to A-6
B - GLOSSARY	B-1 to B-24
C - FLOW NETS	C-1 to C-4
D - STRESS CONCENTRATION FACTORS	D-1 to D-15
E - BEAM FORMULAE	E-1 to E-4
F - DIMENSIONAL ANALYSIS AND SIMILITUDE	F-1 to F-5

CHAPTER 1

MECHANICAL NATURE OF ROCKS

INTRODUCTION

In this first chapter a review is given of some important tools in mechanics. This is necessary so that the conventions and symbols used throughout the book will be readily understood. Then some of the more common patterns of rock behaviour are described together with theories that may govern failure. A valid failure theory is of practical importance for it permits rock failure under one set of conditions, either in a test or experienced on a project, to be extrapolated to other conditions. At the present time, not enough is known about rock failure to be certain that a valid theory has been established. However, two of the most probable hypotheses are presented and used throughout the book.

The testing of rock properties thus must fall into the category of obtaining test data that can be used to predict the behaviour of rock in various circumstances. Also, the classification of rocks serves to indicate the area within which potential problems will occur and the tests that would be of use to establish whether such problems are likely to occur.

SOME DEFINITIONS

The communication of information requires mutual knowledge of the meaning of the words being used. In everyday use our normal language is more or less adequate. However, in scientific or technical work it is important to use words that have been defined with exact meanings. For example, it can be confusing if the word strain is used as a synonym for either 'stress', 'load', or 'deformation' when all these terms have precise but different meanings. Consequently, definitions of the most commonly used words in the field of rock mechanics are given in a Glossary at the end of the book with a few of the most basic terms repeated below.

<u>Mechanics</u> is the study of the effects of forces on bodies. The effects may be acceleration, velocity, and displacement; or the forces may produce changes in volume and shape; finally, fracture or flow may result.

<u>Rock Mechanics</u> is the study of the effects of forces on rocks. The principal effects of interest for the geologist are the changes in shape that can occur. Those of interest for the geophysicist are probably the dynamic aspects of changes in volume and shape, i.e., seismic waves. The engineer is primarily concerned with the phenomena influencing the predictability of fracture and flow and to some extent with changes in the volume and shape of rocks.

Thus, for the engineer the subject of rock mechanics involves - the analysis of loads or forces being applied to the rocks - the analysis of the internal effects in terms of either stress, strain, or stored energy - and, finally, the analysis of the consequences of these internal effects, i.e., fracture, flow, or simply deformation of the rock.

<u>Stress</u> is the internal force per unit area when the area approaches zero (see Chapter 2 on <u>Free Body Diagrams</u> for the difference between internal and external forces). It is useful to reserve the word 'pressure' for the average external normal force per unit area, even though the pressure at a boundary will equal the normal stress in the material at that point.

<u>Normal Stress</u>, σ, is the component of stress normal or perpendicular to the plane on which it is acting.

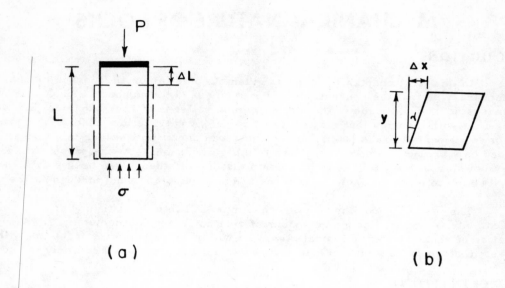

Fig. 1-1 Deformed or Strained Bodies

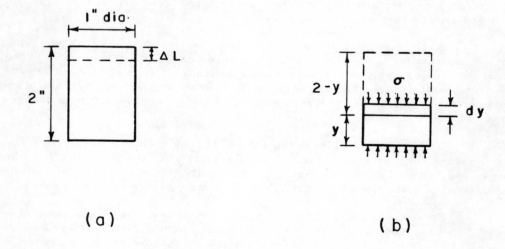

Fig. 1-2 A Cylinder Subjected to Gravitational, Body Forces

Shear Stress, τ, is the component of stress tangential or parallel to the plane on which it is acting.

Deformation, δ, is the absolute or relative movement of a point on a body or the change in a linear dimension, e.g., ΔL in Fig. 1-1.

Strain is the unit deformation or the deformation per unit length or width. For example, in Fig. 1-1(a) the strain resulting from the load, P, is $\Delta L/L$. In engineering work it is common practice to use the original, unstrained length, L, for calculating any strain. To be rigorous, the instantaneous length should be used. However, for small strains the difference arising from using the original length is not important.

Normal Strain, ϵ, is the deformation per unit length in the direction of the deformation.

Shear Strain, γ, is commonly described as the relative change in the angle defining the sides of an infinitesimal element. Or, following the general definition of strain, it can be defined as the deformation per unit length where the length over which the deformation occurs is at right angles to the direction of the deformation (see Fig. 1-1(b), $\gamma = \Delta x/y$).

Modulus of Deformation, E, is the ratio of normal stress to normal strain for a particular material where the increase in strain is caused by the increase in stress. For elastic bodies the term 'modulus of elasticity' is used for this property.

Poisson's Ratio, μ, is the ratio, for a particular material, of the transverse normal strain to the longitudinal normal strain under uniaxial stress in the longitudinal direction. Fig. 1-1(a) shows the reaction of a typical sample to compressive stress.

It should be understood that the transverse strain is caused by the longitudinal stress. In the case of longitudinal compression, lateral extension occurs. The lateral extension or strain in this case is not caused, as some are inclined to assume, by tensile stresses in this direction.

Poisson's Effect is a term sometimes used to describe the lateral deformation resulting from longitudinal stress.

Elastic describes a material or a state of a material where strain or deformation is recoverable, nominally instantaneously but actually within certain tolerances and within some arbitrary time.

Plastic describes a material or a state of a material where strain or deformation is partially irrecoverable within certain tolerances and within some arbitrary time. It can, among other usages, also describe a material that can still resist pressure while producing irrecoverable strain beyond its yield point. (See the Glossary in Appendix B for a more extensive examination of the meaning of the word.)

Viscous describes a material whose strain is dependent on time.

Body Forces arise from conditions that result in every particle of the body experiencing an element of the total body force, e.g., if a body has mass, a gravity force will be acting on every particle in it while it is in a gravitational field; similarly, iron and nickel bodies will be subject to a magnetic body force when placed in a magnetic field.

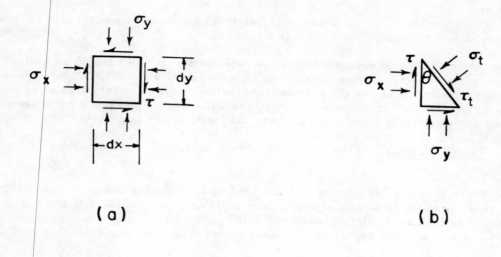

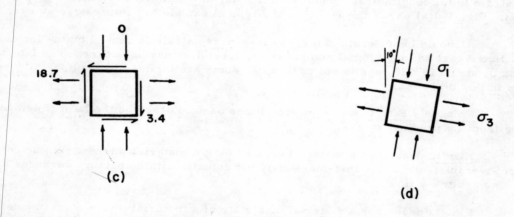

Fig. 1-3 Wedge Analysis of Stresses at a Point

Example - The vertical deformation of the top of the cylinder shown in Fig. 1-2 is to be determined. Its density is 160 pcf and its modulus of deformation is 1×10^6 psi.

The deformation contributed by the elements of thickness dy at a distance y from the bottom of the cylinder (see Fig. 1-2(b)) can be expressed by the following equation:

$$dL = \frac{\sigma \, dy}{E} = \frac{(2-y)(160/1728) \, dy}{1 \times 10^6}$$

The total deformation, L, will be obtained by integrating this equation:

$$L = \int_0^2 \frac{(2-y)(160/1728) \, dy}{1 \times 10^6}$$

$$= \left[\frac{160}{1728 \times 10^6} (2y - y^2/2) \right]_0^2$$

$$= 1.85 \times 10^{-7} \text{ in.}$$

It might be noted here that the deformation resulting from the body force is half the deformation that would occur if the force equal to the weight of the body were acting on the top of the sample.

<u>Strength</u>, in brief, is the maximum stress that a body can withstand without failing (i.e., by rupture or continuous deformation).

STRESSES AT A POINT

By definition, the area over which stress acts is infinitesimal. Hence, stresses at a point can be considered to be acting on elements bounded by infinitesimal areas.

In Fig. 1-3(a) the stresses acting in one plane on an elemental cube are shown. If the stresses on the remaining two planes of the cube are zero, this is a case of plane stress.

It should be recognized that plane stress generally involves triaxial strain; in other words, the element shown in Fig. 1-3(a) will expand in the direction normal to the (x, y)-plane besides straining in the x- and y-directions. Conversely, the case of plane strain usually requires a triaxial stress condition. A selection is generally required for purposes of simple analysis of one of these two cases - plane stress or plane strain.

The shear stresses on the x- and y-faces of the element in Fig. 1-3(a) are shown to be equal in magnitude by virtue of the single symbol, τ. That this must be true can be established by analysing the equilibrium of the element. Let the shear stress on the vertical faces be τ_x and on the horizontal faces be τ_y. Then, using one of the equations of equilibrium, the moments around the lower right hand corner can be taken:

$$\sum M = \sigma_x \, dy^2/2 - \sigma_x \, dy^2/2 + \sigma_y \, dx^2/2 - \sigma_y \, dx^2/2 + \tau_x \, dy \, dx - \tau_y \, dx \, dy = 0$$

then as $dx = dy$

$$\tau_x = \tau_y \qquad \qquad \text{Eq. 1-1}$$

The element in Fig. 1-3(a) can be cut by an oblique plane oriented at θ to the vertical. Until shown to be otherwise, it must be assumed that both normal and shear stresses, σ_t and τ_t, act on this oblique plane. To determine the magnitude and sense of these unknown stresses, σ_t and τ_t, in terms of the known stresses, σ_x, σ_y and τ, two equations of equilibrium can be used:

$$\sum F_x = \sigma_x\, dy + \tau\, dy \tan\theta - (\sigma_t\, dy/\cos\theta)\cos\theta + (\tau_t\, dy/\cos\theta)\sin\theta = 0$$

$$\sum F_y = \sigma_y\, dy \tan\theta + \tau\, dy - (\sigma_t\, dy/\cos\theta)\sin\theta - (\tau_t\, dy/\cos\theta)\cos\theta = 0$$

The two unknowns, σ_t and τ_t, can then be determined, and the following general equations established:

$$\sigma_t = \frac{\sigma_x + \sigma_y}{2} + \frac{(\sigma_x - \sigma_y)}{2} \cos 2\theta + \tau \sin 2\theta \qquad \text{Eq. 1-2(a)}$$

$$\tau_t = -1/2\,(\sigma_x - \sigma_y)\sin 2\theta + \tau \cos 2\theta \qquad \text{Eq. 1-2(b)}$$

These equations can be used for the determination of directions of stresses by assuming the directions in Figs. 1-3(a) and 1-3(b) as positive, i.e., normal stresses compressive, the shear stress on the right-hand face acting downwards, the angle θ measured counter clockwise from the left-hand face, and the shear stress τ_t acting downwards and to the right.

From Equations 1-2(a) and 1-2(b) it can be seen that σ_t and τ_t vary with the angle θ. The variation can be explored by examining σ_t for maxima and minima. If σ_t is differentiated with respect to θ and equated to zero, it is found that:

$$\tan 2\theta = \frac{2\tau}{\sigma_x - \sigma_y} \qquad \text{Eq. 1-3}$$

This equation shows that the maxima and minima in σ_t occur at angles that vary only with σ_x, σ_y and τ, the given stresses on the orthogonal faces of the element. There are two values of 2θ 180 degrees apart that will satisfy this equation. In other words, two values of θ 90 degrees apart will satisfy the equation. By the substitution of these two values of θ into the equation for σ_t it is found that one value produces a maximum and the other a minimum for σ_t.

By substituting the above values of θ into Equation 1-2(b), it is found that τ_t is zero on these oblique planes. In other words, the planes on which the maxima and minima in σ_t occur are planes with no shear stress. The planes are called <u>principal planes</u> and the normal stresses on these planes are called <u>principal stresses</u>. It is useful to remember, and it will be referred to later, that principal stress trajectories must form a pattern of orthogonal curves.

The symbols σ_1 for the major principal stress, σ_2 for the intermediate principal stress, and σ_3 for the minor principal stress are used. In structural work tensile stresses are assumed to be positive, and hence the major principal stress is the maximum tensile or algebraically the minimum compressive stress. In rock mechanics it is useful to assume that compressive stresses are positive. Hence, the major principal stress in this work is the maximum compressive or the minimum tensile stress at a point.

From the above analysis it can be concluded that for any state of stress at a point there will be principal stresses at that point. Then if the x- and y-axes are oriented in the direction of the principal stresses, it can be seen that the stresses at a point, σ_t and τ_t, are determined by the principal stresses and the orientation of the oblique plane θ, e.g.,

$$\sigma_t = \frac{\sigma_1 + \sigma_2}{2} + \frac{\sigma_1 - \sigma_2}{2} \cos 2\theta \qquad \text{Eq. 1-4(a)}$$

$$\tau_t = -1/2 \, (\sigma_1 - \sigma_2) \sin 2\theta \qquad \text{Eq. 1-4(b)}$$

Example - At a point within a body σ_x is 18.7 ksi tension, σ_y is zero and τ is 3.4 ksi acting as shown in Fig. 1-3(c). The principal stresses are to be determined. Hence, using Equations 1-3 and 1-2:

$$\tan 2\theta = \frac{2 \times 3.4}{-18.7 - 0} = -0.364$$

therefore $\qquad 2\theta = -20°$ or $+160°$

$$\sigma_{1,3} = \frac{-18.7 + 0}{2} + \frac{-18.7 - 0}{2} \cos(340) + 3.4 \sin(340)$$

$$= -19.30$$

also $\qquad \sigma_{1,3} = \dfrac{-18.7 + 0}{2} + \dfrac{-18.7 - 0}{2} \cos(160) + 3.4 \sin(160)$

$$= 0.60$$

hence $\qquad \sigma_1 = 0.60$ ksi and $\sigma_3 = -19.30$ ksi as shown in Fig. 1-3(d).

MOHR'S CIRCLE

Equations 1-4(a) and 1-4(b) can be simplified by letting:

$$a = \frac{\sigma_1 + \sigma_2}{2}$$

and $\qquad b = \dfrac{\sigma_1 - \sigma_2}{2}$

Then $\qquad \sigma_t = a + b \cos 2\theta$

or $\qquad \sigma_t - a = b \cos 2\theta$

and $\qquad \tau_t = -b \sin 2\theta$

By squaring and adding:

$$(\sigma_t - a)^2 + \tau_t^2 = b^2 \qquad \text{Eq. 1-5}$$

This equation may be recognized as the equation of a circle, i.e., analogous to the equation $(x-a)^2 + y^2 = r^2$. With σ_t- and τ_t-axes, the centre of the circle would be at $\sigma_t = a$, $\tau_t = 0$, and its radius would be equal to b. Fig. 1-4 shows such a circle.

This circle, known as Mohr's Circle, represents the locus of the stresses, σ_t and τ_t, that exist at a point. In Fig. 1-4 by adding 'a' and 'b' the abscissa of the point on the circle at A can be seen to represent the major principal stress, σ_1, i.e., it is the stress on the plane where the normal stress is the maximum and the shear stress is zero. Point B represents the stresses on the minor principal plane.

Point E in Fig. 1-4 represents the stresses, σ_t and τ_t acting on a plane which is at an angle θ to the major principal plane. By examining the geometry of this point the

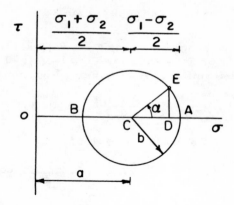

Fig. 1-4 Geometrical Representation of Stresses at a Point by Mohr's Circle

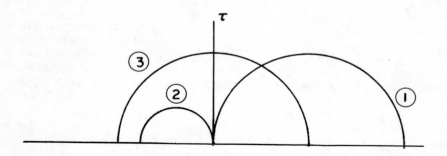

Fig. 1-5 Mohr's Circles for some Common Stress Conditions:
    (1) Uniaxial Compressive Stress
    (2) Uniaxial Tensile Stress and
    (3) Torsion or Pure Shear Stress

value of the angle α can be determined:

$$\overline{OD} = \overline{OC} + \overline{CD}$$

$$= \overline{OC} + \overline{CE} \cos \alpha$$

therefore
$$\sigma_t = \frac{\sigma_1 + \sigma_2}{2} + \frac{\sigma_1 - \sigma_2}{2} \cos \alpha$$

but from **Equation 1-2(a)**

$$\sigma_t = \frac{\sigma_1 + \sigma_2}{2} + \frac{\sigma_1 - \sigma_2}{2} \cos 2\theta$$

thus
$$\alpha = 2\theta.$$

Hence Mohr's Circle also shows the angle between the plane in question and the major principal plane such that α is twice the actual angle.

As mentioned above, it is only necessary to know the principal stresses to define all the stresses at a point. For example, in a uniaxial compression test the major principal stress is equal to the applied compression and the minor principal stress is zero. The stresses at any point in the sample are then represented by Circle ① in Fig. 1-5. From this circle it can be seen that the maximum shear stress is half the major principal stress and that it acts on a plane at 45° to the major principal plane, i.e., $\alpha = 90° = 2\theta$. Similarly, Circle ② represents the stresses at any point within a sample subjected to uniaxial tension.

Circle ③ in Fig. 1-5 can be considered to represent either one of two situations. If a sample is subjected to a compressive major principal stress and a tension of equal magnitude as a minor principal stress, Circle ③ will represent the stresses that exist at any point in the sample. It can be seen from the circle that the maximum shear stress exists on planes that have no normal stress. In other words, the circle also represents the state of stress known as pure shear, i.e., shear stresses acting on planes at 90° to each other with no normal stress acting on these planes. Thus, in the state of so-called pure shear normal stresses act on all the planes through any point in the sample except those of maximum shear stress.

Usually in rock mechanics the signs of the shear stresses and the signs of the angles between the major principal plane and other planes are not important. Hence, it becomes only necessary to draw semi-circles as shown in Fig. 1-5. In the cases where sign is important, rather than attempting to remember the sign convention to use with Mohr's Circle, the solution of the problem by a wedge analysis (as in Fig. 1-3(b)) has much to recommend it.

BEHAVIOUR OF ROCKS

It was stated earlier that the strain in an elastic body is fully recoverable. Many hard rocks can be considered elastic as samples produce stress-strain curves that are not only reversible but also straight, as shown in Fig. 1-6(a).

These same rocks, when including the structural features such as joints, often produce a load-deformation curve in a plate load test as shown in Fig. 1-6(b). The closing of joints accounts for much of the initial curvature in Zone I. In this case the rock is not linearly elastic, but as the curve in Zones I and II is usually reversible the material is still elastic.

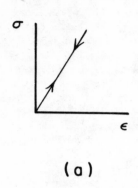
(a)

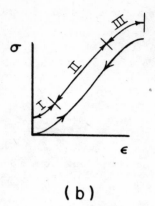
(b)

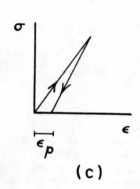
(c)

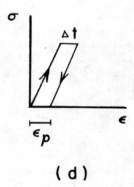
(d)

Fig. 1-6 Some Aspects of the Behaviour of Rocks
- (a) A Linear, Reversible Stress-Strain Curve
- (b) A Curvilinear, Reversible Stress-Strain Curve
- (c) A Stress-Strain Cycle Indicating Some Strain is Plastic or Irreversible
- (d) A Stress-Strain Curve Indicating Plastic Strain is Dependent on Duration of Stress

In Zone III failure is initiated; tests on samples often show that the rate of lateral strain increases and the volume of rock may increase. The maintenance of stress in this zone will usually lead to failure with the time required varying inversely with the increment of stress above the threshold, or yield, value at the transition point between Zones II and III (9). Another interesting aspect is that, when the level of stress exceeds about 50 per cent of the ultimate strength, a sensitive crystal pick-up will detect internal cracking or microseismic activity (1). This activity commonly increases with stress, indicating some breakdown or readjustment in the grain structure of the sample.

Laboratory samples of rock can produce a Zone I curve if the porosity of the samples is significant (9). In these cases the voids probably occur as thin wedge-like openings between grains, which are closed under pressure.

A plastic body, it was stated earlier, under stress produces strains that are partially or completely irrecoverable. Some ground, which compacts under stress, can produce stress-strain curves as shown in Fig. 1-6(c) with ϵ_p being the plastic strain. For most rocks that give plastic strain these lines are curved rather than straight as shown by Cycle B-C in Fig. 1-6(h).

Some rocks that give plastic strain do so regardless of the duration of the stress. In other words, the stress-strain curve of Fig. 1-6(c) would be produced even if the stress were applied and removed quickly. Other rocks require time for the plastic strain to develop.

An increase in plastic strain with time is usually described as creep and could be represented as shown in Fig. 1-6(d) where Δt represents the increment of time required to produce ϵ_p. Such creep may follow a rheological law and be characterized by a coefficient of viscosity. Hence, these rocks are described in this book as being viscous.

It should be recognized, however, that time effects can occur without having a distinctly viscous rock. As an example of actual time effects that occurred in one mine without being caused by a viscous rock, it was observed that the sliding joint sets in a scram drift closed at the rate of 4 in. per month, not because the rock was viscous but because the back was caving and the working rock increased the load on the sets. When the sliding joints in the sets operated, the load was reduced but caused increased working in the rock. The load would thus build up again and another increment of deformation in the set would occur. All of these actions and reactions required time, hence giving a result that was similar to but not the same as that which could occur in viscous rock.

In Fig. 1-6(e) one of the many rheological models, the Maxwell Substance, is shown. The model characterizes a material that, under stress, will produce a strain one part of which is instantaneous and recoverable and the other part time dependent and irrecoverable. The latter, or plastic strain, will thus vary with both time and the magnitude of stress.

The stress-time curve given by this body is also shown in Fig. 1-6(e). If the modulus of elasticity of the spring and the coefficient of viscosity of the dash-pot in the model are constant, the following equation gives the strain in the model:

$$\epsilon = \frac{\sigma}{E} - \frac{\sigma t}{n} \qquad \text{Eq. 1-6}$$

where t is the time during which the stress, σ, has acted, and n is the coefficient of viscosity.

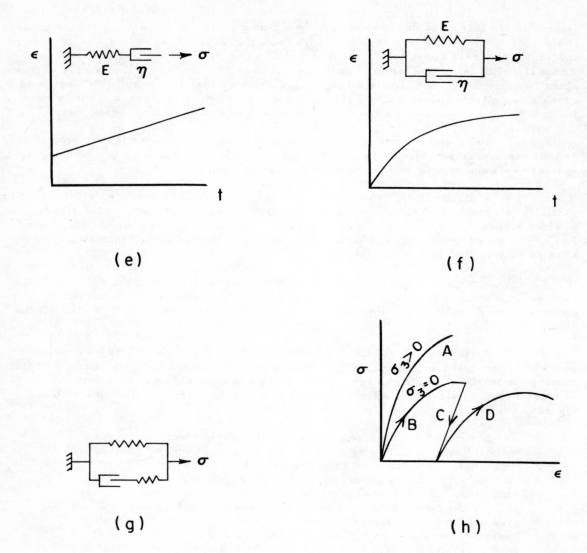

Fig. 1-6 Some Aspects of the Behaviour of Rocks
 (e) A Maxwell Model Showing Stress Producing Instantaneous Strain Plus Time Dependent Strain
 (f) A Kelvin Model Showing Stress Producing Time Dependent Strain Approaching Some Ultimate Value
 (g) A General Linear Substance on which Stress would Produce Instantaneous Strain and a Time Dependent Strain that would Vary Asymptotically towards an Ultimate Value
 (h) A Reaction to Cycling of Stress of Some Geological Materials and the Effect of Confinement ($\sigma_3 < 0$)

In Fig. 1-6(f) another simple rheological model, the Kelvin Substance is shown. The strain that is produced in this model by the application of stress varies with time as shown. However, any strain that occurs is fully recoverable if sufficient time is provided. Consequently, this model represents an elastic body, although from a practical point of view the time required to recover all the strain in a rock acting in this manner might be so great that some of the strain would be considered plastic.

If the characteristics in the Kelvin Model are constant, then the following equation gives the strain resulting from the application of stress:

$$\epsilon = \frac{\sigma}{E}\left(1 - e^{-Et/n}\right) \qquad \text{Eq. 1-7}$$

It can be seen from Equation 1-7 that, as time approaches infinity, the strain approaches the elastic strain corresponding to the stress, σ. On release of stress the following equation applies:

$$\epsilon = \epsilon_o e^{-Et/n} \qquad \text{Eq. 1-8}$$

where ϵ_o is the strain when the time, t, is zero.

Most natural materials, of course, do not act exactly like the above ideal models. To come closer to actual behaviour various combinations can be devised. For example, in Fig. 1-6(g) a General Linear Substance combining both Maxwell and Kelvin elements is shown. Such a material would give both elastic and plastic elements of strain with the plastic elements being time dependent and some of the elastic recovery of stress being also time dependent.

At the present time the only use of rheological models in rock mechanics is in obtaining an increased awareness of the variety of strain-time patterns that might exist in rocks. At this stage some work is being done to solve problems of stress distribution for various types of geometry and loadings assuming a rheological rather than an elastic material. However, it will be some time in the future before such a procedure may be applied to actual field problems.

The behaviour of rock varies with another aspect of time, that is, with stress or strain rate. Generally, the stress at failure and often the modulus of deformation of rock will increase with an increase in the rate of application of load. Table 1 includes the results of dynamic and static testing of several rock types, which shows the comparative effect of stress rate on tensile failure stress and strain (20).

TABLE 1

Dynamic Versus Static Rock Properties (20)

	Marble	Sandstone A	Sandstone B	Granite
Dynamic Tests				
Stress Rate, kg/cm^2/sec	1.7×10^6	1.4×10^6	1.5×10^6	1.5×10^6
Strain Rate, μ/sec	3.6	3.7	3.3	5.5
Failure Stress, kg/cm^2	215	220	190	170
Failure Strain, μ	490	610	460	630
E, kg/cm^2	51×10^4	64×10^4	40×10^4	30×10^4
Static Tests				
Stress Rate, kg/cm^2/sec	1.1	1.8	0.5	2.2
Failure Stress, kg/cm^2	53	80	29	53
Failure Strain, μ	145	410	370	510
E, kg/cm^2	47×10^4	19×10^4	10×10^4	12×10^4

In Fig. 1-6(h) Cycle B-C shows a stress-strain curve in uniaxial compression that is obtained in some geological materials. In this case the application of stress produces a curved stress-strain relationship up to a maximum where yielding occurs. The determination of the representative modulus of deformation in this case is difficult. A tangent modulus, $d\sigma/d\epsilon$, represents the slope of the curve at any specific level, whereas the secant modulus, σ/ϵ, represents the average slope to a specific stress.

In such materials, there are two different concepts of plasticity involved. Below the failure stress some irrecoverable or plastic strain occurs (i.e., the recovery curve from a stress level below failure is commonly a straight line down to some intercept on the x-axis). Hence the material can be described as being plastic or elasto-plastic.

Another type of plasticity occurs when the strength of the material, as in Fig. 1-6(h), is exceeded. Flow rather than fracture, as can happen in soft, relatively incompetent rock, occurs. This flow is similar to the reaction of mild steel to stress above the yield point. A large amount of plastic strain, as indicated by the second cycle, D, can occur before the material breaks down and ruptures. (It is unfortunate that there are so many different concepts of plasticity being used in the various fields of applied mechanics. It is thus particularly important when using this term as a means of communicating important information that the particular meaning implied is either defined or quite clear.)

In Fig. 1-6(h) stress-strain Curve A for the same material, under triaxial stress conditions, as Curves B, C and D is shown. This curve illustrates the common reaction of rocks when subjected to a confining pressure. Increased strength generally results and an increased modulus of deformation may result, both of which vary with the magnitude of the confining pressure. However, for brittle, dense rocks confining pressures can have very little effect on the modulus of deformation.

MOHR'S STRENGTH THEORY

Cohesionless Ground: The strength theory that has been shown to have the most validity at the present time in representing the strength of rocks subjected to compressive stresses is Mohr's Strength Theory. It does not explain, completely satisfactorily, all the observed variations of strength; however, no other theory has as much experimental substantiation.

The basis of this strength theory is that failure requires the overcoming of internal friction on incipient surfaces of failure. Elementary friction mechanics can be examined as shown in Fig. 1-7. Here a block on a plane is subjected to a horizontal force, P. A Free Body Diagram shows three external forces acting on the block; P; the force of gravity, W; and the reaction of the plane, R. R can be replaced by its components N acting normal to the plane and F_r acting tangential to the plane.

Coulomb's Law of Friction has established that:

$F_r \leq N \cdot u$ where u = the coefficient of friction

or $F_r/N \leq u$.

Hence $\tan \beta \leq u$, where β equals the angle of obliquity of the reaction of the plane.

Where $\tan \beta = u$, β is the maximum angle of obliquity to the normal that can exist, or F_r is at its maximum value. Hence, if P is increased, equilibrium cannot be established and motion occurs. It is common to use the Greek letter φ for this maximum value of β and to call it the angle of friction. Just as u, the coefficient of friction, varys for different materials so does the maximum angle of obliquity, or the angle of friction, φ.

In Fig. 1-3(b) the stress on the plane defined by the angle θ is represented by the two components σ_t and τ_t. These stresses could have been represented by an oblique stress inclined to the normal to the plane at an angle β such that $\tan \beta = \tau_t/\sigma_t$.

In Fig. 1-8(a) E represents the state of stress on one of the planes through the point in the material represented by Mohr's Circle ②. The obliquity of stress on the plane of E is $\beta = \tan^{-1} \tau/\sigma$. Hence a line joining E to the origin 0 will give an angle β, the angle of obliquity of the resultant stress, with the x-axis.

As β has a maximum value, φ, a line can be drawn to the x-axis that will be an envelope of all the possible angles of obliquity of resultant stresses for the material. For example, on Stress Circle ① the resultant stress on the plane represented by F is at the maximum angle of obliquity, φ. If an attempt were made to increase the obliquity of this stress, it would be similar to attempting to increase the shear stress τ while keeping the normal stress, σ, constant; or similar to trying to increase F_r in Fig. 1-7 sufficient to counteract an increase in P when β equals φ while keeping N constant. It could not be done in each case and motion would follow.

Thus, as the line in Fig. 1-8(a) at φ to the x-axis represents the maximum angle of obliquity of the resultant stress on any plane in the material, it also represents the envelope above which it is not possible for stress circles to exist. In this case, it would describe the strength characteristics of a material that relies on internal friction only (i.e., a granular mass) to resist the applied forces. The equation of this envelope is:

$$\tau_f = \sigma \tan \varphi \qquad \text{Eq. 1-9}$$

where τ_f = the shear resistance or the shear stress required to produce failure. It might be mentioned here that this strength theory could be explained without using the internal friction concept by just observing empirically that shear strength varies with normal stress on the plane of failure.

Some of the implications of this theory can be seen. First, from Equation 1-9 failure as a result of the application of a compressive stress would actually be failure in shear.

Second, there would be no single value of the major principal stress at failure; the compressive strength, as the failure stress, τ_f, depends on the normal stress acting on

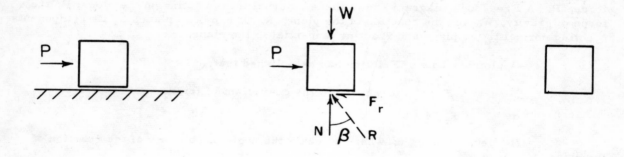

Fig. 1-7 The Forces Involved in Moving a Body Against Frictional Resistance

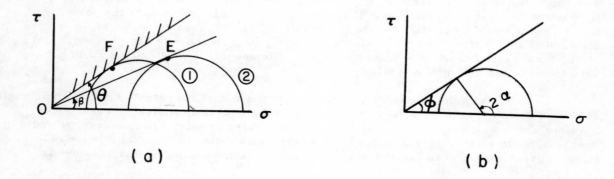

Fig. 1-8 Shear Failure Envelopes

the incipient plane of failure. Thus, in Fig. 1-8(a) Circle ① represents a failure condition with failure occurring on the plane represented by F. However, Circle ② does not represent failure as the maximum angle of obliquity of stress on any plane through this point is β, and β is less than φ. Circle ② can be seen to describe a stress condition with a higher major principal stress, σ_1, than Circle ①. However, as the minor principal stress, σ_3, is more than proportionately higher than the minor principal stress for Circle ①, the stress condition is less critical.

The third observation that can be made is that, from Fig. 1-8(a), as the confining pressure, σ_3, is increased, the normal stress is increased on the incipient plane of failure; hence, the shear stress required to cause failure, τ_f, is increased.

A mathematical relation exists between σ_1 and σ_3 at failure. In Fig. 1-8(b) the angle 2α is twice the angle between the major principal plane and the plane of failure. From the geometry of the diagram:

$$2\alpha = 90 + \varphi$$

or

$$\alpha = 45 + \frac{\varphi}{2} \qquad \text{Eq. 1-10}$$

In other words, failure always takes place theoretically on planes at $(45 + \varphi/2)$ to the major principal plane. There is a considerable amount of test data supporting this conclusion; although many tests, possibly due to inhomogeneities in the material or the stress field, have not produced results consistent with these relations.

Fig. 1-8(c) shows an element with stress conditions producing failure. On the plane at α, or $(45 + \varphi/2)$, to the major principal plane the components of stress, σ and τ are shown. In addition, the resultant stress, f, on the plane is shown. Failure on this plane means that the angle of obliquity of f is φ. Alternatively, it can be said that the relation between the shear stress, τ, and the normal stress σ is:

$$\tau = \sigma \tan \varphi$$

A force polygon as shown in Fig. 1-8(d) can be drawn of the forces acting on the element in Fig. 1-8(c). These forces equal the stresses multiplied by the areas over which they act. The direction of f is obtained by using Equation 1-10. It is assumed that the thickness of the element perpendicular to the page is 1. From this force diagram it follows that:

$$\tan \alpha = \frac{\sigma_1 \, dy / \tan \alpha}{\sigma_3 \, dy}$$

or

$$\sigma_1 / \sigma_3 = \tan^2 \alpha$$

$$= \tan^2 (45 + \varphi/2) \qquad \text{Eq. 1-11(a)}$$

This equation states that for failure to occur the ratio of the major principal stress, σ_1, and the minor principal stress, σ_3, must be equal to $\tan^2 (45 + \varphi/2)$. If $\varphi = 30°$, then σ_1/σ_3 at failure is approximately 3; if $\varphi = 45°$ σ_1/σ_3 at failure is approximately 6. Hence, σ_1 required for failure in these cases would be either 3 or 6 times σ_3.

A classical theory for calculating the pressure exerted by a granular mass against a body is based on Equation 1-11(a) and represented by the element in Fig. 1-8(c). The major principal stress, σ_1, is assumed to be equal to the overburden pressure, i.e., $\sigma_1 = \gamma z$. The minor principal stress, σ_3, is then the reaction of the body to the pressure, p_a, acting on it. As the element must be on the point of failure, incipient motion must be towards the body.

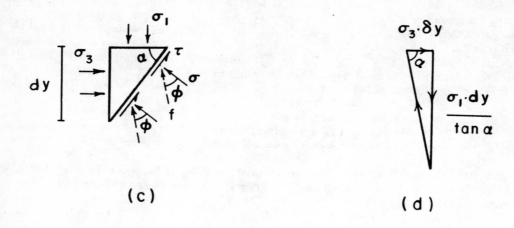

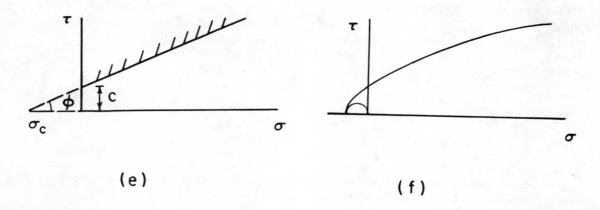

Fig. 1-8 Shear Failure Envelopes

This condition is known as the case of active earth pressure:

$$p_a = \gamma z / \tan^2(45 + \varphi/2) = \gamma z \tan^2(45 - \varphi/2) \qquad \text{Eq. 1-11(b)}$$

This equation, when all the conditions to make it valid are fulfilled (e.g., plane, horizontal, unloaded ground surface and a vertical, smooth wall), provides a fairly accurate prediction of the average pressure exerted by broken rock, sand or gravel against retaining walls, sets, and structures.

When the incipient motion is from the body towards the granular mass, as occurs for example when a thin metal liner tends to deflect laterally out from the centre of the tunnel, the major principal stress, σ_1, is then equal to the reaction of the granular mass, p_p. If the element in Fig. 1-8(c) is turned through 90°, σ_1 or p_p acts horizontally and σ_3 equals γz. This condition is known as the case of passive earth pressure:

$$p_p = \gamma z \tan^2(45 + \varphi/2) \qquad \text{Eq. 1-11(c)}$$

<u>Cohesive Ground</u>: Most rocks resist failure by cohesion between their particles as well as by the internal friction that can be mobilized. A model for this type of strength could be the block shown in Fig. 1-7 when it is glued to the plane. In this case, the force, P, required to produce motion along the plane would be:

$$P \leq K + u.N$$

where K = the lateral force resisting motion due to the glue. This equation assumes that the resistance of the glue and friction are mobilized at the same time or at the same deformation.

The similar Coulomb equation for internal resistance arising from cohesion and internal friction is:

$$\tau_f = c + \sigma \tan \varphi \qquad \text{Eq. 1-12}$$

where c = the cohesion or the resistance to shear along internal planes that is independent of normal stress. The curve of this equation is shown in Fig. 1-8(e). In other words, when the normal stress on an internal plane is zero, the remaining shear resistance is equal to the cohesion, c.

It can be seen that, if a rock had no internal friction, Equation 1-12 would reduce to $\tau_f = c$ = constant, regardless of the value of the normal stress on the plane of failure. In this case, the material would be obeying the Maximum Shear Stress strength theory. The strength envelope in a Mohr diagram would appear as a horizontal line at the level where $\tau = c$.

The strength equation, 1-12, could be written using the intercept on the x-axis, as shown in Fig. 1-8(e), rather than that on the y-axis:

$$\tau_f = (\sigma_c + \sigma) \tan \varphi$$

where

$$\sigma_c = c/\tan \varphi \qquad \text{Eq. 1-13}$$

This equation gives rise to the concept of an intrinsic stress, σ_c, that exists as a result of inter-molecular attraction. This concept has some use in mathematical analyses, but the deduction that it defines the failure envelope on the tension or left side of the origin in the Mohr diagram would not be valid as it is known from tests that the tensile strength of rocks is much less than would be indicated by the extrapolation of the compressive strength envelope.

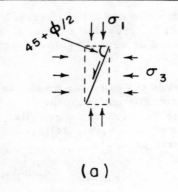
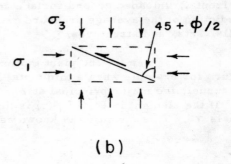

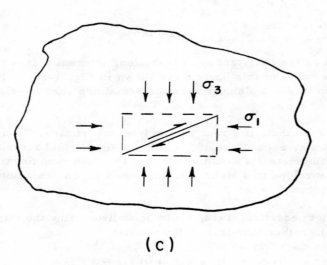

Fig. 1-9 (a) A Normal Fault
(b) A Thrust Fault and
(c) A Transcurrent Fault

Following Equations 1-11 and 1-12, the functional relations between the principal stresses can be established for cohesive materials as follows:

$$\frac{\sigma_1 + \sigma_c}{\sigma_3 + \sigma_c} = \tan^2(45 + \varphi/2) \qquad \text{Eq. 1-14(a)}$$

or $\quad \sigma_1 = \sigma_3 \tan^2(45 + \varphi/2) + \sigma_c(\tan^2(45 + \varphi/2)) - \sigma_c$

$$\sigma_1 = \sigma_3 \tan^2(45 + \varphi/2) + c(\tan^2(45 + \varphi/2) - 1)/\tan\varphi \qquad \text{Eq. 1-14(b)}$$

Actual strength envelopes based on triaxial tests commonly have the shape shown in Fig. 1-8(f). The circle representing a uniaxial tension test then touches the envelope at the point representing the minor principal plane. This is consistent with the mode of failure in tension, since it occurs as a fracture perpendicular to the maximum tensile stress (i.e., the minimum principal stress). The particles are pulled apart without any sliding action. For tensile failure it is possible that the Maximum Principal Stress strength theory applies to rocks, i.e., failure occurs when the minor principal stress or tensile stress reaches some constant value regardless of the value of the other principal stresses.

At high levels of confining pressure the envelope as shown in Fig. 1-8(f) also departs from the linear relations included in Equation 1-13, and the increase in strength with normal stress is less than at the lower levels of confining pressure. Possibly some internal crushing action at these high normal stresses accounts for this change in the strength relations.

<u>Faults</u> - Mohr's Strength Theory can be applied with seeming validity to faulting. For example, in Fig. 1-9(a) at some depth below the surface of the earth the major principal stress, σ_1, could be vertical with the minor or horizontal principal stress either being compression or tension of low magnitude (so that failure is not by tension cracking). Failure along a single plane would then produce what would be later identified as a Normal Fault with its dip at $(45 + \alpha/2)$ to what was at the time the horizontal plane.

In Fig. 1-9(b) the major principal stress is horizontal and the minor principal stress is vertical. Failure would thus be along planes at $(45 + \varphi/2)$ to the vertical and produce Thrust Faults.

The final combination could be where the intermediate principal stress is vertical with the major and minor principal stresses in the horizontal plane as shown in Fig. 1-9(c). Failure would occur in this case on vertical planes in the two directions at $(45 + \varphi/2)$ to the major principal plane.

<u>Effective Stresses</u>: The concept of effective stress must now be examined. If the block of Fig. 1-7 were placed under water as shown in Fig. 1-10(a), the Free Body Diagram would be as shown in Fig. 1-10(b). Here the additional force, U, the uplift on the block, has been added. With the presence of U the reaction from the plane need only be N' to maintain equilibrium in the vertical direction. It can be appreciated that, with the reduction in the normal force, the maximum frictional reaction, F_r', will be less than previously. In other words, at the point of incipient motion:

$$P = F_r' = u \cdot N' = u(W-U).$$

Similarly, it can be stated that for internal stress relations at failure:

$$\tau_f = \sigma' \tan\varphi$$
$$= (\sigma - u)\tan\varphi$$

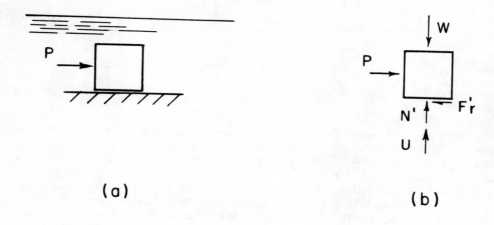

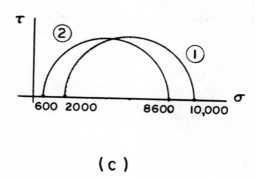

Fig. 1-10 Effective Forces and Stresses

where σ' = the effective inter-granular stress on the plane of failure, u = the pore pressure in the water or air tending to force the particle apart, and σ = the apparent inter-granular stress on the failure plane ignoring pore pressures.

The more general effective stress strength equation is:

$$\tau_f = c + (\sigma - u) \tan\varphi. \qquad \text{Eq. 1-15}$$

From Equation 1-15 it can be seen that if $u = \sigma$ then $\tau_f = c$. Under these conditions the material would appear to have no internal friction and hence would obey the Maximum Shear Stress theory mentioned above. However, such a deduction could be misleading if the pore pressure changed. Although it is not always possible, analyses of failure stresses should include the determination of not only the principal stresses but also the pore pressures appropriate to those conditions.

Example - In a triaxial compression test the confining pressure, or the fluid pressure outside the membrane enclosing the sample, is 2000 psi. The total axial stress at failure is 10,000 psi. The sample contains water in its interstices, and the pressure in this water at failure is measured as 1400 psi. Plot the Mohr Circles for the stress conditions at failure in terms of total stresses and effective stresses.

The total stresses at failure are:

$$\sigma_3 = 2000 \text{ psi}$$
$$\sigma_1 = 10,000 \text{ psi}$$

The Mohr Circle ① is plotted as shown in Fig. 1-10(c).

The effective stresses at failure are:

$$\sigma_3' = 2000 - 1400 = 600 \text{ psi}$$
$$\sigma_1' = 10,000 - 1400 = 8600 \text{ psi}$$

The Mohr Circle ② is plotted as shown in Fig. 1-10(c).

ULTIMATE BEARING STRESSES

For the theoretical determination of strength the case has been used above of a sample subjected to an axial stress that acts over the entire cross-section of the sample. Another case that is of interest for many problems is where only a small part of the surface is loaded as shown in Fig. 1-11(a). This system can be considered to be a prototype of a dam bearing on its foundation, a pillar bearing on the floor, or the stylus in a hardness test bearing on the surface of a rock sample.

The solution of this case is very simple if the assumptions are made - that the failure surface is composed of two plane surfaces as shown in Fig. 1-11(a) - that the loading, q_f, is sufficiently long that the resistance of the ends parallel to the page can be ignored - that no shear stress exists on the bearing plane, i.e., the plane on which q_f is acting - and that average body forces can be used for each wedge.

In Fig. 1-11(b) the stresses on Wedge X are shown. This wedge is analogous to that in Fig. 1-8(c) from which Equation 1-14(b) was derived. Hence for Wedge X:

$$\sigma = \sigma_3 \tan^2(45 + \varphi/2) + c(\tan^2(45 + \varphi/2) - 1)/\tan\varphi$$

where σ_3 is actually a body stress due to gravity acting on the wedge and its average is

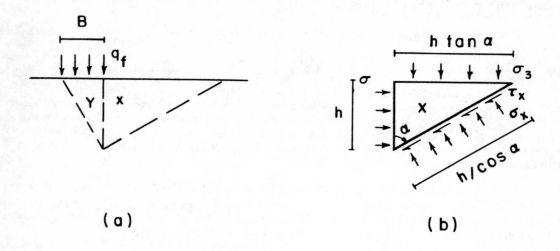

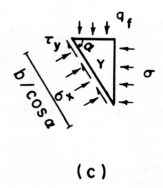

Fig. 1-11 A Wedge Analysis of Ultimate Bearing Stresses

equal to $\gamma h/2$, where γ = the density of the material.

The stresses on Wedge Y are shown in Fig. 1-11(c). In this case the horizontal stress, σ, is the minor principal stress, thus:

$$q_f + \gamma h/2 = \sigma \tan^2(45 + \varphi/2) + c(\tan^2(45 + \varphi/2) - 1)/\tan\varphi$$

$$= 0.5\gamma h \tan^4(45 + \varphi/2) + c(\tan^2(45 + \varphi/2) - 1)/\tan\varphi$$

$$\cdot \tan^2(45 + \varphi/2) + c(\tan^2(45 + \varphi/2) - 1)/\tan\varphi.$$

It follows from

$$h = B\tan(45 + \varphi/2) \text{ that:}$$

$$q_f = 0.5\gamma B \tan^5(45 + \varphi/2) + c(\tan^2(45 + \varphi/2) - 1)$$

$$\cdot (\tan^2(45 + \varphi/2) + 1)/\tan\varphi - 0.5\gamma B \tan(45 + \varphi/2)$$

$$= 0.5\gamma B \tan^5(45 + \varphi/2) + c(\tan^4(45 + \varphi/2) - 1)/\tan\varphi$$

$$- 0.5\gamma B(45 + \varphi/2).$$

From this equation it can be calculated that if $\varphi = 45°$, $\tan^5(45 + \varphi/2) = 225$, $\tan^4(45 + \varphi/2) = 45$ and $\tan(45 + \varphi/2) = 5$. Consequently, there is some justification for dropping the factor $0.5\gamma B \tan(45 + \varphi/2)$ as not being significant with respect to the other factors.

If a surcharge pressure, q, exists on the surface adjacent to the bearing pressure, q_f, the analysis would be the same except for Wedge X where $\sigma_3 = \gamma h/2 + q$ and thus

$$q_f = 0.5\gamma B \tan^5(45 + \varphi/2) + c(\tan^4(45 + \varphi/2) - 1)/\tan\varphi$$

$$+ q \tan^4(45 + \varphi/2) \qquad \text{Eq. 1-16}$$

A more rigorous bearing capacity theory has been worked out which produces an equation of the following form (2):

$$q_f = 0.5\gamma B \cdot N_\gamma + c N_c + q N_q \qquad \text{Eq. 1-17}$$

where N_γ, N_c and N_q are called bearing capacity factors and are all functions of φ. This equation is of the same form as Equation 1-16.

The actual values of these N-factors are larger than those corresponding to the coefficients in Equation 1-16 as this more rigorous theory recognizes that the surface of failure is curved and that shear stresses exist on the boundary between Wedges X and Y and on the bearing surface. The following equations give simple expressions for determining the bearing capacity factors:

$$N_\gamma = \tan^6(45 + \varphi/2) + 1 \qquad \text{Eq. 1-18(a)}$$

$$N_c = 5 \tan^4(45 + \varphi/2) \qquad \text{Eq. 1-18(b)}$$

$$N_q = \tan^6(45 + \varphi/2) \qquad \text{Eq. 1-18(c)}$$

These equations give values that are fairly close to the rigorous solutions for the range of φ from 0° to 45°.

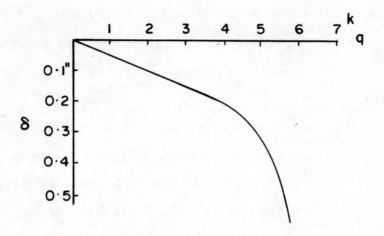

Fig. 1-12 A Bearing Pressure-Settlement Curve from a Plate Load Test

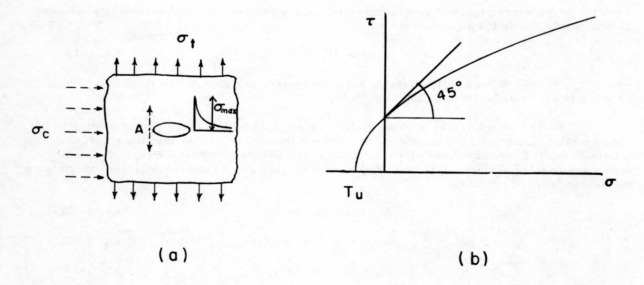

Fig. 1-13 Griffith's Strength Theory

There is evidence that this bearing capacity theory applies to some rocks. However, complications arise in brittle materials like hard rocks when internal concentrations of stress on grain boundaries produce local fracture. The assumption that average stresses act on incipient planes of failure is not valid for these rocks.

These equations were derived for long loadings. Where the length is the same order as the width, B, the only significant change for rocks is in the coefficient of c, and for a square or round bearing area Equation 1-18(b) can be changed to:

$$N_c = 7 \tan^4(45 + \varphi/2) \qquad \text{Eq. 1-19}$$

It is now possible to compare the pressures required for failure of samples where the entire cross-section is loaded with that which apply where only a small part of the cross-section is loaded. From Equation 1-14(b) for a uniaxial compression test:

$$\sigma_1 = c(\tan^2(45 + \varphi/2) - 1)/\tan \varphi$$

From Equation 1-17 for the bearing case with no surcharge:

$$\sigma_1 = 0.5 \gamma B (\tan^6(45 + \varphi/2) - 1) + 7c \tan^4(45 + \varphi/2).$$

In the second case it can be seen that a much higher stress is required for failure, e.g., if $\varphi = 40°$, $\gamma = 170$ pcf, $c = 5000$ psf, $B = 1$ ft:

for the uniaxial compression test

$$\sigma_1 = 5000(4.62-1)/0.838 = 21,600 \text{ psf} = 150 \text{ psi}.$$

For the bearing case

$$\sigma_1 = 0.5 \times 170 \times 1(97-1) + 7 \times 5000 \times 21 = 743,000 \text{ psf} = 5160 \text{ psi}.$$

In weak or soft rocks this theory can be used to determine one of the in situ strength parameters by applying through a plate a bearing pressure sufficient to cause failure of the rock. It has to be restricted to weak rocks as the required loading becomes excessive for other rocks.

Example - A granular material is loaded through a 1 ft x 1 ft plate. The load-deformation curve is as shown in Fig. 1-12. It is assumed that the failure process started in this test at a bearing pressure of 4000 psf. The granular material had a density of 125 pcf. Using Equation 1-17 we can solve for φ.

$$4000 = 0.5 \times 125 \times 1(\tan^6(45 + \varphi/2) - 1) + 7 \times 0 \times \tan^4(45 + \varphi/2)$$
$$+ 0 \times \tan^6(45 + \varphi/2)$$

hence $\varphi = 37°$.

It should be appreciated that the accuracy of this value of φ is affected by, among other factors, the approximations involved in establishing the coefficients in Equation 1-17. These approximations would not be significant if the purpose of the test was to establish the bearing capacity of the ground, as the same approximations would apply in the test as in the system being designed and their influence would be largely cancelled. If, however, the strength parameter was to be applied to a different mechanics system, e.g., for a slope stability analysis, then the error in the approximation could be significant.

In the case of rock with both cohesion and internal friction, only one in situ strength parameter could be obtained from a plate load test. Thus, if the internal friction angle were otherwise determined, the plate load test could be used for determining the in situ cohesion.

GRIFFITH'S STRENGTH THEORY

Griffith's Theory postulates the presence of microscopic cracks within the material. The effect of these cracks is to produce concentrations of stress around their boundaries. If the principal stress in the rock is tension and is normal to the crack, as shown in Fig. 1-13(a) by σ_t, then a tensile stress many times this stress will be created at the ends of the crack.

If the principal stress is compression, σ_c in Fig. 1-13(a), then tension can be induced at the point A on the boundary of the crack. When these stress concentrations are equal to the tensile strength of the material, the cracks will be extended. As the length of the crack transverse to the tensile field stress increases, the stress concentrations become greater; consequently it is visualized that, once initiated, the propagation of the crack will lead to failure of the material. The theory has been substantiated by experimental work on glass.

By assuming that the cracks are elliptical in shape and that they are randomly oriented, the following criteria for failure have been established:

$$\sigma_3 = -T_s \text{ when } \sigma_1 + 3\sigma_3 < 0 \qquad \text{Eq. 1-20}$$

where T_s is the uniaxial tensile strength (and a positive number although compressive stresses are positive) of the rock:

and

$$\frac{(\sigma_1 - \sigma_3)^2}{\sigma_1 + \sigma_3} = 8 T_s \text{ when } \sigma_1 + 3\sigma_3 > 0 \qquad \text{Eq. 1-21}$$

$$\cos 2\theta = \frac{\sigma_1 - \sigma_3}{2(\sigma_1 + \sigma_3)} \qquad \text{Eq. 1-22}$$

where θ = the angle between the minor principal plane and the plane of failure.

By combining Equations 1-21 and 1-22 a failure equation can be obtained for comparison with Mohr's strength Equation 1-12:

$$\tau_f = 2\sqrt{T_s \cdot \sigma + T_s^2} \qquad \text{Eq. 1-23}$$

where τ_f is the shear stress on the plane of failure and σ is the normal stress on the plane of failure.

In Fig. 1-13(b) the failure criteria according to Griffith is plotted. It can be seen that this curve has a shape that is similar to the one in Fig. 1-9(b). The envelope is curved and indicates a much lower tensile strength than would be deduced from a linear envelope; it is consequently more in accordance with experience. Also, the envelope on the compression side of the origin has a decreasing slope starting at 45° at the y-axis. This also is in general agreement with the results of triaxial testing on rocks.

If a wedge analysis, as shown in Fig. 1-11, is made using Griffith's Strength Theory, an alternate equation can be obtained for the bearing capacity of rock. As such an equation would only be applicable to hard rocks, it has been assumed that the contribution to the resistance of the rock in Wedge X of Fig. 1-11(b) of the force of gravity, $\gamma h/2$, is negligible. The resultant expression for a long load is:

$$q_f = 24 \, T_s \qquad \text{Eq. 1-24(a)}$$

$$q_f = 3 \, Q_u \qquad \text{Eq. 1-24(b)}$$

where T_s is the uniaxial tensile strength and Q_u is the uniaxial compressive strength of the rock. In this case, the bearing capacity does not vary with the width, B, of the loaded area.

The effect of any surcharge, q, adjacent to the loaded area would contribute little to q_f if it was small (i.e., less than 1/20) with respect to Q_u. This is likely to be the case in most situations.

Equations 1-24 are based on the assumption that the strength of the rock will be mobilized along the entire failure surface at the same time. When more is known about failure of brittle materials, it is probable that we shall find that failure is initiated at a point due to a concentration of stress (as shown in Fig. 8-1(a)) and propagates into a progressive failure. Hence, theoretical expressions such as Equations 1-16, 1-17, and 1-24 must be used with this possibility kept in mind.

Whereas experimental work on glass has substantiated Griffith's theory very well, insufficient work has been done on rocks to determine whether this theory would predict compression failure by knowing the tensile strength of the rock.

OTHER STRENGTH THEORIES

The classical strength theories that have been considered for various materials have the following criteria: maximum principal stress, maximum shear stress, maximum strain energy, maximum energy of distortion (i.e., maximum octahedral shear stress), and maximum principal strain.

The maximum principal stress criterion is shown in Fig. 1-14(a). The envelope C in this figure would represent a compression failure. Failure according to this criterion would occur under any stress condition that produced a Mohr Circle, such as A, that just touched the envelope, C. This, of course, is contrary to experience as it is known that, by measuring the confining pressure or the minor principal stress, the major principal stress at failure can be increased without any known limit being encountered.

If the maximum principal stress criterion was applied to tensile stresses, the envelope would be shown as T in Fig. 1-14(a). In this case failure would occur with any stress condition that produced a stress circle, such as B, just touching T. Little is known about the parameters governing tension failure, and it is possible that this criterion might be valid for brittle rocks.

The maximum shear stress criterion is represented by the envelope, S, in Fig. 1-14(b). Failure, according to this theory, would occur under stress conditions that produced, as shown by Circles A and B, a maximum shear stress equal to the envelope value. This theory is clearly invalid for rocks, as it does not reflect the increased shear stress that can be sustained as the normal stress on the critical plane is increased nor does it show the much lower strength that would apply to conditions that included tensile stresses.

The total strain energy theory has been rejected for most materials on the basis that it does not fit the facts for a hydrostatic stress condition. In other words, failure that should occur at a critical strain energy level does not occur when this strain energy is created by equal, or near equal, principal stresses.

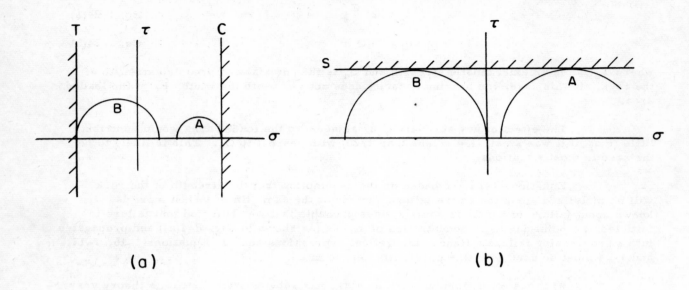

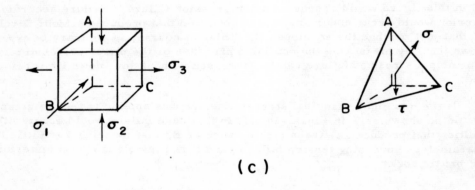

Fig. 1-14 (a) Maximum Principal Stress Theory
(b) Maximum Shear Stress Theory
(c) Octahedral Stresses

The energy of distortion, or the equivalent maximum octahedral shear stress, theory may provide a better basis for extrapolating results for different states of stress taking into account the variation of all principal stresses - not just the major and minor. The concepts behind these two equivalent theories are that failure will occur when either the energy of distortion, or the strain energy due to the shear stresses, reaches a critical value; or failure will occur when the shear stress, τ_o, taking into account the three principal stresses at a point, reaches a critical value on the plane ABC as shown in Fig. 1-14(c), i.e., the octahedral plane. The equation representing both of these concepts is:

$$\tau_o = \frac{1}{3}((\sigma_1 - \sigma_2)^2 + (\sigma_2 - \sigma_3)^2 + (\sigma_3 - \sigma_1)^2)^{\frac{1}{2}} \qquad \text{Eq. 1-25}$$

If the critical value of τ_o is obtained from a uniaxial compression test, Equation 1-25 reduces to:

$$\tau_o = \sqrt{2}\, Q_u / 3 = 0.47\, Q_u \qquad \text{Eq. 1-26}$$

For rocks it would be necessary to modify the original form of this theory for it predicts equal uniaxial compression and uniaxial tension strengths. This modification can be done by postulating that the octahedral shear stress at failure is a function of the normal stress, σ_o, on the octahedral plane. The failure criteria, then, is a function of the intermediate principal stress as well as the major and minor principal stresses.

The maximum principal strain theory postulates failure when a critical strain is reached. This theory has been eliminated for most materials as it is not consistent with the results of hydrostatic stress conditions that would produce strain but not failure. However, it is conceivable that such a criterion could be applied to tensile failure in brittle rocks. More research work is required to investigate this possibility.

A maximum shear strain theory has been suggested for sensitive soils (13). Although it might have some applicability to rocks, no empirical studies have been made to indicate its usefulness.

TESTING

There are two aspects of rocks that must be recognized when formulating any testing program. One aspect is the nature of the rock substance or combination of minerals that comprise the basic material. The other aspect is the nature of the rock mass or formation, which includes not only the rock substance but all of the structural features such as joints, faults, bedding, warping and other discontinuities. Paradoxically, whereas for most problems we would like to know the strength characteristics of the rock mass, the majority of our testing techniques are concerned with the rock substance.

Uniaxial Compression Test: The purpose of the uniaxial compression test has been to determine the compressive strength of the ground for engineering purposes. (Some values obtained for particular rocks are shown in Table 2). It has been assumed that failure in compression could then be predicted by comparing the calculated stresses around the opening with the strength that has been tested. This procedure has been based on the assumption that the strength of laboratory samples would be representative of the rock mass. It is being increasingly recognized, aside from the factors discussed below, that structural geological factors normally make this an invalid assumption.

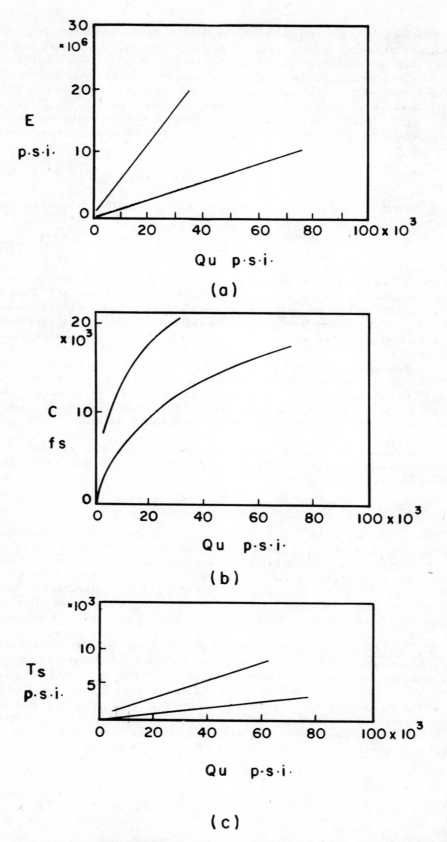

Fig. 1-15 Correlations of Uniaxial Compressive Strength, Q_u, with Modulus of Deformation, E, Seismic Velocity, C_p, and Uniaxial Tensile Strength, T_s

TABLE 2

Uniaxial Compression Test Results on Some Typical Rock Substances

Rock	Uniaxial Compressive Strength (average) - psi	Coefficient of Variation (%)	Modulus of Deformation - psi
Limestone (7)	15,500	21	8.8×10^6
Limestone (7)	4,960	40	5.4
Sandstone (16)	31,100	26	4.3
Shale (16)	5,200	37	3.9
Shale (7)	5,220	53	1.8
Siltstone (7)	3,500	66	1.9
Conglomerate (6)	24,000	30.4	10.7
Quartzite (6)	28,500	29.6	9.6
Granite (7)	21,580	33	5.2
Granite (7)	6,190	27	3.0
Tuff (6)	38,100	30.1	11.1
Tuff (7)	530	21	0.2
Lava (6)	14,700	31.5	9.0
Hornblende Schist (6)	35,400	48.5	12.7
Jasperoid	64,000	7.5	12.8

In addition to this direct use of results, the compression test has also sometimes been considered to provide an indirect measure of other properties. As in concrete work, it is sometimes assumed that the tensile strength, the shear strength, and the modulus of deformation of rocks are proportional to the compression strength. Some studies suggest that these relationships might actually exist (12).

Fig. 1-15(a) shows the possible variations of the modulus of deformation, E, with the uniaxial compression strength, Q_u. Fig. 1-15(b) shows a possible range of the P-wave velocity, C_p, with Q_u. Fig. 1-15(c) shows the type of relationship that seems to apply to the variation of uniaxial tensile strength, T_s, with Q_u. Should these relationships be valid, it would then be possible to determine these other strength properties by uniaxial compression testing (or if more convenient vice versa).

Alternately, compression testing can be used (rather than for design purposes) simply to classify rocks for both strength and deformation properties as well as to provide a rough index of drilling and grinding properties. It is this general purpose that is likely to provide most justification for this type of testing in the future.

As most test specimens come from diamond drill core they are cylindrical in shape. Fortunately, the axial symmetry of the cylinder usually makes it the preferable shape for a test specimen. The remaining variable to be determined, then, is the length to be cut from the core.

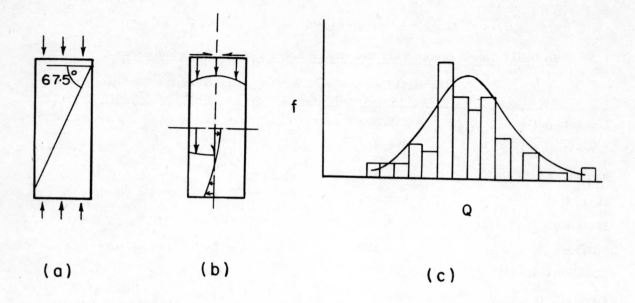

Fig. 1-16 Factors Affecting the Uniaxial Compressive Strength
   (a) Length of Specimen
   (b) Variation of Stresses Throughout the Sample and
   (c) Dispersion of Strengths About a Mean

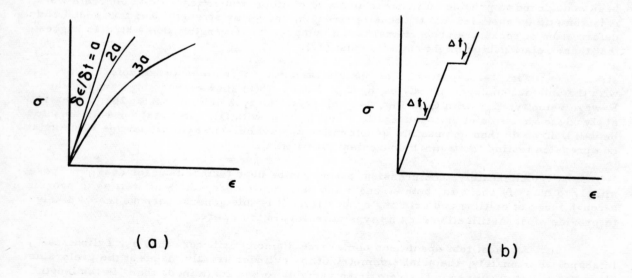

Fig. 1-17 Stress-Strain Curves for Time Dependent Rocks

The ratio of length to diameter of the specimen, L/D, influences the strength of the test sample. Long specimens may fail by elastic instability. Short specimens can eliminate the otherwise preferred plane of failure (see Fig. 1-16(a)). There is a medium range that eliminates both of these possibilities. Samples with an L/D of 2.5 to 3.5 are generally the optimum size for compression testing.

It is reasonable to assume that a uniform compressive stress would exist throughout a sample in a uniaxial compression test. Furthermore, it can be expected that the horizontal stress in such a specimen would be equal to zero. However, after detailed examination it is found that a uniform stress condition very rarely exists in these specimens (3).

The major cause of non-uniformity in the stress field arises from the end conditions of the specimen. For example, if the sample is tested with plane ends between plane steel platens, recent work shows that the compressive stress will vary across the ends with the maximum stress existing at the edges (see Fig. 1-16(b))(3). In addition, horizontal stresses will exist at the ends that may be compressive at the centre of the specimen and may change to tensile towards the edges.

At the mid-height of the specimen the maximum vertical compressive stress is in the centre of the specimen with the horizontal stress often being tensile across the entire width of the specimen. Vertical tension cracking of compression specimens, particularly in the mid-section, is consistent with this stress distribution, although some of the splitting that is observed may be due to wedge action after failure has been initiated at one point. Furthermore, any cracks in the rocks would induce horizontal tension at their boundaries, which could propagate into vertical failure planes (see Fig. 1-13(a)).

As the non-uniform stress distribution arising at the ends of the specimen occurs because of the frictional resistance to lateral expansion of the specimen under longitudinal stress, attempts have been made to eliminate this constraint by use of a lubricant. So far no satisfactory technique of this nature has been established. Lubrication of the ends of the specimen produces horizontal tensile stresses across the entire width of the ends of the specimen as a result of the lubricant being squeezed out towards the edges. In these cases, almost invariably, the method of failure is by vertical splitting.

Another aspect that is important for brittle materials is the effect of minor deviations of the sample ends from being parallel. For example, if a deviation of 0.002 in. exists in a 2 in. long sample, the difference in strain from one side of the sample to the other can be 0.001 in./in. In a hard, brittle rock this could represent a difference in stress of as much as 15,000 psi. Failure could thus be initiated on the side with the higher local stress, but at an average stress less than the strength of the material.

The above example can also be used to point out the effects on the determination of the elastic properties of the sample. For example, if strain were measured on only one side of the sample, the modulus of deformation thus determined could easily be 100 per cent in error. In addition, if two strain gauges at different locations, one vertical and one horizontal, were used to determine Poisson's ratio, the number that would be calculated could easily be greater than 0.5, which has occurred and puzzled many workers.

As mentioned above, it has been shown by experimental work that brittle specimens often decrease in strength with an increase in volume. The increasing probability of large flaws in the specimen with increasing volume probably explains this type of variation in strength. Thus, it could be expected that with sufficiently large samples, so that flaw distribution would not be affected (i.e., flaw size would have a limit), strength would not vary with volume. This has been found to be the case in some studies. In addition, the coefficient of variation for the distribution of test results has been found to decrease with an increase in size of the specimen; this also supports the concept that strength is affected by flaws in the material and that these flaws have a maximum size.

In testing the strength of any material, it must be recognized that the stress at failure is not likely to be as reproducible as either the modulus of deformation or the yield point (where a distinctly ductile material such as steel is being tested). It can be easily imagined that slight imperfections, weaknesses, or inhomogeneities can initiate failure; then, with the shifting of load to other parts of the material, failure progresses throughout the sample. Thus, a dispersion of values is obtained, which is found even for processed, homogeneous materials such as concrete and steel (see Fig. 1-16(c)). It is to be expected, although it is not always so, that greater dispersion would occur for a material such as rock. Consequently, it is important that a series of tests be run on any one rock with the results expressed in the statistical terms of mean, standard deviation (or coefficient of variation, which is the standard deviation expressed as a percentage of the mean), and possibly the range of values.

When safety factors are calculated, it should be clear whether these apply to a mean value of strength or to some other version of strength. For example, one concrete specification in use in heavy construction requires that 80 per cent of the test cylinders should provide strengths greater than the required specified strength (11). The safety factor is then related to the specified strength to obtain the permissible stress.

The relationship of this 'specified strength' to the average strength, assuming a normal frequency distribution of results, depends upon the coefficient of variation of this distribution. Based on the mathematics of probability, if the coefficient of variation were 20 per cent, then the 'specified strength' would be 83 per cent of the average strength (see Fig. 4-12(b)); if the coefficient of variation were 30 per cent, then the specified strength would be 75 per cent of the average strength. Thus, a safety factor that was computed to be equal to 4 based on the specified strength would be equal to about 5, for the above coefficients of variation, when related to the average strength.

By measuring the longitudinal deformation of the compression sample, the modulus of deformation can be determined. In some rock substances the amount of strain produced is not only dependent on the stress but also on time. In these cases the modulus of deformation that would be determined would vary with the rate of stress or strain applied during the test (see Fig. 1-17(a)).

Expressed mathematically, the constant stress-rate test on a Kelvin-type body would have a stress-strain equation as follows (4):

$$\epsilon = \frac{\sigma}{E} - \frac{n\sigma'}{E^2}\left(1 - e^{-Et/n}\right) \qquad \text{Eq. 1-27}$$

where ϵ is strain, σ is stress, σ' is the rate of change of stress, t is the duration of the test from the start until the stress equals σ, n is the coefficient of viscosity and E is the modulus of deformation. It can be seen from this equation that such a material would not produce a straight line unless t were equal to zero.

When a rock substance is encountered with distinct creep properties, it may be of value to conduct a test for the determination of the coefficient of viscosity at various stress levels. This can be done by using a stress controlled test and by applying a certain increment of stress for a certain time interval. A strain-time curve such as shown in Fig. 1-6(f) would then result. With such a curve it is then possible to extrapolate for the asymptote from which E can be obtained and then Equation 1-7 can be solved for n. The stress-strain curve in such a test then looks as shown in Fig. 1-17(b).

<u>Triaxial Compression Test</u>: To determine the strength parameters according to Mohr's strength theory and to suppress any horizontal tensile stresses that occur in the uniaxial compression test, the triaxial compression test can be used. A typical apparatus is shown in Fig. 1-18(a). Here the sample is enclosed in a heavy rubber membrane. A lateral stress can be applied through fluid under pressure acting on the outside of the

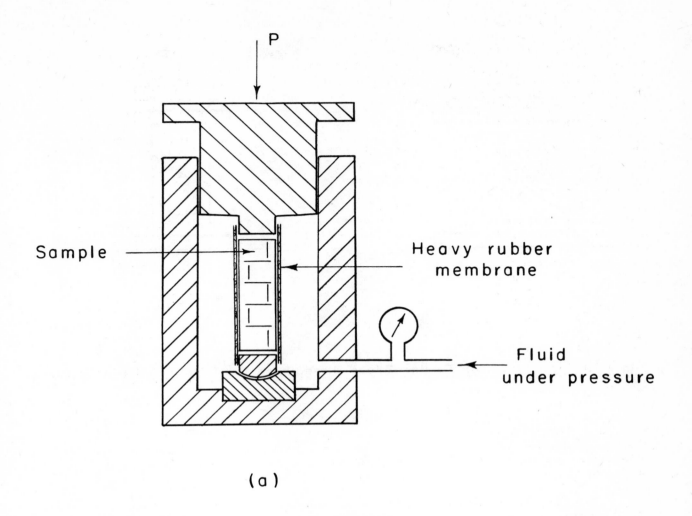

Fig. 1-18 Triaxial Compression Test

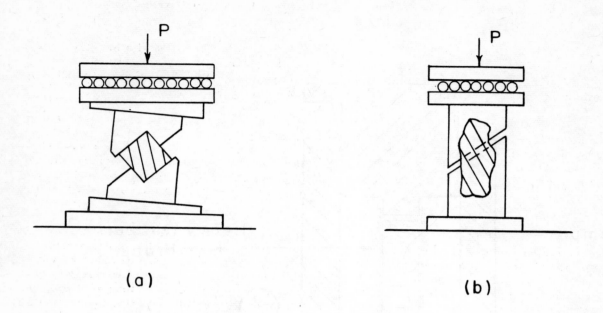

Fig. 1-19 Apparatus for Testing Shear Strength on Pre-Determined Planes (5)

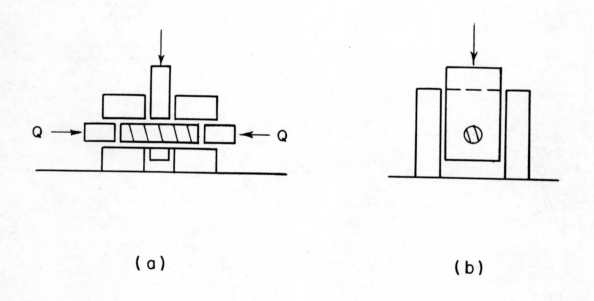

Fig. 1-20 Direct Shear Test

membrane and inside a heavy cylindrical vessel. The longitudinal stress is then applied directly to the specimen through a piston.

Failures in triaxial samples are generally in accordance with Mohr's theory, i.e., the angle of fracture is at $(45 + \varphi/2)$ to the major principal plane as shown in Fig. 1-18(b). However, preferential planes of weakness, such as in-filled joints and laminations of various types, can also influence the actual failure plane.

Where such planes are a distinct part of the rock substance, triaxial tests can be run for the purpose of determining the strength properties (c and φ) that apply to these planes. In this case, as shown in Fig. 1-18(c), a wedge analysis can be made, by knowing the major principal stresses and the angle of the weak plane, to determine the normal and shear stresses acting on this plane. A series of (σ, τ) points are in this way obtained, which can be plotted on a Mohr diagram to produce a failure envelope applicable to these weak planes.

Shear Tests: Other tests have been devised for the examination of the strength properties on specific planes. In Fig. 1-19(a) an apparatus is shown that predetermines the angle of failure that is to be produced in the specimen (5). By using different wedges under the jig holding the sample, different combinations of normal and shear stresses on the plane of failure can be obtained. With this series of (σ, τ) points a Mohr envelope can be plotted for the samples. This test can be used for examining the strength properties on specific planes of interest or simply as an alternative to the normal triaxial test for samples without particular planes of weakness.

In Fig. 1-19(b) another testing technique is shown for achieving the same results (5). In this case a gasket is placed around the specimen either at a plane of weakness that is to be examined or simply in a plane with a predetermined angle to the major principal plane. The sample is then encased in cement, which is placed in the compression testing machine with rollers between the upper platten of the machine and the top plate on the specimen to accommodate the lateral deformation that will occur on failure of the specimen. In this test, by varying the orientation of the predetermined plane of failure, a series of (σ, τ) points can be obtained so that a Mohr envelope can be plotted.

The direct shear test shown in Fig. 1-20 is another technique that can be used to determine the Mohr failure envelope. In this test the normal stress on the plane of failure is controlled by the axial force Q. The shear stress is controlled by the transverse force P. By varying the normal stress and determining the shear stress at failure, a series of (σ, τ) points can be obtained for plotting a Mohr envelope. The results of this test have been found to be more reproducible than those of the triaxial test (6). This is probably due to the elimination of the effects of any planes of weakness that might exist in the samples.

There are two unsatisfactory aspects of the direct shear test. The machining of the samples is important but time consuming. In addition, the actual, as opposed to the assumed, stress distribution on the plane of failure is complex and indeterminant. However, in the present state of rock mechanics the simplicity and cheapness of the method as well as the reproducibility of the results provide good reasons for continuing to use this test.

Tension Test: The uniaxial tension test has been used to determine tensile strengths of rock samples. This test is not very satisfactory, particularly for brittle rocks, owing to the difficulty of eliminating any eccentricity of loading. In addition, the concept of the weak link probably has more applicability to this test than in most of the other tests. For these reasons, it is not used very much either for research purposes or for practical applications.

Other tests have been used for obtaining a measure of the tensile strength of the rock substance. The common test, which has been used for materials similar to rocks such as concrete, is the beam test. The tensile strength that is deduced from this test

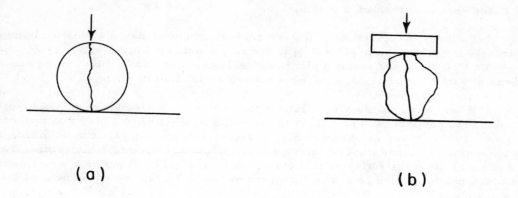

Fig. 1-21 Tensile Splitting from Compressive Forces

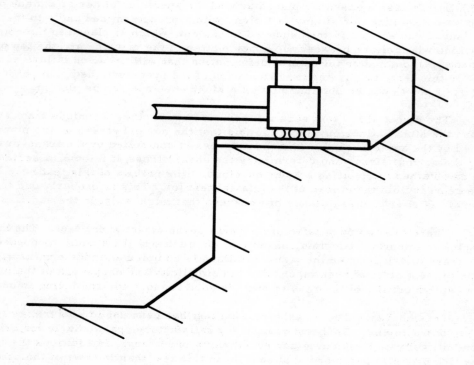

Fig. 1-22 In situ Shear Test

often is from two to three times greater than the average strength determined from the uniaxial tension test on the same rock.

There is some evidence that, when a four point beam test is used (i.e., the load applied at two points with reactions at the two extremities of the beam), the length along the beam of the extreme fibre subjected to the maximum stress can influence the maximum stress at failure. Again, the weak link concept suggests that, as the length of extreme fibre subject to the maximum bending stress increases, the stress at failure should decrease.

Other indirect tests of tensile strength have been devised. The Brazilian test, as shown in Fig. 1-21(a), consists in loading diametrally a disc of rock (5). The result of such a loading is to produce a more or less uniform tensile stress over the major part of the vertical diameter. When this stress produces failure, it becomes a measure of the tensile strength of the rock. The equation for calculating the average tensile stress at failure is:

$$T_s = \frac{2P}{\pi DL} \qquad \text{Eq. 1-28(a)}$$

where P is the external load, D the diameter of the sample and L its length.

In the Brazilian test, before failure occurs it is not uncommon for some crushing to have taken place at the points of application of the load, in which case the elastic stress distribution no longer necessarily applies. Wedging action from the crushed zones can then be an important mechanism in determining the magnitude of the horizontal tensile stress. It has been suggested that this complication can be eliminated by applying the load on an arc of $\tan^{-1} 0.125$ (18), but without inducing any tangential stresses such as could result from using a loading pad of yielding material such as copper.

Alternatively, an empirical equation can be used to take into account this crushing action (19):

$$T_s = \frac{0.79 P}{(D - 1.7 P/Q_u)^2} \qquad \text{Eq. 1-28(b)}$$

where Q_u must be in psi, P in lbs and D in inches.

A test similar to the Brazilian test, as shown in Fig. 1-21(b), is conducted on rough chunks of rock with the only specification being that the L/D should be equal or less than 1.5 (5). A strength number is obtained by dividing the failure load by the volume raised to the 2/3 power.

Although the coefficient of variation of the dispersion of the test results on these rough samples is higher than that for carefully prepared samples, it is felt by those who use the test that the increased number of tests required to obtain a satisfactory mean value is still cheaper than the effort required to shape the smaller number of samples that would otherwise be required for alternate tests.

In Situ Shear Test: Recognizing that the geological structural features of the rock mass are likely to make the in situ strength characteristics different than those obtained on small samples in the laboratory, attempts have been made to measure the strength of a rock mass by in situ shear testing. Such tests normally require deep trenches around exposed slices of rock, which are then loaded to provide both normal and shear stresses on the ultimate plane of failure.

If such a test is done underground, a configuration as shown in Fig. 1-22 can be used. In this case the reaction that is provided by the roof of the opening eliminates the otherwise significant expense of providing a very large mass for such a reaction when

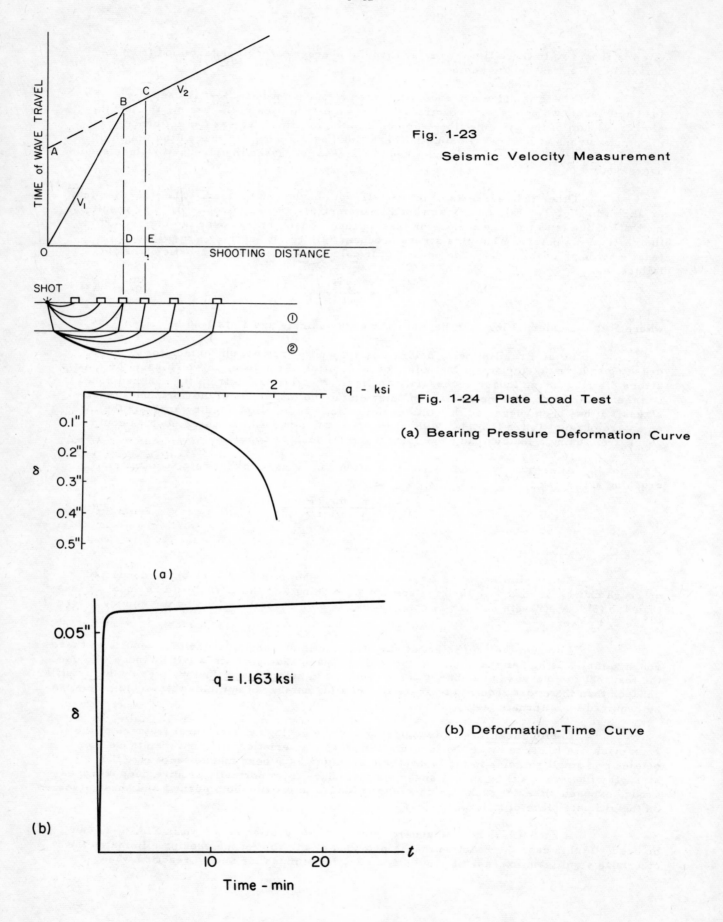

Fig. 1-23 Seismic Velocity Measurement

Fig. 1-24 Plate Load Test

(a) Bearing Pressure Deformation Curve

(b) Deformation-Time Curve

conducting the test on the surface. The configuration of Fig. 1-22 also has the advantage that both normal and shear stresses are applied on the ultimate plane of failure with only one jack being used.

In certain circumstances these tests are quite valuable for obtaining some information on the in situ strength properties. However, each test is very expensive, many tests should be run to obtain mean values and dispersions, and in many circumstances the results are still not a completely valid measurement of the in situ strength properties of larger rock masses.

In Situ Seismic Velocity Test: The seismic method of geophysical exploration requires the measurement of seismic velocities through the various layers involved in the exploration work. These velocities for rocks generally vary between 3000 and 20,000 fs and are theoretically related to the modulus of deformation of the rock mass. The theory of this relationship will be examined in Chapter 8; however, the resulting equations are:

$$C_p = \left\{ \frac{E g (1 - \mu)}{\gamma (1 - \mu - 2\mu^2)} \right\}^{\frac{1}{2}} \qquad \text{Eq. 1-29}$$

$$C_s = \left\{ \frac{G g}{\gamma} \right\}^{\frac{1}{2}} \qquad \text{Eq. 1-30}$$

where C_p is the longitudinal or compression wave velocity, C_s is the transverse or shear wave velocity, E is the modulus of deformation, G is the modulus of rigidity, μ is Poisson's ratio, γ is the unit weight of the rock, and g is the acceleration due to gravity.

The apparatus used in the refraction exploration technique consists of geophone pickups with an oscillograph recorder. A disturbance is created by either exploding a charge of dynamite or, for shallow depths, by hitting the ground with a hammer in a layout as shown in Fig. 1-23. A pick-up at the disturbance records the starting time of the ground wave. The detectors then pick up the arrival time of the waves, which are shown on the oscillograph and are photographically recorded.

These arrival times and the horizontal distance of the detectors from the shot are plotted as shown in Fig. 1-23. The slopes of the resulting curve are then a measure of the velocities through the various layers of ground. The detectors are normally placed at distances up to four times the depth to which the information is required.

By extrapolating the velocity line, V_2, back to the y-axis, or to Point A in Fig. 1-23, an intercept is obtained that is a measure of the depth of layer (1). The Point A is theoretically the time of arrival for a detector placed at the position of the shot. In this case the travel distance is two times the depth of the layer (1). Consequently, the thickness of the layer can be calculated:

$$H_1 = V_1 \times \overline{OA}/2.$$

The above relationships are based on the assumption that the path of the seismic wave is vertically down from the shot point, horizontally through the second or third layer and then vertically up to the detector. More accurate equations can be used; however, the differences in results are normally not significant.

Both for exploration and structural purposes the depth to which this method can give useful information is limited. Consequently, the investigation of the ground for underground openings by this method is generally impractical. Proposed dam-sites and open pit mines are projects on which this type of test might be useful.

Aside from determining the depths to various layers, the determination of the in situ seismic velocities provides some measure of the strength characteristics of the rock mass as opposed to that of individual samples. For example, if the rock substance is hard but is located in a badly jointed formation with possibly some weathering in the joints, the in situ seismic velocity will tend to be low (Table 3). This introduces the ambiguity of whether the material is a weak homogeneous rock or whether the formation is composed of a hard rock substance in a highly jointed mass. Furthermore, if the fissures of a highly jointed formation are filled with water, the velocity will be only slightly affected by the joints. The following table gives some actual, typical values of seismic velocities illustrating the above points (8).

TABLE 3

Some Typical Seismic Velocities

Rock Type	Seismic Velocity
Shale	2900 - 12,800 fs
Sandstone	4700 - 14,000 fs
Granite, Massive	18,500 fs
Granite, Partly Decomposed and Slightly Seamed	10,500 fs
Granite, Highly Decomposed and Badly Fractured	2,200 fs
Granite, Highly Decomposed and Friable	1,500 fs
Limestone	16,400 - 20,200 fs

Some studies have shown that there may be a linear relationship between the uniaxial compressive strength and the modulus of deformation (see Fig. 1-15(a)) similar to that which is assumed to exist in concrete where the compressive strength is often assumed to be approximately 1/1000 of the modulus of deformation (6, 12). There is also the indication that this relationship might vary for different types of rocks, e.g., the uniaxial compressive strength of igneous rocks might be one fraction of their modulus of deformation, whereas sedimentary rocks would be another fraction. If such relationships could be more firmly established, then the measurement of in situ seismic velocities (which could be converted to modulii of deformation) would be a very simple method of obtaining in situ strength characteristics. Already contractors are using seismic velocity as a measure of the ripability without blasting of surface rock.

Laboratory Seismic Velocity Test: The longitudinal seismic velocity of core samples can be measured in the laboratory. The velocity can then be used to calculate the modulus of deformation applicable to dynamic stresses of low magnitude using the formula:

$$E = \rho C_p^2 \qquad \text{Eq. 1-31}$$

where ρ is the mass density, and C_p is the longitudinal seismic velocity.

Alternatively, the velocity determined in the laboratory can be compared with that obtained in the field. If the field value is much lower than the laboratory value, it is possible that jointing, or other structure features, accounts for the reduction.

The laboratory test can be conducted in two ways. One technique is to vibrate longitudinally a sample and measure the frequency, f, at which resonance occurs. Then the longitudinal velocity equals 2fL, assuming the length of the sample, L, at resonance is half the wavelength of the oscillation.

The second technique is to measure the time, t, required for a pulse to travel the length of the sample, L, and calculate the velocity as L/t.

Plate-Load Testing: A plate-load test, as mentioned above if sufficient force can be applied, is another way of obtaining some information on the strength of the rock mass as opposed to the strength of small samples of the substance. The basis of the test is that, if a plate is placed on the surface of the rock and the contact pressure is increased, shear failure of the material will ultimately occur. Failure may be recognized, depending upon the material, either by a definite fracture, by a sudden yielding, or by a gradual continuous yielding. All types of failure are assumed to be governed by the same strength theory as expressed in Equation 1-17.

Other assumptions in this theory are that the bearing pressure is uniformly distributed over the plate and that the contact between the plate and subgrade is rough, giving rise to a frictional resistance to lateral movement. The assumption of roughness is generally valid, and the assumption of a uniform pressure can be obtained by placing a rubber, water-filled bag between the plate and the rock.

Equation 1-17 is valid for a test that is run on a floor or horizontal surface. If the test is run on a wall or vertical ground surface, then the first and third terms no longer contribute to the ultimate bearing capacity. Consequently, the equation reduces to one term, which modified for a square instead of long bearing area is:

$$q_f = 1.3 \, c N_c \qquad \text{Eq. 1-32}$$

Besides simply applying an increasing load to the plate, it is important to control the rate of application of load to insure that all viscous components of deformation are obtained for each increment of load. Failure can be in some rocks by continuous plastic flow at a relatively slow rate. In this case, if a fast rate of load application were used, the test results would over-state the actual strength of the rock mass.

Example - In Fig. 1-24 the results of a plate-load test conducted on the wall of a drift in altered rock are shown (10). Owing to the limitation of the capacity of the jack being used, the size of plate was only 4 in. in diameter. Failure in this test was by plastic yielding, although in other tests in the same drift many of the failures were by sudden, brittle fracture. In this test small pieces of rock started flying off the adjacent face at a bearing pressure of 1.9 ksi with the ultimate bearing pressure being 2.1 ksi. The angle of internal friction had been found from laboratory studies to be 37°.

From Equation 1-19, N_c was determined:

$$N_c = 7 \tan^4(45 + 37/2) = 113$$

From Equation 1-32, c was calculated:

$$c = 2,100/113 = 19 \text{ psi}$$

CLASSIFICATION OF ROCKS

Efficient communication between the various groups concerned with applied problems (e.g., physicists, geologists, engineers, operators, and contractors) is important. For example, a rock may be described as a Cretacous shale with minor quantities of quartz and pyrite. However, some of the principal factors of interest would be: (a) Is it jointed, badly fractured, distinctly layered or massive? (b) Does it fail in a brittle or plastic manner? (c) Is it viscous before failure? (d) Can information be interpolated between bore holes? In other words, the significant properties should be recognized, and the appropriate information passed on to those concerned so that an initial appraisal can be made of any potential problem.

"Classification is the arrangement of things in classes according to the characteristics they have in common." (Encyclopaedia Britannica). Classification can be on the basis of natural, inherent characteristics (e.g., the classification of animals into families, etc.) or on the basis of purpose (e.g., the classification of books into large and small sizes for library purposes). Clearly, the classification of rocks for applied problems should be functional, and the words used for communication between groups should be defined, widely recognized, and related to the use of the information.

In selecting classification properties, most of the applications of rock mechanics should be examined. The principal areas where the subject might be of use are: in the design of slopes; in the excavation of tunnels, shafts, drifts, cross-cuts, and underground power houses; in the design of mine workings (e.g., stopes, rooms and pillars); in the control of ground water; in the design of tunnel and power house linings, rock bolting systems, drift and cross-cut sets, stope supports, and other systems of support; in the design of caving operations (i.e., to achieve the initial cave and efficient drawing, and to minimize subsidence effects); in the design of underground defence installations to resist nuclear explosions (i.e., in analysing the stress and ground motion effects of the dynamic strain pulse together with the design of the geometry of the openings and the required linings); in the design of foundations; in the appraisal of rockburst dangers; in the design of optimum drilling patterns and explosive charges; and in the study of comminution processes.

From this list of potential applications, the most common and easily determinable properties of importance in most of these areas are:

(i) The strength of the rock substance.
(ii) The deformation characteristics of the rock substance both before and after failure.
(iii) The gross homogeneity and isotropy of the formation.

Many other properties will be important for specific problems. However, most projects will require additional test and engineering data, which would not be usefully included in a general classification. A classification system should simply provide information which indicates what types of problem may occur and what types of special investigations or tests should be conducted.

For example, one part of a site (mine, dam, etc.) may be faulted. Then the faults may be - clean, sharp, with no contortion or alteration of the formations - with contortion and/or fracturing of the adjacent formations - with alteration of the adjacent rock - braided or sheared over a considerable width - brecciated in a matrix of gouge or secondary mineralization. Obviously, all such structural details may be important for stability, but could not be included in a simple classification system.

Hitherto the geological classification of rocks according to origin has been the most common method of labelling rocks. It is interesting to examine this classification method from a functional point of view with respect to rock mechanics. Briefly the classification system is as follows:

      (i) Sedimentary: a. mechanical, b. chemical, c. organic.

      (ii) Igneous: a. acid, b. intermediate, c. basic, and
                 i. plutonic, ii. hypabyssal, iii. extrusive.

    (iii) Metamorphic.

A geological classification contains some, although not sufficient, information on the above important rock properties for engineering purposes. If structural information is recorded when the mapping is done, then additional useful information is obtained from this traditional practice.

A more detailed petrological description of the rock substance, as is often required for the geologist's purposes, which would include chemical, mineralogical and petrographic data, does add useful information. The chemical analysis seldom provides information on strength, deformation or continuity. However, the mineralogical data might be of use in judging the probable effects of exposure to the atmosphere, the effect of percolating water, the possibility of wet strength being different from dry strength, and the possibility of swelling.

Similarly, petrographic studies can also give other useful information such as the reason for the deterioration of minerals. For example, the determination of whether alteration has resulted from surface weathering or from hydrothermal action would assist in judging whether the alteration products should be associated with an erosion surface or otherwise. The detection of alteration adjacent to joints could be very significant with respect to the strength properties of the formation. However, such detailed information could never be included in a simple classification system. It would be determined, when warranted, by special testing.

Petrographic analyses, which include information on texture and origin as well as on the minerals, could provide some pertinent information. In an indirect way the magnitude of strength and the nature of deformation properties might be deduced from such analyses, e.g., microfractures detected in quartz crystals in a granite would be significant with respect to the strength of the granite. However, this would normally be an expensive and indirect way to obtain this information when, for example, the direct testing of strength would probably give less expensive and more practical strength information. When it is shown that a rock like granite can have compressive strengths as low as 6000 psi or as high as 50,000 psi, then it must be recognized that a geological classification is incomplete for engineering purposes.

Another rock classification system that was proposed in the past was for the determination of the appropriate kind and amount of support in tunnels (14). This system divided all rocks into seven categories as follows:

  i) Intact rock, i.e., rock containing no joints.

  ii) Stratified rock, i.e., rock that has little strength between the beds.

  iii) Moderately jointed rock, i.e., a rock mass that is jointed but cemented or strongly interlocked so that a vertical wall requires no support.

  iv) Blocky and seamy rock, i.e., a jointed rock mass without any cementing action in the joints and weakly interlocked so that a vertical wall requires support.

  v) Crushed rock, i.e., rock that has been reduced to sand-like particles without having undergone any chemical change.

  vi) Squeezing rock, i.e., rock containing minerals with low swelling capacity.

vii) Swelling rock, i.e., rock containing minerals of high swelling capacity.

This classification system is useful in the design and construction of tunnels, but it has deficiencies for more general use. For example, it gives little information on strength and no information on the failure characteristics.

Another system that has been used for describing in mechanics' terms the rock around underground openings is as follows (15):

(i) Competent rock, i.e., rock that will sustain an opening without artificial support.

  (a) Massive-elastic, i.e., homogeneous and isotropic

  (b) Bedded-elastic, i.e., homogeneous, isotropic beds with the bed thickness less than the span of the opening and little cohesion between the beds

  (c) Massive-plastic, i.e., rock that will flow under low stress.

(ii) Incompetent rock, i.e., rock which requires artificial supports to sustain an opening.

This classification system gives some information on the relative strength of the rocks, but the information is related to the size and shape of the particular opening. This means that different classifications can exist for the same rock mass. No information is included on the failure characteristics or on the structural aspects, e.g., jointing within the formation. It does contain some information on layering, but this is restricted to sedimentary rocks rather than including the banding of metamorphic rocks and dikes and sills. In addition, this type of strength classification has little pertinence for slopes, foundations, or drilling.

Another system of classifying different types of ground uses a series of deformation models (16). The purpose of this classification system is for idealizing various types of ground for the special theoretical studies in ground dynamics. Some of the models included in this work are as follows:

(i) Linear elastic body, i.e., a body with a straight line, reversible stress-strain curve.

(ii) Curvilinear elastic body, i.e., a body with a curved, reversible stress-strain curve.

(iii) Bi-linear elastic body, i.e., a body with a reversible stress-strain curve comprising two straight lines.

(iv) Elasto-plastic body, i.e., a body with a stress-strain curve composed of a sloping straight line connecting with a horizontal line indicating plastic deformation.

(v) Plasto-elastic body, i.e., a body with a stress-strain curve composed of an initial horizontal line connecting to a straight, sloping line.

(vi) Visco-elastic body, i.e., a body whose deformation varies both with the level of stress and the duration of the stress but the deformation is fully recoverable.

(vii) Visco-plastic body, i.e., a body whose deformation varies with stress level and duration and is not fully recoverable.

(viii) Locking medium, i.e., a mass with stress-strain curve composed of an initial, horizontal straight line connecting to a vertical straight line.

This system is very useful for advanced stress distribution studies and could provide a basis for differentiating patterns of deformation. However, it contains no information on strength or structural aspects. Furthermore, the concepts involved in this system are well in advance of any simple testing techniques that could be used to supply the required empirical information.

As can be seen, none of these systems of classifications is satisfactory for engineering purposes. Ideally, such a system should indicate for the in situ rock mass its strength, deformation characteristics and continuity. At the present time, it is very difficult to test for the strength and deformation properties of the rock formation. Thus, the best that can be done on a routine basis is to establish for the substance, generally from drill cores, its strength and deformation properties.

By knowing these properties for the rock substance some pertinent information is obtained for the formation. For example, if the substance is weak then certainly the formation will be weak. On the other hand, if the substance is strong the formation may be strong or the formation may be weak, but it would be known that this weakness arises from structural causes.

A practical system of classification at the present time would be as follows:

Substance: STRONG or WEAK
ELASTIC or PLASTIC

Formation: MASSIVE, LAYERED, BLOCKY, BROKEN.

If a geological name can be easily established, it should be used; otherwise the classification would just be describing a rock. For example, a classification of one rock could be as follows:

Substance: STRONG, ELASTIC LIMESTONE
Formation: LAYERED and BLOCKY.

Another case would be:

Substance: WEAK, PLASTIC ROCK
Formation: MASSIVE.

The strength of the rock substance could be defined as STRONG when the uniaxial compression strength is greater than 10,000 psi, or some other test might be used with an appropriate division line. Below this value the rock substance would be described as WEAK. In view of the limited use for analytical purposes that can be made of the strength of the substance, the two divisions of STRONG and WEAK should be adequate for classification. Where useful, the system could be expanded to form additional categories by calling rock substances VERY STRONG when their compressive strengths are greater than 25,000 psi and, similarly, describing as VERY WEAK those rock substances with compressive strengths less than 5,000 psi.

Other terms have been used to describe strength properties, but they are somewhat ambiguous. For example, the terms "hard" and "soft" not only imply relative strengths but also durability and drillability. Similarly, the terms "competent" and "incompetent" are often used; however, these terms also imply certain characteristics of stiffness and brittleness.

The classification of strength could also be in terms of the basic strength parameters of cohesion and angle of internal friction. In this case, classification would require triaxial testing, which, considering the initial use of the information, would be too expensive simply for the purposes of classification. Furthermore, many of the practical problems that are concerned with strength involve a configuration where the rock is in an unconfined condition; consequently, the uniaxial compression strength is of most concern.

The classification of the deformation properties of the rock substance should provide some information on the possibilities of creep, swelling, the amount of stored strain energy, and the nature of failure. An elastic rock could be expected not to creep or swell; it would also be expected to store a relatively large amount of the strain energy so that on failure, which would probably be by brittle fracture, this energy would be released explosively. The term PLASTIC could be used to describe those rock substances where more than 25 per cent of the total strain at any stress level is irrecoverable. Among these materials some would creep, some would swell on exposure, some would fail by yielding rather than rupturing, and even those that ruptured would not do so with the explosive violence associated with the previously described group. A subdivision in this group might be for those rock substances that give increases in strain that vary with time while the stress remains constant, i.e., they could be called VISCOUS.

It should be appreciated, however, that the classifying of a rock as ELASTIC or PLASTIC would not establish the post-failure deformation characteristics under all circumstances. For example, the properties of rock after yielding or rupture can be different in the uniaxial test from that experienced in the triaxial test. Furthermore, the formation containing a brittle rock substance could give a plastic reaction. For example, in a brittle rock mass a relaxed zone can be created around an opening, which can supply some back pressure. As the rock in this relaxed zone produces some permanent strain, it can be described as a plastic ring inside the highly stressed elastic zone.

The continuity of the formation can be described as MASSIVE when the distances between layers and joints are more than six feet. Layering in a mechanic's sense should mean that the bonding between layers is less than within any one layer.

The structure of the rock mass is described as BLOCKY when the joint spacing is greater than one foot and less than six feet and as BROKEN when the individual blocks are smaller than one foot. An additional category of VERY BROKEN could be added for those cases where the fragments are smaller than three inches.

There would, of course, be additional information that would be required for any specific project. For example, the details of the geometry of the formation - strike and dip of layers, joints and faults, as well as the existence of anticlines or synclines - would often be a minimum requirement. Porosity and permeability would be required for gas and oil work. The presence of altered zones, the nature of the joint infilling material, and the shear strength of the in situ rock could be important for other projects. However, these items must be obtained when the study of the project is being made and not when the initial classification is being established.

REFERENCES

1. Obert, L. and Duvall, W., "The Microseismic Method of Predicting Rock Failure in Underground Mining", USBM RI, 3803 (1945).

2. Terzaghi, K., "Theoretical Soil Mechanics", Wiley (1943).

3. Trumbachev, V., "Distribution of Stress in Inter-Room Pillars and Immediate Roofs", Gosgortekahizdat Moscow (1961).

4. Coates, D., Burn, K. and McRostie, G., "Strain-Time-Strain-Strength Relationships in a Marine Clay", Trans EIC, Vol. 6 No. A-11 (1963).

5. Protodyakonov, M., "Methods of Studying the Strength of Rocks, Used in the U.S.S.R.", Internat. Symp. on Mining Res., Vol. 2, Univ. of Missouri, Pergamon (1961).

6. Coates, D., Udd, D. and Morrison, R., "Some Physical Properties of Rocks and their Relationships to Uniaxial Compressive Strength", Proc. Rock Mech. Symp., McGill University, Mines Branch (1962).

7. U.S. Bureau of Reclamation, "Physical Properties of Some Typical Foundation Rocks", Concrete Lab. Rpt. No. SP-39 (1953).

8. U.S. Corps of Engineers, "Sub-surface Investigation Geophysical Explorations", Engineering Manual, Civil Works Construction, Part 118, Chap. 2 (1948).

9. Brace, W., "Brittle Fracture of Rocks", GSA Proc. Internat. Cfce. State of Stress in the Earth's Crust, Rand Corp., Santa Monica (1963).

10. Coates, D., Helliwell, J. and Gyenge, M., unpublished work.

11. "Concrete Manual", U.S. Bureau of Reclamation, U.S. Govt. Printing Office, Washington, D.C. (1956).

12. Judd, W. and Huber, C., "Correlation of Rock Properties by Statistical Methods", Internat. Symp. Mining Res., Univ. of Missouri, Pergamon (1962).

13. Coates, D. and McRostie, G., "Some Deficiencies in Testing Leda Clay", ASTM Spec. Tech. Pub. No. 361 (1964).

14. Terzaghi, K., "Introduction to Tunnel Geology" in "Rock Tunnelling with Steel Supports", by Proctor, R. and White, T., Youngstown Printing, Youngstown, Ohio (1946).

15. "A Study of Mining Examination Techniques for Detecting and Identifying Underground Nuclear Explosions", US Bureau of Mines IC 8091 (1962).

16. Hardy, H.R., unpublished work.

17. Morrison, R. and Coates, D., "Soil Mechanics Applied to Rock Failure in Mines", Trans CIMM, Vol. 58, p. 401 (1955).

18. Fairhurst, C., "On the Validity of the Brazilian Test for Brittle Materials", Internat. J. Rock Mech. and Mining Sc., Vol. 1, No. 4, p. 535 (1964).

19. Paone, J. and Bruce, W., "Drillability Studies", US Bureau of Mines RI 6324 (1963).

20. Ito, I. et al., "Rock Behaviour for Tension under Impulsive Load using Detonator's Attack", Trans. MM Alumni Assoc. Kyoto Univ., Vol. 15, No. 2, p. 61 (1963).

CHAPTER 2

ELASTIC PROTOTYPES

INTRODUCTION

The common procedure in any applied mechanics subject is to make a sufficient number of assumptions so that the system that is being examined is simple enough to be able to be analysed mathematically. For example, in fluid mechanics it is common to assume for some systems that liquids are incompressible and then simply to apply Newton's equations of motion. By making this assumption, a wide range of flow problems can be solved quite satisfactorily. On the other hand, the actual compressibility of the fluid is of importance when sudden closures occur that produce water hammer. For this latter problem another set of simplifying assumptions is made to permit the problem to be analysed.

Similarly, in structural engineering it is common practice to assume that beams are either simply supported or fixed, when actually they are most often partially fixed, owing to the nature of the real connections as opposed to the theoretically assumed connections. However, simple theory in this field has been eminently useful for over a hundred years and is only now being refined to include some aspects of actual behaviour not formerly recognized in the theoretical solutions.

Of course, good engineering practice has always taken into account the factors that have not been analysed by exercising judgment on their possible effects when making final decisions. The theoretical analysis, together with empirical data and intuitive appraisal of intangible factors, are all components entering into the solution of a technical problem.

In rock mechanics there are several simple models or analogues of common problems that can be used as a first approximation to determine the consequences of creating certain excavations or of applying certain forces. By using these simple theoretical models, the functional relations between loads, material properties, and stress or deformation can be established. The consequences of actual deviations from the simple representation can then be either determined by judgment or by gathering empirical data.

In most cases it is useful to make different sets of assumptions so that more than one theoretical model of the actual situation can be analysed. For example, the roof of an underground opening in stratified rock can be considered as a beam and approximate calculations made for either stress or deformation. Alternatively, the roof can be considered to be the inside surface of a thick-walled cylinder with an infinite outside radius. Another set of calculations can then be made for either stress or deformation. It is possible that, when two analogues are used, one is more valid for a stress at one location, whereas the second is more valid for the stress at another location.

The time spent on any such analyses is not great, and the benefit, even if the only result is to educate one's judgment, is usually far greater than the expense involved. Consequently, anyone concerned with analysing rock mechanics problems should be familiar with the mathematics of the following simple theories.

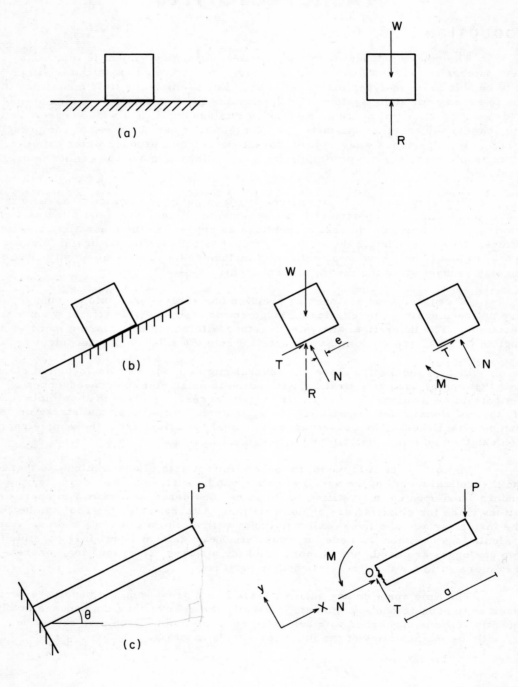

Fig. 2-1 Construction of Free Body Diagrams

EQUILIBRIUM

All rigid bodies at rest are in a state of equilibrium and thus must obey the three equations of equilibrium:

$$\sum F_x = 0 \qquad \text{Eq. 2-1(a)}$$

$$\sum F_y = 0 \qquad \text{Eq. 2-1(b)}$$

$$\sum M = 0 \qquad \text{Eq. 2-1(c)}$$

These equations provide one source of information for the determination of unknown forces. The equations simply state that the algebriac sum of the external forces on a body in equilibrium in any direction must be equal to zero and that the sum of the moments of the external forces on this body about any point must also be equal to zero.

In mechanics of materials, internal forces are generally of predominant interest. To be able to use the equations of equilibrium, these internal forces must be converted into external forces. A construction called the drawing of a Free Body Diagram is very useful in achieving this purpose.

The rule in drawing a Free Body Diagram is that, whenever a contact with another body is eliminated, the contact must be replaced by a force. If nothing is known about the force, it must be drawn at an unknown angle and at an unknown point, i.e., it is drawn at an angle θ, which is unknown, and at a distance x from a known point. However, the direction and the point of application of the internal force are often known from recognizing certain obvious requirements.

In addition to the forces on a body arising from contacts with other bodies, there are body forces that must be included in the Free Body Diagram, e.g., if a body has mass, a gravity force will be acting on it while it is in a gravitational field; similarly, iron and nickel bodies will be subjected to a magnetic body force when placed in a magnetic field.

In Fig. 2-1(a) a block is sitting on a table. To draw a Free Body Diagram of the block, the contact with the table is removed and replaced by a force. In this case, it is known that the force of the table on the block must act vertically upwards through the centre of gravity of the block; hence, R is drawn. Then, as the block has mass and is in a gravitational field, there will be a force W acting through the centre of gravity of the body.

If the table is not horizontal, as shown in Fig. 2-1(b), the Free Body Diagram of the block will have a force R acting on it at some unknown location and possibly (depending upon the analyst's perception) in an unknown direction. It is often convenient to replace the force R by orthogonal components, T and N, acting tangential and normal to the surface of the body. Similarly, it is often convenient to apply these components at the centre of gravity of the face on which they are acting. In this case the effects of the actual force acting at some distance e from this centre point is shown by a moment M.

In Fig. 2-1(c) the internal forces acting in this beam can be determined by drawing a Free Body Diagram of any part of the beam. When this Free Body Diagram has been drawn, the internal forces must balance the external forces and thus are calculated

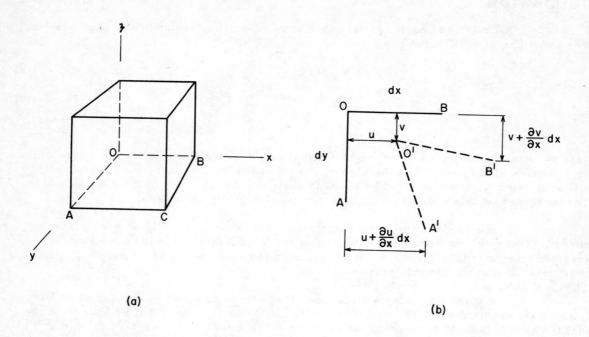

Fig. 2-2 Strain at a Point

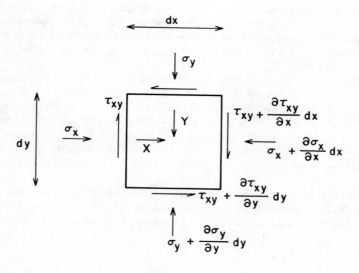

Fig. 2-3 Stress at a Point

by applying the three equations of equilibrium:

$$\sum F_x = N - P \sin \theta = 0$$

$$\sum F_y = T - P \cos \theta = 0$$

$$\sum M_o = M - Pa \cos \theta = 0$$

Hence,

$$N = P \sin \theta$$

$$T = P \cos \theta$$

$$M = Pa \cos \theta$$

When these internal forces have been calculated, the stresses and deformation resulting from them can be analysed.

PLANE STRESS AND PLANE STRAIN

These simple laws can be used to examine deformation relations in a stress field where stress gradients exist. In Fig. 2-2, AOB refers to an angle of an infinitesimal element in a mass before the stress is applied to the mass. A'O'B' represents this angle after the stress field is applied. The displacement of O in the x-direction is $\delta_{ox} = u$, in the y-direction is $\delta_{oy} = v$, and in the z-direction is $\delta_{oz} = w$. It then follows that:

$$\delta_{Ax} = u + (\partial u / \partial x) dx$$

and

$$\delta_{By} = v + (\partial v / \partial y) dy$$

The change in length of OB to O'B' is thus $(\partial u / \partial x) dx$, from which it follows that the strain equations are:

$$\epsilon_x = \partial u / \partial x$$

$$\epsilon_y = \partial v / \partial y$$

$$\epsilon_w = \partial w / \partial z.$$

As the change in the angle AOB is equal to the shear strain, γ_{xy}, it follows that:

$$\gamma_{xy} = \{v + (\partial v / \partial x) dx - v\} / dx + \{u + (\partial u / \partial y) dy - u\} / dy$$

$$\gamma_{xy} = \partial v / \partial x + \partial u / \partial y$$

Similarly,

$$\gamma_{xy} = \partial u / \partial z + \partial w / \partial x$$

$$\gamma_{yz} = \partial v / \partial z + \partial w / \partial y$$

The general three-dimensional stress field can then be limited to the special cases of plane stress and plane strain. By plane stress we mean that the forces that are applied to a body are all within one plane, as would be the case for forces applied to the edges of the plate and in the plane of the plate; it follows that σ_x, τ_{xz} and τ_{yz} are all zero and that the stress field is characterized only by σ_x, σ_y and τ_{xy}. By the term plane strain, we mean the case of a long body in the z-direction with the ends fixed between smooth, rigid abutments so that the deformation in the z-direction is zero and the forces that are applied to the body are all perpendicular to the z-direction and do not vary with z so that each section is the same; it follows that the ϵ_z, γ_{yz}, γ_{xz}, τ_{yz} and τ_{xz} are all equal to zero.

For the case of plane stress, the general stress equations can be established by summing the forces in the x-direction in Fig. 2-3, which include the body forces X and Y.

$$\sum F_x = \sigma_x dy - \{\sigma_x + (\partial \sigma_x/\partial x) dx\} dy + \tau_{xy} dx -$$

$$\{\tau_{xy} + (\partial \tau_{xy}/\partial y) dy\} dx + X dx dy = 0$$

$$= (\partial \sigma_x/\partial x) dx dy - (\partial \tau_{xy}/\partial y) dx dy + X dx dy = 0$$

$$\partial \sigma_x/\partial x + \partial \tau_{xy}/\partial y - X = 0 \qquad \text{Eq. 2-2(a)}$$

Similarly,

$$\partial \sigma_y/\partial y + \partial \tau_{xy}/\partial x - Y = 0. \qquad \text{Eq. 2-2(b)}$$

From the strain equations above, the following differentiation can be made:

$$\partial^2 \epsilon_x/\partial y^2 = \partial^3 u/\partial x \partial y^2$$

$$\partial^2 \epsilon_y/\partial x^2 = \partial^3 v/\partial x^2 \partial y$$

$$\partial^2 \gamma_{xy}/\partial x \partial y = \partial^3 u/\partial x \partial y^2 + \partial^3 v/\partial x^2 \partial y$$

$$\therefore \partial^2 \epsilon_x/\partial d^2 + \partial^2 \epsilon_y/\partial x^2 = \partial^2 \gamma_{xy}/\partial x \partial y.$$

In the case of plane stress the following relations exist between stress and strain:

$$\epsilon_x = (\sigma_x - u \sigma_y)/E$$

$$\epsilon_y = (\sigma_y - u \sigma_x)/E$$

$$\gamma_{xy} = \tau_{xy}/G = 2(1+u)\tau_{xy}/E$$

$$\partial^2(\sigma_x - u\sigma_y)/\partial y^2 + \partial^2(\sigma_y - u\sigma_x)\partial x^2 = 2(1+u)\partial^2 \tau_{xy}/\partial x.\partial y$$

From Equations 2-2(a) and 2-2(b) it follows:

$$2\partial^2 \tau_{xy}/\partial x \partial y = -\partial^2 \sigma_x/\partial x^2 - \partial^2 \sigma_y/\partial y^2 + \partial X/\partial x + \partial Y/\partial y \qquad \text{Eq. 2-3}$$

By inserting into Equation 2-3 it follows:

$$(\partial^2/\partial x^2 + \partial^2/\partial y^2)(\sigma_x + \sigma_y) = (1+u)(\partial X/\partial x + \partial Y/\partial y) \qquad \text{Eq. 2-4}$$

This is known as the compatibility equation in terms of stresses for the case of plane stress.

For the case of plane strain the relationships between stresses and strains are as follows:

$$\epsilon_z = \left(\sigma_z - \mu(\sigma_x + \sigma_y)\right)/E = 0$$

$$\sigma_z = \mu(\sigma_x + \sigma_y)$$

$$\epsilon_x = \left\{\sigma_x - \mu(\sigma_y + \sigma_z)\right\}/E$$

$$= \left\{\sigma_x - \mu\left[\sigma_y + \mu(\sigma_x + \sigma_y)\right]\right\}/E$$

$$= \left\{\sigma_x(1-\mu^2) - \mu\sigma_y(1+\mu)\right\}/E$$

Similarly,

$$\epsilon_y = \left\{\sigma_y(1-\mu^2) - \mu\sigma_x(1+\mu)\right\}/E$$

Then, proceeding as above for the case of plane stress, it follows that the compatibility equation for plane strain is:

$$(\partial^2/\partial x^2 + \partial^2/\partial y^2)(\sigma_x + \sigma_y) = (\partial X/\partial x + \partial Y/\partial y)/(1-\mu) \qquad \text{Eq. 2-5}$$

By comparing the two compatibility equations, it can be seen that, with the presence of body forces, Poisson's ratio, μ, influences the stress distributions. Without body forces, none of the elastic constants influence the stress distribution and, furthermore, the compatibility equations for both plane stress and plane strain are equal.

Another useful observation can be made by examining the equations relating stress and strain for the two cases. It can be seen that, if in the equations for the case of plane stress E is converted into $E/(1-\mu^2)$ and μ is converted into $\mu/(1-\mu)$, the equations applicable to plane strain are obtained.

THICK-WALLED CYLINDERS

To approach the subject of stresses around underground openings, the relatively simple case of the thick-walled cylinder can be analysed. First, the analysis of stress in a thin-walled cylinder for plane stress as shown in Fig. 2-4(a) is as follows:

$$2t\sigma = p_i d$$

or

$$\sigma = \frac{p_i d}{2t}$$

where σ is the average stress in the walls of the cylinder and t is the thickness of the material in the cylinder. The underlying assumption in this analysis is that the thickness of the cylinder is relatively small compared to its diameter; consequently, the stress in the wall of the cylinder is very nearly uniform from the inside surface to the outside surface.

In the case of the thick-walled cylinder where the thickness is either of the same order of magnitude or much larger than the inside diameter, the assumption of uniformity of stress would be grossly inaccurate. Fig. 2-4(b) shows the stress distribution in the walls of a thick-walled cylinder subjected only to internal pressure (i.e., the outside pressure, p_o, is zero). The tangential stress varies from a maximum at the inside surface to a minimum

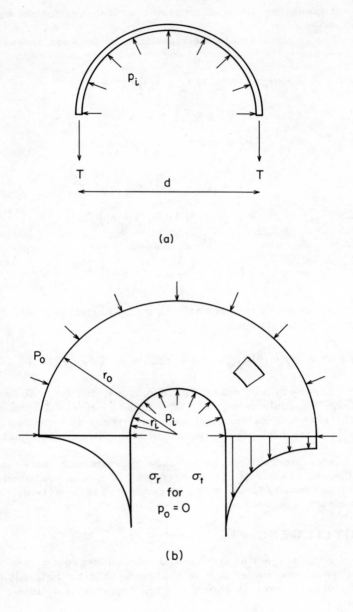

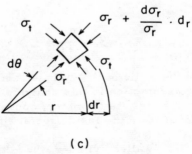

Fig. 2-4 Thin- and Thick-walled Cylinders

at the outside surface. The radial stress varies from a maximum at the inside surface to zero at the outside surface.

By considering the equilibrium of the element taken from the thick-walled cylinder shown in Fig. 2-4(c) and assuming zero axial stress, i.e., a case of plane stress, the equilibrium equation can be written:

$$\sum F_r = \sigma_r \, r \, d\theta + 2\sigma_t \, dr \, d\theta/2 - \left(\sigma_r + \frac{d\sigma_r}{dr} \, dr\right)(r + dr) \, d\theta = 0$$

Thus,

$$\sigma_t - \sigma_r - r \, d\sigma_r/dr = 0 \qquad \text{Eq. I}$$

This equation is the equilibrium equation for the element.

In addition to fulfilling the equilibrium requirements, the strains of the element must be compatible with the stress relations. If u is the displacement of the material at a radius r, then $(u + (du/dr) dr)$ is the displacement at a radius $(r + dr)$. Thus, for the element the elastic strain equations are:

$$\epsilon_r = \frac{u + (du/dr) dr - u}{dr} = \frac{du}{dr}$$

$$\epsilon_t = \frac{(r + u) d\theta - r d\theta}{r d\theta} = \frac{u}{r}$$

where ϵ_r and ϵ_t are unit strains of the element in the radial and tangential directions. These equations can be expressed in terms of stresses using Poisson's Number, m, the reciprocal of Poisson's Ratio:

$$\epsilon_r = \frac{\sigma_r - \sigma_t/m}{E}$$

$$\epsilon_t = \frac{\sigma_t - \sigma_r/m}{E}$$

thus

$$\sigma_r = \frac{m^2 E}{m^2 - 1} \left\{\frac{du}{dr} + \frac{u}{mr}\right\} \qquad \text{Eq. II(a)}$$

and

$$\sigma_t = \frac{m^2 E}{m^2 - 1} \left\{\frac{u}{r} + \frac{du}{m \, dr}\right\} \qquad \text{Eq. II(b)}$$

Equations II are the compatibility equations.

The third source of information regarding the stress field is obtained from the boundary conditions. Equations I and II must satisfy the known conditions of:

$$\text{when } r = r_i \quad \sigma_r = p_i$$

$$\text{when } r = r_o \quad \sigma_r = p_o \qquad \text{Eq. III}$$

These Equations III are known as the boundary equations.

By combining the three sets of equations, I, II, III, as is the general procedure in solving cases in the theory of elasticity, and solving for σ_r and σ_t, it follows that

$$\sigma_r = \frac{r_o^2 p_o - r_i^2 p_i}{r_o^2 - r_i^2} - \frac{(p_o - p_i)r_i^2 r_o^2}{(r_o^2 - r_i^2)r^2} \qquad \text{Eq. 2-6(a)}$$

$$\sigma_t = \frac{r_o^2 p_o - r_i^2 p_i}{r_o^2 - r_i^2} + \frac{(p_o - p_i)r_i^2 r_o^2}{(r_o^2 - r_i^2)r^2} \qquad \text{Eq. 2-6(b)}$$

Several interesting observations can be made from these equations regarding the stress distribution in a thick-walled cylinder:

1. The sum of σ_t and σ_r is a constant regardless of position of the point in the cylinder, e.g.,

$$\frac{\sigma_r + \sigma_t}{2} = \frac{r_o^2 p_o - r_i^2 p_i}{r_o^2 - r_i^2}$$

2. σ_t can be seen to be always greater than σ_r.

3. In the case of a cylinder with an infinite external radius, r_o, the equations become

$$\sigma_r = p_o + (p_i - p_o) r_i^2/r^2 \qquad \text{Eq. 2-7(a)}$$

$$\sigma_t = p_o - (p_i - p_o) r_i^2/r^2 \qquad \text{Eq. 2-7(b)}$$

4. In the case where the internal pressure, p_i, is zero, the equations in item 3 become

$$\sigma_r = p_o(1 - r_i^2/r^2) \qquad \text{Eq. 2-8(a)}$$

$$\sigma_t = p_o(1 + r_i^2/r^2) \qquad \text{Eq. 2-8(b)}$$

From this it can be seen that, at the inside surface of a thick-walled cylinder of infinite radius with external pressure only (in other words in a stressed infinite mass), the tangential stress is equal to $2p_o$.

Example - Calculate the tangential stress, σ_t, on the inside surface of a concrete shaft lining with an inside diameter of 10 ft and an outside diameter of 14 ft, the ground pressure around the lining being 40 psi.

From Equation 2-6(b):

$$\sigma_t = \frac{(7 \times 12)^2 40 - (5 \times 12)^2 \, 0}{(7 \times 12)^2 - (5 \times 12)^2} + \frac{(40 - 0)(5 \times 12)^2 (7 \times 12)^2}{\{(7 \times 12)^2 - (5 \times 12)^2\}(5 \times 12)^2}$$

$$= 163.3 \text{ psi.}$$

The relationship between deformation and stress can now be established. The change in the circumference of the inside of the cylinder, C, is examined:

$$C = 2\pi r_i$$

and

$$dC = 2\pi \, dr_i$$

also

$$= C\epsilon_t$$

Thus

$$dr_i = r_i \epsilon_t$$

$$= r_i(\sigma_t - u\sigma_r)/E \qquad \text{Eq. 2-9}$$

Positive values of the displacement or change in radius, dr, indicate inward movement.

HOLE IN AN INFINITE ELASTIC SOLID

The case of a circular hole in an infinite plate, as shown in Fig. 2-5, is analogous to the thick-walled cylinder. Where the pressures on the edges of the plate, p_x and p_y, are equal, the stresses at any point within the plate will be equal to p_x in any direction. Consequently, if an imaginary circular cut were made in the plate to produce the thick-walled cylinder as shown by the dashed line in Fig. 2-5(b), the solution of the stress distribution around the hole would be exactly the same as for the thick-walled cylinder as derived above. In other words, the tangential stresses around the hole would be equal to twice the external pressure p_x decreasing exponentially away from the hole. The radial stress at the surface of the hole would, of course, be zero, increasing exponentially to the value of the external pressure at the external radius.

For a plate where p_x and p_y are not equal, the stress distributions out from the hole in the plate are given by the following equations(2):

$$\sigma_r = \frac{1}{2}(p_x + p_y)(1 - a^2/r^2) + \frac{1}{2}(p_x - p_y)(1 - 4a^2/r^2 + 3a^4/r^4)\cos 2\theta \qquad \text{Eq. 2-10(a)}$$

$$\sigma_t = \frac{1}{2}(p_x + p_y)(1 + a^2/r^2) - \frac{1}{2}(p_x - p_y)(1 + 3a^4/r^4)\cos 2\theta \qquad \text{Eq. 2-10(b)}$$

If the plate is only loaded in one direction, i.e., $p_x = 0$, as shown in Fig. 2-5(c), these equations can, of course, still be used. In this case, at the end of the diameter normal to the uniaxial stress field, as shown in Fig. 2-5(c), the tangential stress will be equal to three times the field stress p_y. At the ends of the diameter that is parallel to the stress field, the tangential stress will be equal in magnitude to the uniaxial stress p_y but will be opposite in sign.

In Fig. 2-5(c) the vertical shear stresses are examined along a line CD. The distribution will be as shown in Fig. 2-5(d). This shear diagram is similar to that of the more familiar case of the beam as shown in Fig. 2-9(a). In both cases the load is being transferred into abutments through shear stresses.

By combining the general polar equations relating displacements and strain (as used to produce Equation II above for the thick-walled cylinder) with Hooke's Law and Equations 2-6, it can be shown that for plane stress the displacement of the inside of a hole in a plate is (4):

$$dr = r_i(p_x + p_y - 2(p_x - p_y))\cos 2\theta/E \qquad \text{Eq. 2-11}$$

Positive values of dr indicate inward movement.

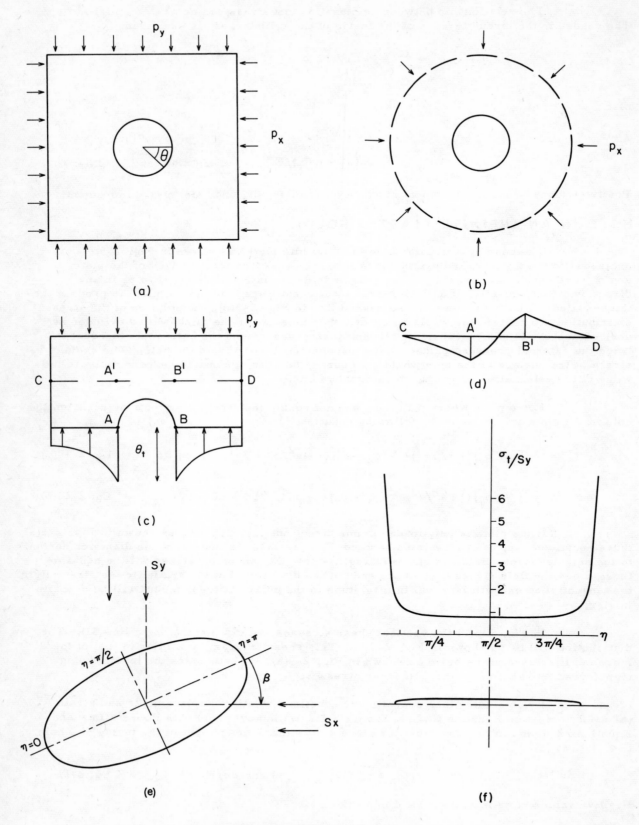

Fig. 2-5 A Hole in an Infinite Plate

A more general solution, in the sense that a circle is a special case of an ellipse, has been derived that is more applicable to stope or room geometry (5). The following equation for the tangential stress along the boundary of the ellipse, σ_t, is for an infinite plate, or in a case of plane stress:

$$\sigma_t = p_y \frac{2v(1+k)+(1-k)(1-v^2)\cos 2\beta + (1-k)(1+v)^2 \cos 2(\beta - n)}{(1+v^2) + (1-v^2)\cos 2n} \qquad \text{Eq. 2-12}$$

where p_y is the vertical field stress, v is the ratio of the major axis to the minor axis, k is the ratio p_x/p_y, β is the angle measured clockwise from the major axis to the direction of p_x and n is the elliptic coordinate measured clockwise from the major axis. It is probable that the geometry of many mine openings can be approximated by ellipses sufficiently closely that the limiting errors will not be the geometry but rather the uncertainties in the values of k and the constancy of the elastic properties of the rocks.

Fig. 2-5(e) shows the general case of an elliptical hole (5). Fig. 2-5(f) then shows the variation of the boundary tangential stress for the particular case of $v = 20$, $k = 1/3$ and $\beta = 0$. A series of solutions is shown in Appendix D.

SEMI-INFINITE ELASTIC SOLID

When a semi-infinite elastic solid is only loaded by its own weight, the deformation of any element of ground can only be in the vertical direction. In other words, horizontal deformation and strain must be equal to zero. This deduction can be used to relate the horizontal stresses, σ_x and σ_y, which will be equal in isotropic ground, to the vertical stress, σ_z, as follows:

$$\epsilon_x = \sigma_x/E - \mu \sigma_z/E - \mu \sigma_y/E = 0$$

where μ is Poisson's Ratio. Then as $\sigma_x = \sigma_y$, it follows:

$$\sigma_x = \mu \sigma_z/(1-\mu)$$

or
$$= \sigma_z/(m-1) \qquad \text{Eq. 2-13}$$

where m is Poisson's Number.

When a point load, Q, is applied on the surface of a semi-infinite elastic solid, as shown in Fig. 2-6(a), stresses are created. By following the same procedure as for the thick-walled cylinder, the stress distribution resulting from this load can be determined by solving the equilibrium, compatibility, and boundary equations.

The solution produces Boussinesq's equations (2):

$$\sigma_z = \frac{3Q \cos^5 \psi}{2\pi z^2} \qquad \text{Eq. 2-14(a)}$$

$$\sigma_x = \frac{Q}{2\pi y^2}(3\sin^4\psi \cos\psi - (1-2\mu)(1-\cos\psi)) \qquad \text{Eq. 2-14(b)}$$

$$\tau = \frac{3Q \cos^5\psi}{2\pi z^3} \qquad \text{Eq. 2-14(c)}$$

$$\sigma_r = \frac{2Q \cos^3\psi}{2\pi z^2} \qquad \text{Eq. 2-14(d)}$$

$$\sigma_\theta = \frac{Q}{2\pi y^2}(1-2\mu)(1-\cos\psi - \sin^2\psi \cos\psi) \qquad \text{Eq. 2-14(e)}$$

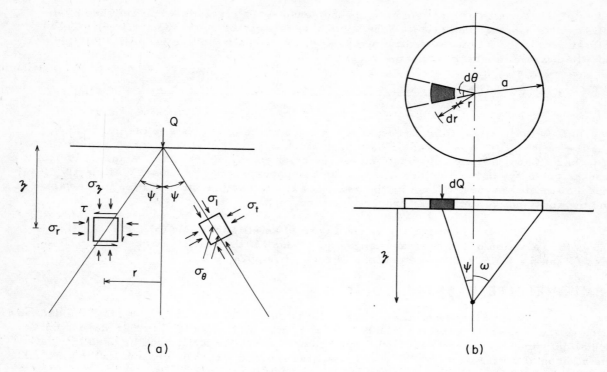

Fig. 2-6 Surface Load on a Semi-infinite Body

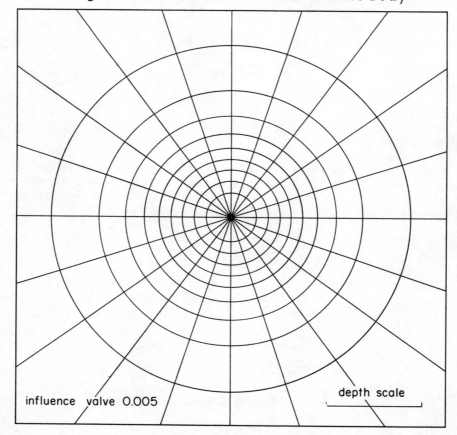

Fig. 2-7 A Newmark Chart

where σ_z is the vertical normal stress at the depth z at a point defined by the angle, ψ, between the vertical and the radius vector, σ_x is the horizontal stress in the plane of the radius vector, τ is the shear stress on vertical, tangential planes, σ_r is the major principal stress and is in the direction of the radius vector, σ_θ is the intermediate principal stress and is in a horizontal plane, and σ_t is the minor principal stress and is in a vertical plane including the radius vector.

Similarly, the equations for a line load, or a two-dimensional case, are (2):

$$\sigma_z = \frac{2Q}{\pi z} \cos^4\psi \quad \text{Eq. 2-15(a)}$$

$$\sigma_x = \frac{2Q}{\pi z} \sin^2\psi \cos^2\psi \quad \text{Eq. 2-15(b)}$$

$$\tau = \frac{2Q}{\pi z} \sin^2\psi \cos^3\psi \quad \text{Eq. 2-15(c)}$$

$$\sigma_r = \frac{2Q}{\pi z} \cos^2\psi \quad \text{Eq. 2-15(d)}$$

$$\sigma_t = 0 \quad \text{Eq. 2-15(e)}$$

With these equations it is possible to determine the significant depths that are influenced by various surface loads on material that can be assumed to behave elastically. Bulbs of pressure, or lines of equal increase in vertical pressure, can be drawn for a surface loading. It would be seen by doing this that there is a depth, which varies with the width of the loaded area and the material, beyond which the increase in stress in the body is insignificant.

<u>Example</u> - Determine the vertical stress at a depth of 3 ft at an angle of 45° out from under a surface point load of 1000 lb.

From Equation 2-14(a) the vertical stress is obtained:

$$\sigma_z = \frac{3 \times 1000 \cos^5 45}{2\pi \times 3^2} = 9.4 \text{ psf}$$

To be able to determine the increase in stress within a semi-infinite elastic body as a result of an area, not a point, loading on the surface, a useful method has been devised (3) based on the following derivation:

$$d\sigma_v = 3 dQ \cos^5\psi$$

$$= \frac{3q}{2\pi z^2} \cos^5\psi \, dA$$

$$= \frac{3q}{2\pi z^2} \cos^5\psi \, r \, d\theta \, dr$$

where q = surface pressure, and dA = the elemental area as shown in Fig. 2-6(b).

Thus,

$$\sigma_v = \frac{3q}{2\pi z^2} \int_0^{2\pi} \int_0^R r \cos^5\psi \, d\theta \, dr$$

$$= q(1 - \cos^3\omega), \text{ where } \omega = \tan^{-1} a/z$$

$$= q\left(1 - (1 + (a/z)^2)^{-1.5}\right)$$

When $\sigma_v/q = 1$, $a = \infty$, and when $\sigma_v/q = 0.9$, $a/z = 1.92$, which follows from the above equation. Thus, it can be seen that the increase in stress, σ_v, resulting from a

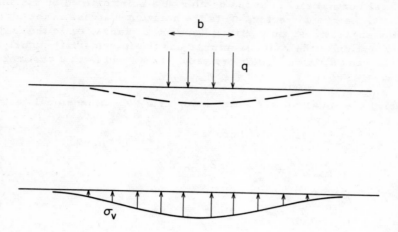

Fig. 2-8 The Deflection Crater from a Surface Load on a Semi-infinite Body

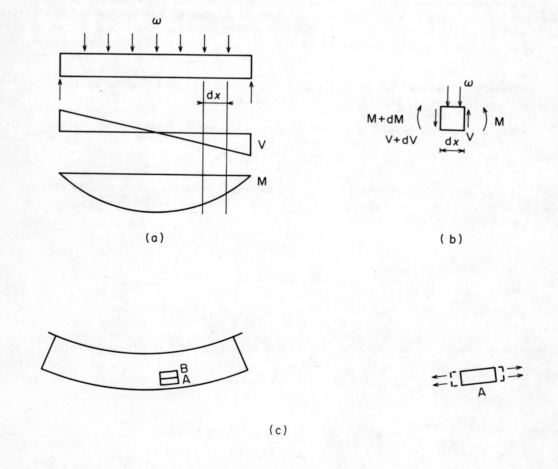

Fig. 2-9 The Inter-relationship Between Load, Shear and Moment in a Beam

surface loading with a pressure q is a function of the ratio of the width of the loaded area to the depth at which the stress is being calculated. In other words, if the radius of the loaded area, a, is equal to 1.92 times the depth, z, to which the stress is being calculated, then the increase in vertical stress will be equal to 0.9 q. Similarly, it can be calculated that, if a = 1.38 z, the increase in vertical stress will be equal to 0.8 q at the depth z under the centre of the loaded area.

If we give z any arbitrary value of, say 1 in., we could draw circular areas with radii of 1.92 in. and 1.38 in. that would represent respectively the areas that would be required to be loaded to produce at a depth of 1 in. below the centre of the loaded area increases of stress of 0.9 q and 0.8 q respectively.

Then, if only the annular ring between these two radii is loaded, the increase in vertical stress under the centre of these circles would be:

$$\sigma_v = 0.9q - 0.8q = 0.1q$$

If only 1/10 of this annular ring is loaded, similar to the loaded area shown in Fig. 2-6(b), the increase in vertical stress would be 0.01 q.

Following this procedure, a Newmark chart can be drawn as shown in Fig. 2-7. The influence value of each element will depend on the number of radii that are used corresponding to pressures up to 0.9 q, as well as on the number of radial divisions that are made. In the chart shown in Fig. 2-7, the influence value is 0.005, which means that the increase in stress at the depth z is:

$$\sigma_v = 0.005 q \, N \qquad \text{Eq. 2-16}$$

where N is the number of elements or squares covered by the actual loaded area.

To determine N, the actual loaded area is drawn on transparent paper to the scale of 1 in. = the depth to which the stress is to be calculated. This drawing is then placed on the Newmark chart with the horizontal position at which the stress is to be calculated placed at the centre of the chart. The number of squares covered by the loaded area drawn to scale is then counted to give the value of N.

In Fig. 2-8 the surface of a semi-infinite elastic solid is loaded with a pressure q over a width b. As a result of this surface loading, the increase in vertical stress at depth will follow a pattern as shown in the figure. Because of this dish-shaped increase in vertical stress, there will be a corresponding increase in strain and deformation. The result will be that the surface of the semi-infinite elastic solid will deflect into a dish-shaped crater with vertical deflection taking place beyond the loaded area.

If in Fig. 2-8 the loaded area has been applied through a rigid body, e.g., either a reinforced concrete footing, a massive concrete dam, or a pillar of hard rock on a floor of soft rock, this rigid body would resist the curvature associated with the dish-shaped surface crater. The result of this resistance would be to produce concentrations of pressures towards the edges of the loaded area so that the deformation of the surface of the semi-infinite body under the loaded area would be uniform. Such concentrations of pressure could, of course, be significant with respect to the strength of the materials involved, as mentioned in Chapter 1 regarding the use of plate load testing. This topic is treated in more detail in Chapter 7.

BEAMS

Beams are structural elements that support loads applied transversely to their length and, as such, are similar to the cases of roof rock over extensive workings or at shallow depths. As a consequence of supporting loads in this manner, beams are subjected to bending stresses and shear stresses. The internal forces producing these stresses can be

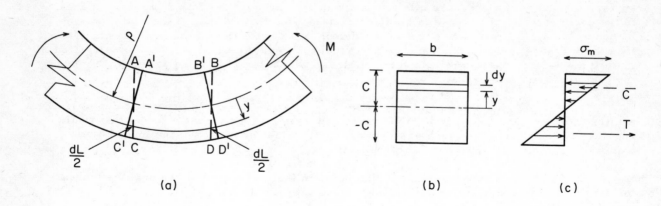

Fig. 2-10 Pure Bending of a Beam

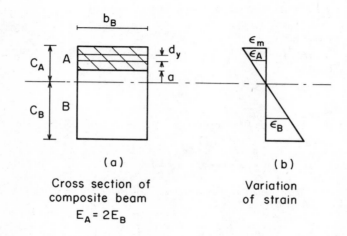

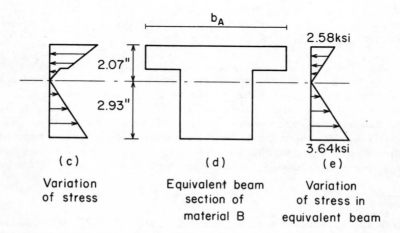

Fig. 2-11 The Analysis of a Composite Beam

determined by creating free body diagrams and using the equations of equilibrium as explained earlier. The resultant stresses are then determined, as explained below, by examining the internal behaviour of the beam material.

It was mentioned above that the equations of equilibrium are one source of information for the determination of unknown forces. By cutting a section out of a beam that is supporting a distributed load, as shown in Figs. 2-9(a) and 2-9(b), a Free Body Diagram of this section can be drawn. By analysing the equilibrium of this section it can be established (by ignoring infinitesimals of higher order) from the sum of the vertical forces and from the sum of the moments about any corner respectively that:

$$\omega = dV/dx \qquad \text{Eq. 2-17}$$

$$V = dM/dx \qquad \text{Eq. 2-18}$$

These equations provide useful relationships between the loading, ω, the resisting shear, V, and the resisting moment, M. For example, if we know the point in a beam where the shear force, V, is zero, then dM/dx is zero and hence the resisting moment, M, is a maximum (i.e., to determine where M is a maximum, one would find dM/dx, equate it to zero and then solve for x).

A second source of information for determining the internal effects of the external loading is the requirement of compatibility. It is obvious on reflection that the deformation of any element of a stressed body must be compatible with the deformation of the adjacent element. For example, in the beam shown in Fig. 2-9(c), as a result of bending Element A will be elongated. Unless the material were to rupture, it would be necessary for Element B, which is adjacent to Element A, to elongate.

To fulfil this requirement of compatibility, the assumption is made in the theory of pure bending that plane, parallel, transverse sections of unloaded or unbent beams remain plane after bending. Consequently, as shown in Fig. 2-10(a), the elongation, dL, in any layer of a beam will be proportional to the distance of that layer from the neutral axis (i.e., the neutral axis is at the fibre of the beam, which is subjected to neither compressive nor tensile forces).

The length of such sections between parallel planes would be equal before bending; thus, normal strain and hence normal stress (for a constant modulus of elasticity) resulting from bending are also proportional to the distance from the neutral axis as shown in Fig. 2-10(c).

By drawing a Free Body Diagram of part of the beam from either end up to section BD in Fig. 2-10(a), one of the equilibrium equations can be applied by taking the horizontal forces acting externally and on the face BD; i.e., there will be a compressive force, C, and a tensile force, T, acting as shown in Fig. 2-10(b) to produce the following equation:

$$\sum F_n = \int_{+c}^{-c} \sigma_m(y/c)\, dA$$

where dA is the elemental area equal in this case to (b dy), c is the distance from the neutral axis to the extreme fibre, and σ_m is the maximum or extreme fibre stress.

Then
$$\sum F_n = \frac{\sigma_m}{c} \int_{-c}^{+c} y\, dA = \frac{\sigma_m}{c} \cdot \bar{y} \cdot A = 0$$

where $\bar{y}$ is the distance to the centroid of the area A. In other words, the neutral axis of this beam coincides with the centroidal axis of the cross-sectional area. This only happens to be the case for straight, not curved, beams.

By equating the moment external to the section at BD with the couple resulting from the tension and compression acting normal to the section BD, a second equation can be written by summing the moments about the neutral axis of the forces on the elements as shown in Fig. 2-10(c):

$$M_{na} = \int_{+c}^{-c} y \cdot \sigma_m(y/c) \, A - M = 0$$

$$\therefore M = \sigma_m/c \int_{+c}^{-c} y^2 \, A$$

Then, as
$$I = \int_{+c}^{-c} y^2 \, A$$

$$\sigma_m = Mc/I \qquad\qquad \text{Eq. 2-19(a)}$$

where M is the sum of the moments of the external forces applied to the beam to one side of the section and taken about the neutral axis at the section, and I equals the moment of inertia of the cross-sectional area. This bending or flexure equation can be used for determining the normal stresses caused by bending in a beam.

Another useful relation that is inherent in the flexure formula follows from recognizing similar triangles:

$$\frac{\rho}{CD/2} = \frac{y}{dL/2}$$

or
$$\frac{y}{\rho} = \frac{dL}{CD} = \epsilon = \frac{\sigma}{E} = \frac{My}{EI}$$

Thus,
$$\frac{1}{\rho} = \frac{M}{EI} \qquad\qquad \text{Eq. 2-19(b)}$$

$\underline{\text{Example}}$ - Apply the theory of pure bending to the composite beam shown in Fig. 2-11 to determine the extreme fibre stresses in Materials A and B. The moment at this section is 100-kip-inches (the word "kip" is an abbreviation for kilopound or 1000 pounds). The width of the beam is 3 in., the total depth is 5 in., the thickness of Material A is 2 in., the modulus of elasticity of A, E_A, is 10×10^6 psi and E_B equals 5×10^6 psi.

As plane transverse sections before bending must remain plane after bending, the strain in the beam will still be proportional to the distance from the neutral axis as shown in Fig. 2-11(b). However, stress will not only vary with strain but also with the modulus of elasticity of the material as shown in Fig. 2-11(c).

If we examine the force, dF, acting on any elemental layer, dy, we can write the following equations:

$$dF = \epsilon_A E_A b_B \, dy$$

Then, if we let $E_A = E_B n$

$$dF = \epsilon_A E_B n b_B \, dy$$

Then, if we let $b_A = n b_B$ \qquad\qquad Eq. 2-20

we can construct an equivalent beam all of Material B with the part originally composed of Material A stretched out laterally to equal ($n\, b_B$), thus providing the same resisting force as for the actual beam. At this point we can apply our normal beam analysis to determine the extreme fibre stresses.

The neutral axis is found first:

$$\bar{y} = \frac{3 \times 3^2/2 + 2 \times 6 \times 4}{9 + 12} = 2.93 \text{ in.}$$

Hence,
$$I = 6 \times 2^3/12 + 12 \times 2.07^2 + 3 \times 3^3/12 + 9 \times 1.43^2 = 80.4 \text{ in.}^4$$

Therefore, at the top of the beam

$$\sigma_m' = 100 \times 2.07/80.4 = 2.58 \text{ ksi (cpn)}$$

and at the bottom of the beam

$$\sigma_m'' = 100 \times 2.93/80.4 = 3.64 \text{ ksi (tsn)}$$

Now these stresses apply to the equivalent beam in Material B. Consequently, when the actual beam is considered, it is necessary to shrink the width b_A back to the width b_B and change the stresses in the Material A to provide the same total force as in the equivalent beam; in other words the stresses are multiplied by n. Hence, the extreme fibre stress in Material A is 2.58 ksi x 2 = 5.16 ksi. Material B requires no conversion; therefore, the extreme fibre stress in this material is 2.64 ksi as calculated above.

In addition to analysing the stresses caused by the internal moment, M, we are interested in calculating the stresses caused by the shear force, V. For this purpose we isolate a section of the beam as shown in Figs. 2-9 and 2-12(a). As the moment varies along the length of the beam, there will be a moment, M, acting on one face and a moment (M + dM), acting on the other face. Consequently, the normal stresses on the left face in Fig. 2-12(a) will be less than those acting on the right face.

We then examine the equilibrium conditions of a slice of the beam between these two sections of thickness dy as shown in Fig. 2-12(b). The sum of the horizontal forces acting on this part of the beam can be written as follows:

$$\sum F = \int_y^c \left[(M + dM)y/I \right] dA - \int_y^c (My/I) dA - \tau b\, dx = 0$$

Hence
$$\tau = \frac{dM}{dx} \frac{\bar{y} A}{I\, b}$$

where $\bar{y}$ is the distance from the neutral axis to the centre of gravity of the area beyond the distance, y, A is the area beyond y, I is the moment of inertia of the entire cross-sectional area of the beam, and τ is the horizontal shear stress at y.

As it was previously established that V = dM/dx the equation can be written:

$$\tau = V \frac{\bar{y} A}{I\, b} \qquad \text{Eq. 2-21}$$

This gives us an equation for the calculation of the horizontal shear stress at any distance y from the neutral axis.

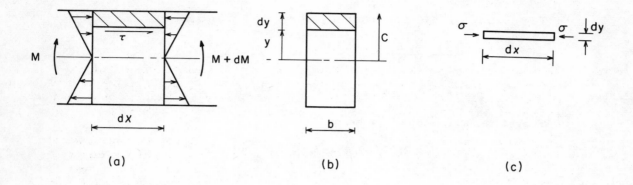

Fig. 2-12 The Analysis of Shear Stress in a Beam

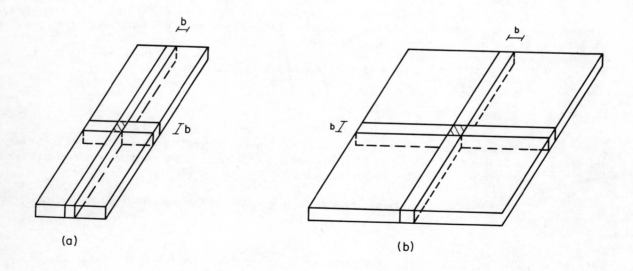

Fig. 2-13 Two-way Bending in a Plate

Example - The shear force, V, at a section of a beam is 100 lb. The moment of inertia of the beam is 500 in.^{4}, its width is 6 in. and its depth is 10 in. Calculate the shear stress at a point 3 in. above the neutral axis using Equation 2-21. $\bar{y}$ is 4 in. and A is 12 in.^{2}; hence, the shear stress is:

$$\tau = 100 \ \frac{4 \times 12}{500 \times 6} = 1.6 \ \text{psi}$$

By examining the rotational equilibrium requirements for an element of a stressed body, it can be established that the shear stresses on faces at right angles to each other of an infinitesimal element must be equal in magnitude. Consequently, the above equation can be used for the calculation of the vertical shear stresses at any point in a beam.

Another consequence of applying external loads to a beam that is sometimes of interest is the amount of strain energy stored within the beam. In Fig. 2-12(c) an element with a length dx, a height dy and a width dz is subjected to an axial stress σ. The stress produces a strain ϵ and a deformation of ϵ dx. Thus the amount of work done during the application of the stress, which started from zero and built up to its ultimate value, is as follows:

$$1/2 \ \sigma \, dy \, dz \, \epsilon \, dx = dU$$

where dU is the stored, or strain, energy in the element. This equation can then be written:

$$dU = \frac{\sigma^2}{2E} \, dV \qquad \text{Eq. 2-22(a)}$$

where dV is the volume of the element (dx dy dz). Alternatively:

$$U = \int \frac{\sigma^2}{2E} \, dV \qquad \text{Eq. 2-22(b)}$$

In the case of beams this equation can be re-written as

$$U = \iint \frac{(My/I)^2}{(2E)} \, dx \, dA$$

where dx is the length of the infinitesimal element and dA is the cross-sectional area of the infinitesimal element. The equation can be simplified to:

$$U = \int \frac{M^2 \, dx}{2 \, EI} \qquad \text{Eq. 2-23}$$

When this equation is integrated over the full length of the beam, the total strain energy stored in the beam due to bending is obtained. This stored energy must be equal to the external work done by the external forces in causing deflection and thus in causing the beam to bend; hence the deflection, d, of a beam can be calculated as follows:

$$Pd/2 = \int \frac{M^2 \, dx}{2 \, EI} \qquad \text{Eq. 2-24}$$

where P is the external force on the beam acting through the deflection, d.

The factor of 1/2 enters the left side of the equation as P must be applied gradually so that in travelling the distance d it varies from zero to its full value of P. As this variation is linear for an elastic body, the average force P/2 multiplied by the distance through which it travels will give the external work done on the beam. If the force P were applied with its full value immediately to the beam, a dynamic problem would arise, the beam would oscillate and hence be a different prototype case.

Example - A beam has a span of 100 in., a moment of inertia of 500 in.^{4} and a modulus of elasticity of 10^6 psi. It is loaded at the centre with 120 lb. Calculate the deflection of the beam at the centre.

Using Equation 2-24, $M = Px/2$, hence the deflection is:

$$d = \frac{2}{P} \int_0^{L/2} \frac{2(Px/2)^2 \, dx}{2 EI}$$

$$= \frac{120}{2 \times 10^6 \times 500} \int_0^{50} x^2 \, dx$$

$$= 1.2 \times 10^{-7} \times 50^3/3 = 0.005 \text{ in.}$$

PLATES

If the effect of bending in narrow beams (i.e., defined in this context as beams where the width is the same order of magnitude as the depth) is examined in more detail, it is found that not only is there curvature of the beam in the plane of the moment but curvature also develops in the plane transverse to the axis of the beam. This curvature results from the case of plane stress, as mentioned in Chapter 1, actually producing triaxial strain. In the zone of tensile bending stresses there will be transverse strains leading to a contraction of the section with the deformation varying with the distance from the normal axis. Similarly, the compressive bending stresses produce transverse strains leading to an expansion of the section.

For wide beams this distortion of the cross-sectional shape of the beam is resisted except at the edges. In the central part of the beam the circumstances are such as to produce a case of plane strain and hence triaxial stress. In other words, if the stresses resulting from bending are called σ_x, then at right angles to these stresses and transversely in the beam there will be stresses σ_y resisting the distortion of the cross-sectional shape of the beam. The stresses σ_y will have the same sign as σ_x, i.e., in the tension zones σ_y will be tension and thus act to prevent the transverse contraction of the beam.

As a result of this triaxial stress condition the strain ϵ_x will be less than σ_x/E. The actual strain will be:

$$\epsilon_x = \frac{\sigma_x}{E} - \frac{\mu \sigma_y}{E}$$

However,

$$\epsilon_y = \sigma_y/E - \mu \sigma_x/E = 0.$$

Consequently $\sigma_y = \mu \sigma_x$. In other words, the transverse stress will be about 1/4 to 1/3 of the bending stress.

The presence of this transverse stress produces the strain equation:

$$\epsilon_x = \sigma_x/E - \mu^2 \sigma_x/E = \sigma_x(1-\mu^2)/E$$

It can be considered that the equivalent modulus of elasticity of the material is:

$$E' = \frac{E}{1 - \mu^2} \qquad \text{Eq. 2-25}$$

It follows that E' is greater than E and that σ_x for a plate will be less than for a beam.

Hence, it can be seen that, for equal deformations, a plate can be expected to carry more load per unit width than a beam, even though it is only supported on two edges like the beam.

Example - A beam has a span of 100 in., a depth of 10 in., a moment of inertia of 83.3 in.^{4}/in.-width, a modulus of elasticity of 10^6 psi and $\mu = 0.3$. When it is 6 in. wide, a load of 20 lb/in. of width at the centre causes a deflection of 0.005 in. Calculate the deflection when it is 100 in. wide and loaded at the centre by 20 lb/in. of width. Also, calculate the maximum transverse stress.

From Equation 2-25 the equivalent modulus of elasticity for a wide beam is:

$$E' = 10^6/(1-0.3^2) = 1.1 \times 10^6 \text{ psi}$$

$$\therefore d = 0.005 \times 10^6/(1.1 \times 10^6) = 0.00454 \text{ in.}$$

From Equation 2-14(c) the maximum longitudinal stress is:

$$\sigma_m = (20 \times 100/4) \, 5/83.3 = \pm 30 \text{ psi}$$

The maximum transverse stress is thus:

$$\sigma = 0.3 \times 30 = \pm 9 \text{ psi}$$

The case of the plate supported on four edges means that bending takes place in the x- and y-directions. In this case, every element of load acting on the beam can be considered to be supported by two individual beams acting at right angles to each other, as shown in Fig. 2-13. When we consider the relationship between bending moment and radius of curvature the following equations can be written $M_x = E'IR_x$ and $M_y = E'I/R_y$ where I is the moment of inertia of the element of beam of width b and R equals the radius of curvature of the individual beams supporting the one element of load.

It can be seen intuitively, and established mathematically, that the support given to the element of load on the area b x b in Fig. 2-13(a) by the elemental beam (of width b) in the long direction will be very small compared to the support given by the short transverse beam (of width b). In other words, the radius of curvature of the long beam will be much larger than the radius of curvature of the short beam, and hence the induced moments will have the inverse comparison.

On the other hand, for the plate in Fig. 2-13(b) where the long span is equal to the short span, it can be visualized that the moment from bending on either span will be less than half (curvature in the second direction by itself decreases the moment in the first direction) what it would be if the plate were only supported on two sides rather than on four.

For practical considerations it can be shown that, when the long span of the plate is three times the short span, the moment is only reduced by about 5 per cent in the short span (1). When the long span of the plate is about twice the short span of the plate the moment is reduced by about 20 per cent. These figures indicate the relative dimensions of plates for which the two-way action in bending is significant in reducing bending stresses.

ARCHES

Simple arches can be considered as curved beams. For example, the arch in Fig. 2-14 can be analysed in the same way as a simply supported beam. The equations of equilibrium are used to determine the reactions at the supports. Then the arch can be cut at any section where the internal forces (shear force, axial force and bending moment) are to be calculated.

(Note: When the radius of curvature of the beam becomes of the same order of magnitude as the depth of the beam, then the usual flexure formula can no longer be applied for determining the maximum bending stresses. However, when the radius of curvature of

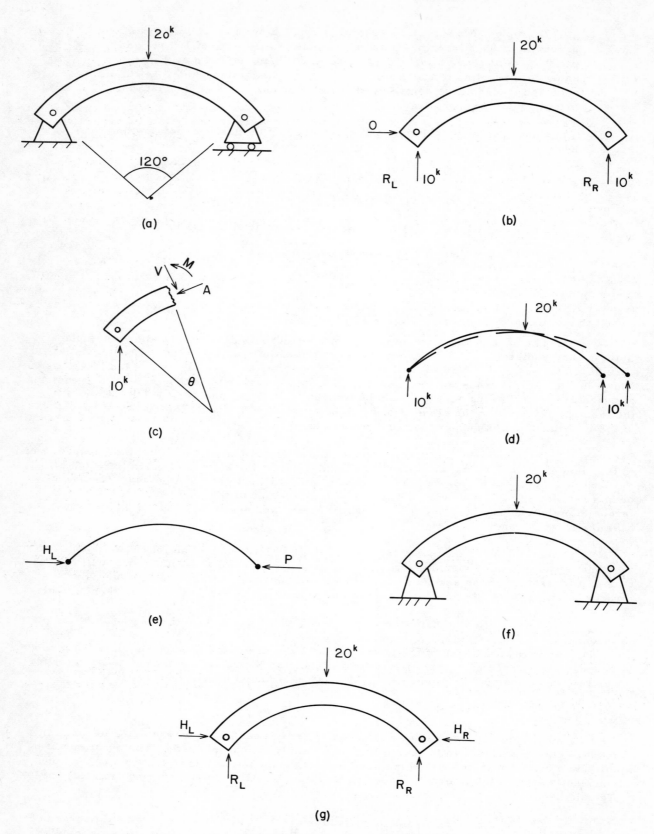

Fig. 2-14 The Analysis of a Hinged Arch

the inside surface is four times the depth of the beam, the maximum stress is only about 10 per cent higher than that calculated using the normal flexure formula. Consequently, for any arches with larger radii of curvature than this figure, the error introduced by using the simple flexure formula can be ignored.)

For the arch shown in Fig. 2-14 the supports are pin-connected and are restrained from moving transversely. Consequently, to prevent the horizontal deflection that occurred as shown in Fig. 2-14(d), a horizontal reaction is required at each support.

By examining the Free Body Diagram of the arch, as shown in Fig. 2-14(g), it can be seen that there are four unknown forces. In this case the reactions cannot be solved by simply using the three equations of equilibrium (as it is only possible to solve for one unknown per equation). It is thus necessary to obtain an equation describing a relationship between these forces and the curved beam that incorporates some other necessary condition aside from equilibrium.

If we consider the horizontal deformation as shown in Fig. 2-14(d), we can visualize the horizontal reaction that occurs in Fig. 2-14(g) as being that required to decrease this horizontal deformation to zero. Consequently, if we can relate the force required to cause this reverse deformation to the properties of the beam, a fourth equation can be obtained so that the four unknowns in the system can be solved.

To relate load and deflection, it is necessary first to establish an equation for the deflection of beams at points other than at the applications of the actual forces. For this purpose, consider the artificial construction of applying a unit load of one kip at the point, A, on a beam where the deflection is to be determined. Let the resultant moments that occur within the beam be represented by m. Then, imagine the actual loads being applied to the beam. As a result of these actual loads being applied to the beam, the force of one kip moves through a deflection d_a.

The work done by the load of one kip moving through the deflection d_a is then $1^k \times d_a$. (This is similar to applying a force of one kip on a spring and then loading the spring with ten kips, which produces an additional deflection of 1 in. The total work done after the ten kips was applied is equal to 11 x 1 or 11 kip-inches; the work done during this interval by the load of one kip equals 1 x 1 or one kip-inch.)

The strain energy arising from the work done during the application of the actual loads on the beam by the moments will be equal to $\int m \cdot d\theta$. In other words, the work done by these moments is the sum of the individual moments multiplied by the angle through which they rotate.

It was shown in Equation 2-14(b) that $1/\rho = M/EI$ and we know that $\rho \cdot d\theta = ds$, where ds is an element of length of the curved beam. Therefore, the internal work done by the moments induced by the one kip force on the beam equals $\int m \frac{M \, ds}{EI}$.

Now, as the external work done by the one kip force during the time when the actual loads are being applied to the beam must equal the internal work stored by the moments induced by the one kip force, we can say:

$$1 \times d_a = \int m \frac{M \, ds}{EI}$$

or,

$$d_a = \int m \frac{M \, ds}{EI} \qquad \text{Eq. 2-26}$$

This general procedure is known as the Theory of Virtual Work.

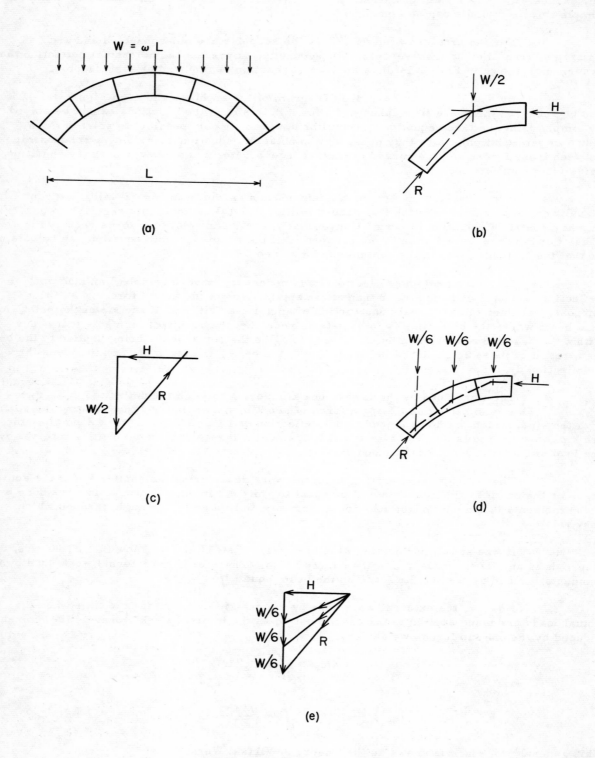

Fig. 2-15 The Analysis of a Fixed Arch

Example - Determine the deflection at mid-span of a cantilever loaded at the end by 120 lb. The span is 100 in., $I = 500$ in.^{4} and $E = 10^6$ psi.

Using the loaded end of the beam as the origin for x, the moments due to the load are:

$$M = Px$$

Applying a unit load at mid-span, the moments are

$$m = 0 \qquad\qquad 0 < x < L/2$$

$$m = (x - L/2) \qquad\qquad L/2 < x < L$$

Using Equation 2-26, the deflection is determined:

$$d_a = \left[\frac{1}{EI} \int_0^{L/2} 0 Px \, \partial x + \int_{L/2}^{L} (x-L/2) Px \, \partial x \right]$$

$$= \frac{P \left[x^3/3 - Lx^2/4 \right]}{EI} \bigg|_{L/2}$$

$$= \frac{5P}{48 EI} = \frac{5 \times 120}{48 \times 10^6 \times 500} = 2.5 \times 10^{-8} \text{ in.}$$

With a method of determining the deformation of any point on a beam due to bending, whether it is straight or curved, it is possible to calculate the amount of horizontal deformation, d_h, shown in Fig. 2-14(d). Thus, the force, P, required to cause the right-hand support of the arch to deflect an amount $-d_h$ can be calculated. The force P of Fig. 2-14(e) is then the same as the reaction H_R of Fig. 2-14(g) in the pinned arch.

Thus for the pinned arch of Figs. 2-14(f) and 2-14(g), where there are four unknown forces, there are now four equations that can be used for their solution:

$$F_v = 0$$

$$\sum F_h = 0$$

$$\sum M = 0$$

$$d_h = \int \frac{mM \, ds}{EI}$$

The deflection equation has supplied additional information about the structural system.

If the abutments of the arch in Fig. 2-14(f) were fixed rather than pin connected, the number of unknown factors would be increased and, hence, more sources of information would be required to obtain the required number of equations.

Fig. 2-15 shows a masonry arch supporting a distributed load. The classical analysis of such an arch includes the assumptions - that the tangential force at the crown, H, is horizontal and acts at the upper third point of the section - and that the reaction, R, acts normal to the springing line and at the lower third point of the section.

With these assumptions the arch becomes a statically determinate body. R can be solved by using the equilibrium equation $\sum F_v = 0$. The internal forces can then also

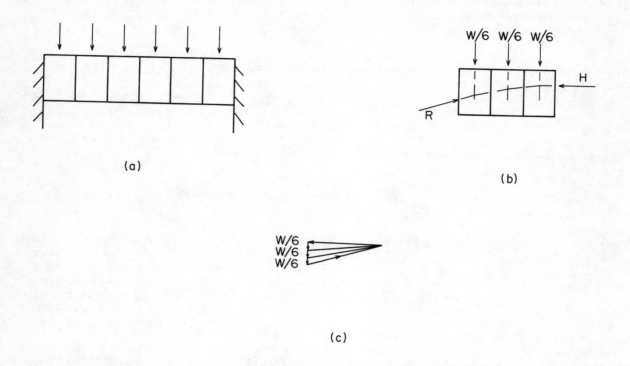

Fig. 2-16 The Analysis of a Flat Arch

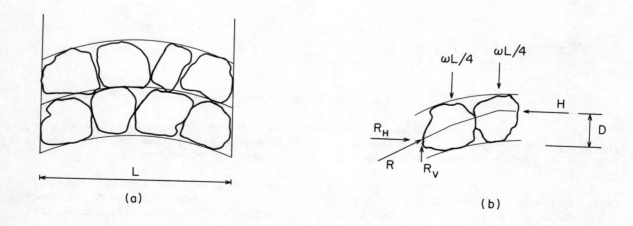

Fig. 2-17 The Analysis of a Boulder Arch

be solved by using the equilibrium equations; e.g., in Fig. 2-15(b), H can be obtained by using $\sum F_h = 0$. Alternatively, the force polygon can be drawn as in Fig. 2-15(c), and the magnitude of R and H obtained from this diagram.

This design analysis was useful for masonry arches even though it was not necessarily an accurate representation of the stresses in the arch, as it represented a limit condition that would be safe. The traditional problem was whether these arches would be stable under a given loading and a preselected rise for a specified span.

It was assumed that the separate masonry blocks were not bonded together. Hence, their contacts could not sustain a tension. Tensile stresses would be avoided if the line of action of the internal force between the blocks was within the middle third of the cross-section of the arch. Fig. 2-15(d) shows the individual force diagrams that can be drawn with the aid of the polar diagram Fig. 2-15(e) to determine the location of the line of action, for the assumptions regarding the reactions stated above, between the blocks. In this way it was learned whether the resultant force between two blocks would be within the middle third and thus whether the arch would be safe.

In Fig. 2-16 a flat or jack arch is shown. The analysis of the line of action of the internal forces between the blocks shown in Figs. 2-16(b) and 2-16(c) indicates that, if the load is low enough, this arch will be stable.

Proceeding with this type of analysis, an arch made up of boulders as shown in Fig. 2-17 can be examined. In this case, the arch could conceivably fail in one of three ways. If the load is high for the span and rise, tensile stresses could open spaces between the blocks permitting them to fall, which would destroy the arching action. A second way in which the arch could fail would be by crushing at the relatively small areas of contact between the blocks. The third way would be by the blocks slipping out of the arch as a result of the frictional resistance at their contacts not being great enough with respect to the shear caused by the loading on the arch.

An analysis can be made on the possibility of such a boulder arch failing by slipping at the abutments. From Fig. 2-17 it can be seen, by taking moments about the point of application of R, where D is the rise of the arch, L is the span, and ω the weight per linear foot, that the horizontal component, R_h, of the reaction R can be determined:

$$R_h D = \frac{\omega L}{2} \times \frac{L}{4}$$

Therefore
$$R_h = \frac{\omega L^2}{8D} \qquad \text{Eq. 2-27}$$

As the resistance at the abutment arises entirely from friction, it can be said that, when slip occurs, the vertical component, R_v, of the reaction is:

$$R_v = R_h \tan\varphi$$
$$= \frac{\omega L^2}{8D} \tan\varphi \qquad \text{Eq. 2-28}$$

If D is $L/8$ and $\tan\varphi$ is 1 and if slip were to occur, R_v would be equal to ωL. But the maximum R_v will only be equal to half the vertical loading $\frac{\omega L}{2}$. Consequently, with these characteristics the arch would be stable against slipping.

The other modes of failure for irregular shaped boulders would be more difficult to analyse.

Example - A boulder arch, as shown in Fig. 2-17, is expected to form over an 8 ft wide accessway through fill underground. If the rise of the arch, D, is expected to be 2 ft and the angle of internal friction 30°, calculate the vertical reaction, R_v, at the abutment in terms of the loading ω psf and the maximum possible vertical reaction.

The actual reaction will be:

$$R_v = 8 \times \omega/2 = 4\omega \text{ lb/ft}$$

The horizontal reaction will be:

$$R_h = \frac{4\omega \times 2}{2} = 4\omega$$

Thus, the maximum vertical reaction will be:

$$R_v \text{ max} = R_h \tan\varphi = 4\omega \tan 30 = 2.31\omega$$

Hence, such an arch would not be stable at the abutments.

REFERENCES

1. Timoshenko, S., "Theory of Plates and Shells", McGraw-Hill (1940).

2. Timoshenko, S. and Goodier, J., "Theory of Elasticity", McGraw-Hill (1951).

3. Newmark, N., "Influence Charts for the Computation of Stresses in Elastic Foundations", Univ. Illinois Eng. Expt. Sta. Bull. 228 (1942).

4. Merrill, R. and Peterson, J., "Deformation of a Borehole in Rock", USBM RI 5881 (1961).

5. Geldart, L. and Udd, J., "Boundary Stresses Around an Elliptical Opening in an Infinite Solid", Proc. Rock Mech. Symp., McGill Univ., Mines Branch, Ottawa (1963).

CHAPTER 3

SHAFTS, DRIFTS AND TUNNELS

INTRODUCTION

The underground openings examined in this chapter are of simple geometry and of more or less permanent nature. For example, tunnels for civil engineering projects, with the exception of diversion tunnels, generally are designed to last forever, e.g., tunnels for water, sewers, highways, subways, railways, and powerhouse intakes and outlets. Also, whereas some shafts may be only for access during construction, the majority of shafts that are sunk must remain open for a considerable period of time.

Drifts and crosscuts are of a more variable nature. They can be for the passage of mine services, in which case they are of a more or less permanent nature; or they can be for the passage of muck out of a mining block where they would normally only be required to remain open during the mining of that zone.

The treatment in this chapter is concerned with possible analyses for various ground reactions that can occur as a result of underground excavations and as a result of placing different types of support. Judgment, of course, in practical work is still required, particularly to account for the deviations of actual ground from the idealizations that can be analysed and also to deal with those situations that have not yet been idealized so that appropriate analyses can be established.

It must also be recognized that in rock mechanics the design methods that have been evolved using steel in structures and soil in foundations and in earth structures will not usually be satisfactory. Both structural steel and most soils have the property of yielding under excessive stress before rupturing (and in some cases, such as foundations on sand, rupture would never occur). This property has made it possible, using these other materials, to ignore stress concentrations and to base design on average stresses.

As most rocks are brittle, stress concentrations can cause local rupture of the rock, which can lead progressively to failure of the entire structure. Consequently, any prediction of failure or any design safety factor should be based on the best possible analysis of actual stress concentrations.

A second element of design of great importance in rock mechanics is the recognition of the variability in strength of all materials. Aside from the great difficulty of measuring the in situ strength of rock masses, the variability is likely to be greater than for other structural materials. Appraisals should ideally be made based on the probability of failure. In an explicit analysis it may be that in one case less than 1 per cent probability of failure would be required, whereas in another case a 50 per cent probability of failure would be acceptable.

STRESSES AROUND SHAFTS

<u>Circular Shafts in Elastic Ground Subjected to Gravity Stresses</u>: The first case that can be examined is that of a circular shaft in ideally elastic, solid rock. The stress distribution in the horizontal plane around the shaft can be analysed by assuming that the shaft is a thick-walled cylinder with the outside radius equal to infinity. From Equations 2-6 we can write the equations applicable to this case:

$$\sigma_r = S_x + (p_i - S_x) a^2/r^2 \qquad \text{Eq. 3-1(a)}$$

$$\sigma_t = S_x - (p_i - S_x) a^2/r^2 \qquad \text{Eq. 3-1(b)}$$

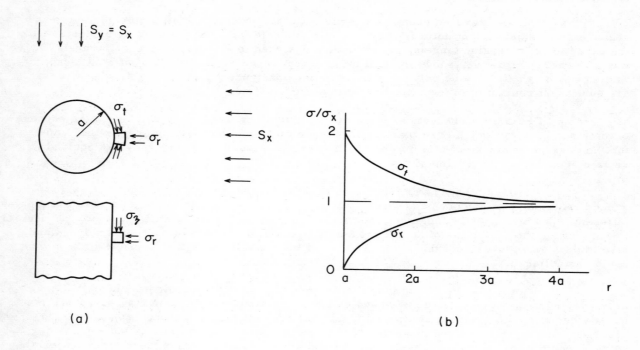

Fig. 3-1 Stresses Around a Circular Shaft in Elastic Ground

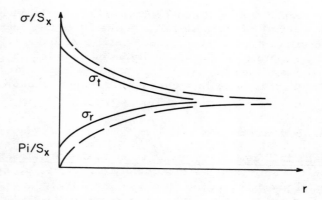

Fig. 3-2 The Effects on the Stresses Around a Shaft from Lining Backpressure

where σ_r is the radial stress, σ_t is the tangential stress, S_x is the horizontal stress, equal in all directions, existing in the rock before any excavation, a is the radius of the shaft, r the radial distance to the stresses, and p_i the internal or lining pressure.

In Fig. 3-1 the case is shown where the internal pressure, p_i, is equal to zero. As shown in Fig. 3-1(b), the horizontal tangential stress at the surface of the shaft is equal to twice the magnitude of the horizontal field stress. The radial stress, of necessity, is zero at the surface of the shaft. Both stresses then approach the field stress, S_x, being within 10 per cent of this value at a distance into the ground equal to the diameter of the shaft, i.e., at 3a from the centre of the shaft.

It can be seen from the equations that the tangential stress at the surface of the shaft does not vary with the size of the shaft but is always (for $S_x = S_y$) twice the magnitude of the horizontal field stress. On the other hand, the amount of ground that is subjected to the increased tangential stresses is proportional to the size of the shaft. It should be kept in mind that, in spite of the stress concentration in the horizontal plane around the shaft, the major principal stress may still be the vertical stress, S_z.

Since many shafts are inclined to the vertical, the analysis is difficult, even for ideally elastic ground. Photoelastic work has shown that, as the angle of inclination from the vertical increases, the stress concentration factors approach those applicable to a tunnel (1). This will be discussed again when examining the stress distribution around tunnels. Also, many shafts are rectangular in shape; however, the mechanics applicable to these cases, although more complex, would be similar.

In Fig. 3-2, the effect of a lining reaction or pressure on the stress distribution around a shaft in ideally elastic ground shows that the radial stress at the surface of the shaft must equal the lining pressure, p_i. The tangential stress at the surface of the shaft is decreased by the amount of the lining pressure. The depth of ground around the shaft subjected to any increased tangential stress is reduced from the case with no lining pressure.

Circular Shafts in Non-Ideal Ground: In Fig. 3-3 the reaction of ground that is not ideally elastic is shown. If the rock has a stress-strain curve with a slope that increases with an increase in stress or if the rock creeps like a viscous material with the strain being a function of time as well as stress, then the tendency will be for the tangential stress to be lower than for ideally elastic ground at the surface of the shaft and somewhat higher in the ground away from the shaft, as shown by the dashed line in Fig. 3-3.

Also, a common pattern that is emerging from current measurements around openings is to have a zone of relaxed, probably fractured ground that sustains very little stress. This relaxed ground remains intact and presumably builds up sufficient back pressure so that a point is reached away from the surface of the shaft where the tangential stress concentration effects are sustained, as shown by the dotted line in Fig. 3-3.

Another variation from the ideal case is shown in Fig. 3-4(a) where a fault passes close and parallel to the ultimate shaft surface. An element of ground around the shaft at the location of the fault would normally be sustaining radial and tangential stresses, as shown in Fig. 3-4(b); this would give rise to shear stresses in the plane of the fault as shown in Fig. 3-4(c). If the magnitude of this shear stress is greater than can be sustained by the fault gouge or breccia, then some displacement of this element will occur. The result will be that less stress will be transmitted across the fault plane than if it were not there. As shown pictorially (i.e., only one set of stress trajectories are shown) in Fig. 3-4(a), this tends to create higher stresses in the neck of ground between the fault and the surface of the shaft. If these stresses are of high enough magnitude, then failure could occur.

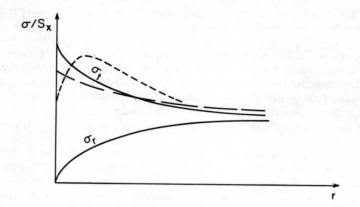

Fig. 3-3 The Effect on Stresses Around a Shaft from Rock Yielding

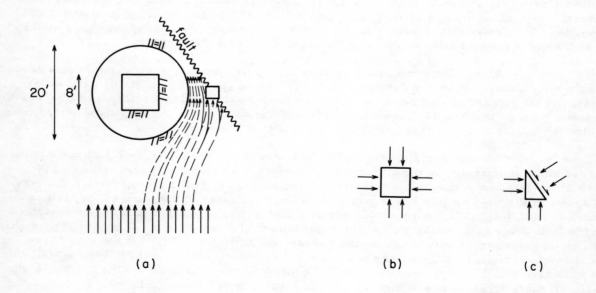

Fig. 3-4 The Effects on Stress Distribution Around a Shaft of an Adjacent Fault

In one case an 8 ft square pilot raise, as shown in Fig. 3-4(a), was driven without mishap. A 20 ft diameter shaft was then sunk on this raise. At one level in this operation a serious rock burst occurred. As several series of strong faults were known to exist in this ground, it is possible that one was located in the critical position shown in Fig. 3-4(a).

<u>Circular Shafts in Residual Stress Fields</u>: It is known that the horizontal stresses in the earth's crust are often greater than those due to the effect of gravity plus confinement. In other words, crustal movements or prestressing from eroded layers of overburden, or whatever the mechanism, can create horizontal stresses greater than those caused by gravity and even greater than the vertical stresses presently existing. Under these circumstances it would be expected that the horizontal stress would not be equal in all directions. Consequently, the tangential stresses at the surface of a shaft at any one elevation in this type of ground would vary around the shaft. Equations 2-10 can be used to determine the stresses around a shaft in such a stress field:

$$\sigma_r = \frac{1}{2}(S_x+S_y)(1-a^2/r^2) + \frac{1}{2}(S_x-S_y)(1-4a^2/r^2+3a^4/r^4)\cos 2\theta \qquad \text{Eq. 3-2(a)}$$

$$\sigma_t = \frac{1}{2}(S_x+S_y)(1+a^2/r^2) - \frac{1}{2}(S_x-S_y)(1+3a^4/r^4)\cos 2\theta \qquad \text{Eq. 3-2(b)}$$

In Fig. 3-5 a case is shown where the horizontal field stress S_y is four times the horizontal field stress S_x. In this case, the tangential stress at the point A would be $2.75 S_y$, whereas the tangential stress at the point B would be $-0.25 S_y$. The variation of these stresses with radial distance is shown in Figs. 3-5(b) and 3-6(c).

It should be recognized that stress concentrations can be increased not only by having σ_x different from σ_y but also by having anisotropic ground. Theoretical solutions, supported by some measurements, indicate that the stress concentrations can be much greater in anisotropic than in isotropic ground (12, 22). For example, in Fig. 3-5(a) if $S_x = 0$, the compressive stress at A would increase from $3 S_y$ to $4.3 S_y$ if $E_y/E_x = 6$, and the tensile stress would increase from $-S_y$ to $-5.1 S_y$ if $E_y/E_x = 0.04$ (25).

It might be reasoned that the distortion of a circular hole created by an excavation in a stress field with different principal stresses would be similar to the generally assumed distorted shape that would occur if a circular hole were created in elastic ground with different moduli of deformation in the x- and y-directions but with a constant horizontal stress in all directions. Consequently, by analogy it could be expected in such anisotropic ground that stress concentrations similar to the case represented in Fig. 3-5 would occur. However, the deformation of a circular hole in anisotropic ground can be conspicuously different from an elliptical shape, and hence such rationalizations should be used with caution (15).

If the stresses at the surface of a shaft in elastic ground in a residual stress field are of sufficient magnitude, failure will occur. With tensile stresses tending to develop at the point B in Fig. 3-5(b), radial cracking might occur. Alternatively, if the stress S_y was particularly large, the concentration of this stress at the point A might cause some crushing or compression failure. As this crushed zone extended into the rock, secondary diagonal tension could cause some tensile cracking at the sides of these crushed zones as shown in Fig. 3-5(d).

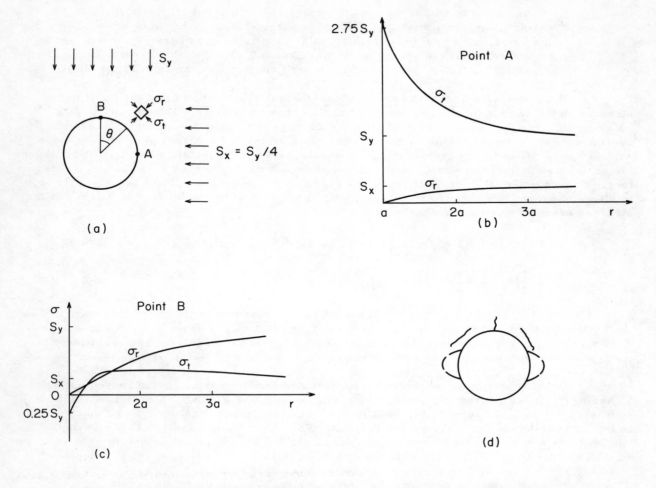

Fig. 3-5 Stress Concentrations Producing Failure Around a Shaft

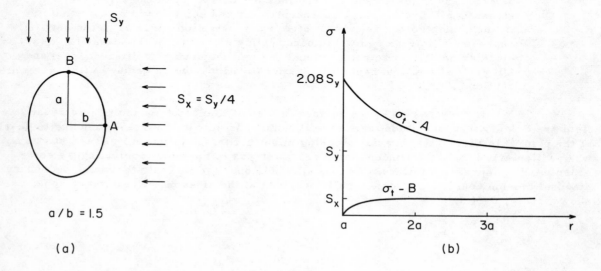

Fig. 3-6 Stress Distribution Around an Elliptical Shaft

Some work has shown that the calculated tangential stress in one case had to exceed the laboratory tested uniaxial compressive strength by about 10 per cent before failure would occur (15); in another case, the strength had to be exceeded by 100 to 200 per cent before failure occurred (17). These results probably were due to relaxation of surface ground with some cohesion being retained. For example, this relaxation of surface material was shown explicitly to occur in a hematite where the exterior 2 feet to 3 feet of the pillars was found by measurements to be sustaining very little tangential stress (18).

It has often been suggested that, where the magnitude of the horizontal field stress varies with direction, shafts should be elliptical with the major axis in the direction of maximum horizontal stress. Fig. 3-6 shows the effects of an elliptical opening in a stress field where $S_y = 4 S_x$. In this case, the tangential stress at the point B is zero as opposed to tensile for the circular opening shown in Fig. 3-5, and the tangential stress at the point A is $2.08 S_y$ as opposed to $2.75 S_y$ for the circular opening. The practical difficulties in this procedure, at the present time, arise from the lack of an established method of determining the field stress characteristics before the shaft is sunk and are due also to the difficulties of determining if the elastic properties in the horizontal plane throughout the length of the shaft are isotropic.

GROUND SUPPORT AROUND SHAFTS

<u>Excavation and Lining to Prevent Failure</u>: To prevent the rock pressures associated with fractured ground from developing, it might be possible to reduce the tangential stresses created by the shaft excavation by applying a pressure to the rock and thus achieving the results as shown in Fig. 3-2. Shafts up to 20 ft in diameter can be drilled by a large machine that cuts a core and withdraws it from the ground.

During the drilling operation a back pressure is applied to the rock surface by having the excavation filled, as in oil drilling, with drilling mud. This mud has a density sufficient to apply a pressure of about 0.6 psi per foot of depth. This pressure at a depth of 1000 feet would amount to 600 psi, which would be sufficient to prevent rock failure as a result of stress concentrations in most rock.

When the entire shaft has been drilled, a precast concrete lining is lowered into place. Grouting behind the lining then provides a pressure against the rock surface to reduce the ratio of major to minor principal stresses even more than the mud pressure.

<u>Uniform Pressure from Granular or Broken Ground</u>: If the rock around a circular, lined shaft fails, the resultant situation might be somewhat as shown in Fig. 3-7(a). In terms of mechanics we can think of this situation as shown in Fig. 3-7(b), where the inside radius of the shaft is 'a' and the radius to the elastic or unfractured zone is 'r_e'. The annular ring of ground between 'a' and 'r_e' is in a plastic state of equilibrium (strain irrecoverable) with a stress relationship being governed by Mohr's Strength Theory.

This annular ring of plastic ground can be analysed following the analysis that was made in Chapter 2 (see Fig. 2-4(c)), for the stress relations in a thick-walled cylinder (3). The equilibrium equation that applies to this ground will be the same as that for the elastic thick-walled cylinder:

$$\sigma_t - \sigma_r - r \cdot d\sigma_r/dr = 0 \qquad \text{Eq. I}$$

The second source of equations for this case, instead of the requirements for compatible elastic strains, comes from the stress conditions for incipient motion in this granular ground. From Mohr's Strength Theory we can write:

$$(\sigma_t + S_c)/(\sigma_r + S_c) = \tan^2(45 + \varphi/2) = f \qquad \text{Eq. II}$$

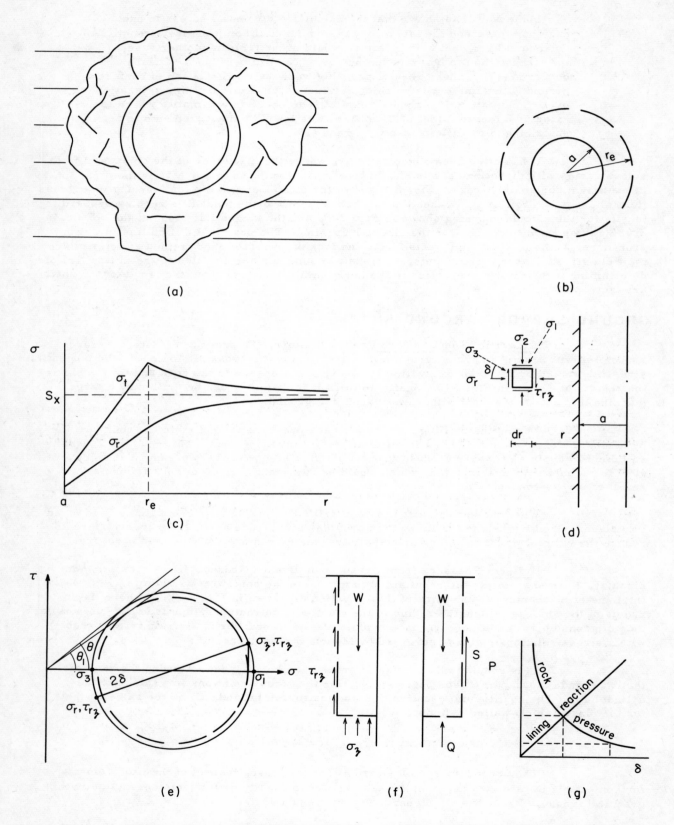

Fig. 3-7 The Analysis of Rock Pressure on a Shaft Lining in Yielding Ground

where φ is the angle of internal friction of the broken ground, and $S_c = c/\tan\varphi$ with c being the cohesion.

The third source of information, as for the elastic case, is obtained from the boundary conditions. These are:

$$\text{when } r = a$$
$$\sigma_r = p_i$$
$$\text{when } r = r_e$$
$$\sigma_r = p_o \qquad \text{Eq. III}$$

where p_i is the internal lining pressure on the thick-walled cylinder of plastic ground and p_o is the pressure on the outside of the annular ring.

By solving the above three sets of equations we can obtain:

$$\sigma_r = (p_i + S_c)(r/a)^{f-1} - S_c \qquad \text{Eq. 3-3(a)}$$

$$\sigma_t = (p_i + S_c)(r/a)^{f-1}\tan^2\alpha - S_c \qquad \text{Eq. 3-3(b)}$$

where f is used for $\tan^2(45 + \varphi/2)$ and α is used for $(45 + \varphi/2)$. It can be seen from these equations that, in the plastic state, only one boundary condition, p_i, is required for the solution of the stresses.

From Equations 3-3, both σ_r and σ_t are shown to increase with the radius, r, raised to some power that will vary between about one and five. However, this cannot continue indefinitely, otherwise there would be extremely large stresses at moderate distances away from the shaft. It is easily imagined that, eventually, at some distance from the shaft elastic conditions would be regained. In other words, the ratio of the principal stresses would become smaller than the critical value as expressed by the flow condition in Equation II.

At the transition point between the elastic and plastic zones, the equations for σ_r and σ_t for both plastic and elastic equilibrium should be valid; in particular, there must be a continuity of σ_r. Solving these equations for σ_r simultaneously (i.e., Equations 2-12 and 3-3), the distance to the transition point, r_e, can be found:

$$r_e = a\left(\frac{2}{f+1} \frac{S_x + S_c}{p_i + S_c}\right)^{\frac{1}{f-1}} \qquad \text{Eq. 3-4}$$

where a is the radius of the shaft, S_x is the horizontal field stress, $S_c = c/\tan\varphi$, p_i is the internal pressure on the rock surface, and $f = \tan^2(45 + \varphi/2)$. Figure 3-7(c) shows the theoretical stress relations for the above case where the horizontal field stress, S_x, is the same in all directions.

<u>Example</u> - Consider the unusual case of a shaft with a radius of 10 ft passing through a soft shale in a horizontal field stress of 1000 psi, i.e., $S_x = S_y = 1000$ psi. Assume that the shale will yield and has a Mohr envelope defined by $\varphi = 30°$ and c = 15 psi. Calculate the radius, r_e, of the plastic zone if no lining resists the expansion of the shale.

For use in Equation 3-4:

$$p_i = 0$$

$$S_c = 15/\tan 30 = 26.0$$

$$f = \tan^2(45 + 30/2) = 3$$

Thus $r_e = 10 \left(\dfrac{2}{3+1} \dfrac{1000 + 26.0}{0 + 26.0} \right)^{\frac{1}{3-1}} = 44.5 \text{ ft}$

The above equations have been derived assuming that the rock would fail by yielding without any reduction in cohesion. Where failure is by fracture with sufficient expansion to make the effective cohesion zero, Equations 3-3 would still describe the stresses in the plastic zone as S_c would simply become zero. However, Equation 3-4 could not be used unless the ground beyond the plastic zone were also granular, that is, with no cohesion. This would be the case for a shaft in sand, but not for one in competent rock. The explicit solution for the thickness of the plastic zone for this brittle rock case can be approximated but not solved(4).

In Fig. 3-7(c), σ_r is shown to have some positive value at the surface of the shaft. This represents the case where there is a lining pressure, p_i, acting against the rock. The maximum value of σ_t is then greater than this pressure by the factor of $\tan^2 \alpha$. In the case of fractured ground without cohesion, no tangential stress, σ_t, could be sustained without a positive lining pressure. On the other hand, where some cohesion has been retained, then a tangential stress, $\sigma_t = S_c (\tan^2 \alpha - 1)$ from Equation 3-3(b), can be sustained and equilibrium established.

However, where the ground around the shaft has failed, the vertical stress for the elastic condition before failure might have been the major principal stress and thus greater (as was mentioned above) than the horizontal tangential stress. In this case, the stress initiating failure would be the vertical stress.

Failure due to the vertical stress would require both downwards and inwards movement of an element on the surface of the shaft. The inward movement for a circular opening would induce a sufficiently high tangential stress, σ_t, to cause yielding in the horizontal as well as the vertical plane. Whereas in the elastic situation the tangential stress, σ_t, would be the intermediate principal stress, under failure conditions it would become equal to the major principal stress. Hence, the above analysis of the stress conditions in the horizontal plane would still have applicability.

The rigorous solution of the three-dimensional problem has not yet been established; however, an engineering type solution with some simplifying assumptions yields interesting results. The following analysis is a modification of work that was done previously (5). In Fig. 3-7(d), the vertical stress, σ_z, and the radial stress, σ_r, are not principal stresses owing to the shear stresses, τ_{rz}, that arise on the tangential planes. As can be seen, the major principal stress, σ_1, will be inclined outwards from the shaft.

The assumption can be made that at the surface of the shaft the angle of inclination of the minor principal stress, σ_3, is close to zero. This would be equivalent to assuming either that there is no friction between the rock and the lining or that the lining moves down with the rock as it sags. By making this assumption the calculated lining pressure, p_i, should be greater than the pressure where the minor principal stress is inclined to the horizontal at the surface of the shaft.

Assuming the inclination of σ_r to be zero at the surface of the shaft would, in the case of a retaining wall, produce a more severe earth pressure than when the inclination was something greater than zero. On the other hand, the thickness of the plastic zone, r_e, would be greater and hence the lining pressure, p_i, could be less as σ_r would have a greater distance to rise to the value appropriate at the transition point from plastic to elastic conditions. In any event, the results of the proximate analysis will indicate that these differences are possibly not significant.

With the above assumption it is then established that at $r = a$, $\sigma_1 = \sigma_z$ and $\sigma_3 = \sigma_r$. Similarly, it is known that when $r = \infty$, $\sigma_1 = \sigma_z$ and $\sigma_3 = \sigma_r$. In between these extreme points, τ_{rz} is greater than zero and thus $\sigma_1 \neq \sigma_z$ and $\sigma_3 \neq \sigma_r$. Hence, it is known that, where $r = a$ and ∞, the maximum possible value of the ratio $\sigma_z/\sigma_r = \tan^2 \alpha$, but in between these two locations the maximum value of the ratio σ_z/σ_r is less than $\tan^2 \alpha$. This ratio can be given the following expression:

$$\sigma_z/\sigma_r = \tan^2(45 + \varphi_1/2)$$

with the significance of φ_1 as shown in Fig. 3-7(e)(5). It is probable that the angle φ_1 will vary with r; however, it may not differ greatly from the angle φ.

If it is assumed that φ_1 is approximately equal to φ, then Equations 3-3 can be used for this three-dimensional case. It also means, since the two circles in Fig. 3-7(e) coincide, that σ_z is approximately equal to σ_1 and σ_t, and that σ_r is approximately equal to σ_3.

With the objective of obtaining a solution for the lining pressure that would result from ground failure around the shaft, it can be recognized that, with Equations 3-3 and the assumption that $\sigma_z = \sigma_t$, there are three equations with four unknowns, which makes it necessary to obtain from some other source an additional equation. This can be done by examining the gross equilibrium of an annular ring around the shaft as shown in Fig. 3-7(f). For this analysis it is necessary to assume that the annular ring of plastic ground is the same thickness from the bottom to the top of the shaft.

If it is assumed that the stresses σ_r and τ_{rz} on the outside of the annular ring shown in Fig. 3-7(f) vary linearly with z, then the following equation can be written (5):

$$W = S + Q$$

$$\pi(r^2 - a^2)\gamma z = \frac{1}{2} \times 2\pi r z \tau_{rz} + \int_a^r 2\pi r \sigma_z \, dr$$

where σ_3 is the stress and Q the force on the bottom of the annular ring, and S is the shear force on the outside vertical surface. It is then convenient to convert the shear stress as follows: $\tau_{rz} = (\tau_{rz}/\sigma_r)\sigma_r = A\sigma_r$. By solving for A using these two equations, Equations 3-3, and recognizing that $\sigma_z = \sigma_t$ as explained above, the following is obtained:

$$A = \frac{(r/a)^2 - 1}{m(r/a)^f} - \frac{2f}{f+1} \frac{(r/a)^{f+1} - 1}{(r/a)^f} \frac{a}{z} \qquad \text{Eq. 3-5(a)}$$

where $m = p_i/\gamma a$ and $f = \tan^2(45 + \varphi/2)$.

This equation can be simplified by taking the extreme case $z = \infty$:

$$A = \frac{(r/a)^2 - 1}{m(r/a)^f}. \qquad \text{Eq. 3-5(b)}$$

It is reasonable to expect that A will increase from zero to a maximum at some distance r_1; hence, at r_1 $dA/dr = 0$. Thus, in the above equation where $z = \infty$, A can be differentiated with respect to r and the value of r_1 determined:

$$r_1 = a\left(\frac{f}{f-2}\right)^{\frac{1}{2}} \qquad \text{Eq. 3-5(c)}$$

With $A = \tau_{rz}/\sigma_r$, it is known that the maximum value that this ratio could have would be $\tan\varphi$ (see Fig. 3-7(e)). This maximum value would apply if the tangential plane at the distance r_1 were a failure plane. If this tangential plane is not a failure plane, then the maximum value of A would be less than $\tan\varphi$ and the computed lining pressure would be greater. However, as there is no obvious way of determining the actual value of A, it will be assumed to be $\tan\varphi$. Then, by inserting this maximum value of A and the value of r_1 for r into the above equation for A, it is possible to solve for the lining pressure, p_i:

$$p_i = \frac{2\gamma a}{\tan\varphi(f-2)} \left(\frac{f-2}{f}\right)^{f/2} \qquad \text{Eq. 3-5(d)}$$

It should be kept in mind that Equation 3-5 may produce an unconservative answer for this case owing to the various assumptions that have been made; on the other hand, the equation applies to an infinite depth, which is an element of conservatism. Furthermore, the deformation properties, as examined below, of both ground and lining are very important.

Example - To illustrate the order of magnitude to be expected for this case, let $\varphi = 40$, $\gamma = 170$ pcf and $a = 10$ ft. Therefore, from Equation 3-5:

$$p_i = \frac{2 \times 170 \times 10}{0.839(4.60-2)} \left(\frac{2.60}{4.60}\right)^{2.30} = 418 \text{ psf} = 2.90 \text{ psi}$$

This calculated lining pressure looks small; however, other factors, as examined below, add to its significance.

By an ingenious numerical analysis it has been suggested that, where φ is greater than 30° and less than 40°, the maximum value of $A = \tan(\varphi-5)$, and thus the average effective ratio $S_z/S_r = \tan^2(45 + (\varphi-5)/2)$. The calculated lining pressure, p_i, is then the minimum to maintain stability (5). Consequently, if it is assumed that these relationships are correct, the same lining pressure as calculated above would be obtained if the angle of internal friction, φ, was 45°, which is a more probable value for broken rock. Consequently, the same conclusion, that p_i looks very small, still applies.

The above analysis, as is the case for any plastic analysis, requires that sufficient strain occurs to mobilize the full internal friction. The amount of deformation of the surface of the shaft should be analysed to determine if this can occur. By combining Equations 2-6 and 2-9 the following equation is obtained:

$$p_c = d E (1-(r_i/r_o)^2)/(2r_i) \qquad \text{Eq. 3-6}$$

where p_c is the pressure on the outside of the lining, d is the radial deformation of the inside surface of the lining, and r_i and r_o are the inside and outside radii. Unfortunately, there is no theoretical method and little empirical information for the determination of the required strain to mobilize internal friction. Thus, when possible, the individual cases should be based upon appropriate test data from the actual ground in question.

To obtain some idea of the order of magnitude of deformation that might be required, typical values of failure strain and Poisson's Ratio for gravel or sand can be tried: $\epsilon_f = 5\%$ and $\mu = 0.3$. For $\varphi = 40°$, r_1 from the equation above is calculated to be $1.33a$. The radial deformation of the surface of the shaft, d_f, required to mobilize the full internal friction of the plastic zone, considering that failure is essentially due to vertical stress, would be:

$$d_f = \epsilon_x \cdot \Delta r = \mu \epsilon_z \Delta r = \mu \epsilon_f \Delta r$$

$$d_f = 0.3 \times 0.05 \times 0.33a = 0.005a.$$

Thus, for a shaft 20 ft in diameter the change in radius would be 0.05 ft and the change in diameter would be 0.1 ft.

Alternatively, if we recognize that sand or gravel has a relatively high porosity, e.g., commonly between 30 and 50 per cent, and that, initially at least, broken rock will have a porosity close to zero, then it might be more valid to consider the failure strain obtained in testing solid rock. The failure strain would be of the order of 0.5 per cent with a Poisson ratio, μ, of possibly 0.3. In this case, the deformation of the surface of the lining would be 0.0005a with a change in radius for a 20 ft shaft of 0.005 ft and a change in diameter of 0.01 ft.

The significance of these values of radial deformation can be examined by taking a typical concrete lining with a thickness of 12 in., $E = 3 \times 10^6$ psi and $\mu = 0.15$ for a shaft with a 20 ft rock diameter.

The external pressure on the lining that would be caused by a deformation of 0.005 ft can be calculated from Equation 3-6:

$$p = 0.005 \times 3 \times 10^6 (1-0.9^2)/(2 \times 9) = 158 \text{ psi.}$$

As this pressure is much greater than the minimum rock pressure calculated above of 2.9 psi, it means that the stiffness of the lining is such that, if it were in intimate contact with the rock, it would not permit the rock failure strain of 0.5 per cent, let alone that of 5 per cent, to occur; thus, the full internal friction resistance of the broken rock could not be mobilized. For the minimum lining pressure of 2.9 psi to apply, either a gap must exist initially between the lining and the rock or the lining must be of less resistant material than concrete. Consequently, it can be seen that the deformation characteristics of the problem are very significant. In addition, the value of timber supports or yielding steel arches can be seen in this context.

We now have a situation whereby, as the deformation of the lining increases, the rock pressure decreases but the lining reaction increases; the functional relations will be as shown in Fig. 3-7(g). For the above case we can write the equation of the lining reaction, p_c, from Equation 3-6 as a function of the deformation of the inside surface, d, with the units being inches:

$$p_c = d \times 3 \times 10^6 (1-0.9^2)/(2 \times 9 \times 12) = 2640 \, d.$$

Also, we can postulate a relationship between A (i.e., τ_{rz}/σ_r) and the deflection of the lining. If we assume, as above, that the maximum value of A is $\tan\varphi$, that the variation with d is linear, and that the radial deformation of 0.005 ft is required to mobilize the full internal friction, then for $\varphi = 40°$:

$$A = k\,d$$

where   k = the constant of proportionality;

when   d = 0.005 ft = 0.06 in.   A = tan 40 = 0.839

hence   k = 0.839/0.06 = 14.0 in.$^{-1}$

and   A = 14.0 d

Using this value of A in Equation 3-5(b) and using Equation 3-5(c) for r, the rock pressure, at $z = \infty$, can be expressed as a function of A:

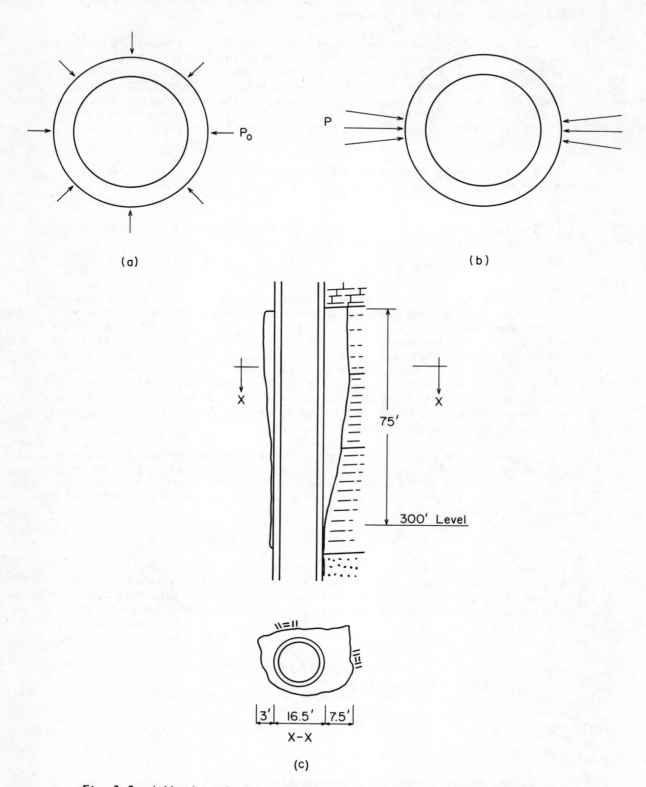

Fig. 3-8 a) Uniformly Distributed Rock Pressure on Shaft Lining
b) Concentrated Rock Pressure on a Shaft Lining
c) Typical Rock Failure Around a Shaft in Sedimentary Rocks

$$p_i = \frac{2\gamma a}{A(f-2)} \left(\frac{f-2}{f}\right)^{f/2}$$

$$= \frac{2 \times 170 \times 10}{144 \times 14.0\,d(4.60-2)} \left(\frac{4.60-2}{4.60}\right)^{4.60/2} = 0.173/d$$

Solving these two simultaneous equations for the rock pressure, p_i, and the concrete reaction, p_c, we can determine the point of intersection of the two curves in Fig. 3-7(g):

$$p_i = \sqrt{0.173 \times 2640} = 21.4 \text{ psi.}$$

Hence, the maximum stress on the inside surface of the lining can be calculated using Equation 2-14:

$$\sigma_t = \frac{2 \times 21.4}{1 - 0.9^2} = 225 \text{ psi}$$

Alternatively, if the deformation required for the full mobilization of the internal friction were as great as 0.05 ft (i.e., for the large failure strain of 5 per cent) the above equations would change as follows:

$$A = 1.4\,dr$$

$$p_i = 1.73/dr$$

Hence, $\quad p_i = 67.4 \text{ psi}$

and $\quad \sigma_t = 710 \text{ psi.}$

<u>Concentrated Pressures from Granular or Broken Ground</u>: The lining stresses calculated above, which are less than normal permissible stresses, are based on the assumption that the rock pressure will act uniformly around the outside of the lining. From experience we know that this is not likely to occur in fractured rock, although it may in sand. Rock pressure can be expected to be concentrated on a few isolated points of contact. The effects of this condition can be analysed using the following equations (6):

$$\sigma_{cpn} = p_o(r_o^2 + r_i^2)/(r_o^2 - r_i^2) + 5\bar{p}r_o^2(3r_o^2 + r_i^2)/(r_o^2 - r_i^2)^2 \qquad \text{Eq. 3-7(a)}$$

$$\sigma_{tsn} = 2p_o r_o^2/(r_o^2 - r_i^2) - 6\bar{p}r_o^2(r_o^2 + 3r_i^2)/(r_o^2 - r_i^2)^2 \qquad \text{Eq. 3-7(b)}$$

The above equations are based on the assumption that a lining is equivalent to a ring that is free to deform under the external pressures. p_o is the component of rock pressure that is uniformly distributed around the outside of the lining; $\bar{p}$ is the pressure obtained by taking the rock force, P, that is concentrated over two diametrically opposite small arcs (less than 10°) (see Fig. 3-8(b)) and dividing by $2\pi r_o$.

With these assumptions, σ_{cpn} is the maximum compressive stress that would exist in such a ring and would occur on the outside surface of the lining; σ_{tsn} is the maximum tension, which would occur on the inside of the lining. These stresses are tangential stresses; the analysis does not include the local effects of the contact pressures.

<u>Example</u> - If we take the previous example and assume that half the rock pressure will be uniformly distributed, i.e., $p_o = 11$ psi, and half the rock pressure is concentrated on diametrically opposite points, i.e., $\bar{p} = 11$ psi, then the maximum stresses are as follows:

$$\sigma_{cpn} = 5,900 \text{ psi}$$

$$\sigma_{tsn} = 6,100 \text{ psi}$$

These stresses, of course, are far in excess of the failure stresses in the concrete lining. However, it should be recognized that, unless the lining were very loose, possibly due to shrinkage of the concrete or possibly due to using pre-cast sections without back grouting, the lining would not be free to deflect outwards at 90° to the pressure points like a ring, and thus the development of these stresses would not be permitted.

A better idealization of the typical case would be to assume that half of the lining is a fixed arch. In other words, in Fig. 3-8(b) the top and bottom points of this ring would be prevented from deflecting outward by the adjacent rock and, for the case of symmetrical loading, these points would not rotate. Consequently, the top and bottom points would act as if they were fixed.

In this case, it can be shown that the computed stresses would be about one-third of those computed above, assuming the lining to be a ring free to deflect. Consequently, the following equations could be used where it is judged that under concentrated loads the lining would behave like a fixed arch:

$$M_o = PR \left[\frac{0.471 + 6.82m}{0.299 + 1.571m} \right] \qquad \text{Eq. 3-8(a)}$$

$$M_1 = PR \left[\frac{0.158 - 0.032m}{0.299 + 1.571m} \right] \qquad \text{Eq. 3-8(b)}$$

where M_o is the moment at the concentration of pressure, M_1 is the moment one-quarter of the way around the shaft from this point, and $m = I/(AR^2)$ (where I is the moment of inertia of the lining section, A is the cross-sectional area and R is the radius to the centroid.)

Several observations can be made as a result of the above analyses. First, the most severe stresses in a lining are likely to arise from local concentrations of rock pressure. Consequently, any device that can reduce these concentrated loads is likely to make a significant difference in the behaviour of the lining; for example, placing some easily deformable material such as timber, sand, or possibly one of the new plastic foam products behind stiff linings could reduce any potential concentrations.

Considering normal preconceptions versus analysis it is interesting to ask: If a high horizontal, residual stress existed in the N-S direction, would the points of concentrated pressure on a lining in an underground opening be on the N-S axis? The high compressive stress concentrations in the rock would be at the E-W points; consequently, if no other factors were involved, compressive failure should occur at these points first with the consequent expansion of the rock. Thus, the areas of concentrated rock pressure would be on the E-W axis and not on the N-S axis as might be initially assumed.

Second, there would seem to be an advantage to select, if possible, a lining material with a low modulus of deformation in relation to the strength; for example, the ratio of the modulus of deformation to the uniaxial compression strength of concrete is commonly assumed to be about 1000. For mild, structural grade steel this ratio for the yield stress would be about 900, but the ratio for the rupture stress would be about 400. Aside from costs and other practical factors, this low ratio of modulus of deformation to strength for steel would be a strong point in its favour.

The third observation that can be made is that, if the original lining were ruptured as a result of rock pressures generated as in the above example, a second lining with the same thickness might be subjected to very low stresses. This would be so if sufficient deformation in the rock occurred to develop the full internal friction of the fractured zone with the average rock pressure reduced to its minimum. Also, the original concentration of rock pressure might be smoothed out as a result of the failure of the original lining.

A fourth observation would be that a ground support method such as rock bolting would have one advantage in that it would not be sensitive to the unequal distribution of rock pressure that is so disruptive to a structural lining. The strain accompanying the fracturing of surface rock might still be a factor to consider in a bolting system, not to mention the destructive effects any fracturing farther back would have on the anchorages of the bolts.

A case where stress measurements were obtained can be examined (6). A shaft, 26 ft in diameter, was excavated in weak, faulted shale. A lining was placed consisting of 10 in. of hollow concrete blocks and 18 in. of poured concrete inside the blocks. The composite lining was made tight with grout. Measurement on the inside of the lining showed that the stresses increased over five to eight months and then generally decreased. Some of the maximum compressive stresses were as follows:

Depth	Tangential Stress
383 ft	2420 psi
415 ft	920 psi
895 ft	710 psi

Unfortunately no information was obtained on the rock characteristics. Assuming the density of the rock was 160 pcf and $\varphi = 30°$, Equation 3-5(d) can be used to calculate the average pressure from the ruptured rock:

$$p = \frac{2 \times 160 \times 13}{0.577(3-2)} \left(\frac{3-2}{3}\right)^{3/2} = 1380 \text{ psf} = 9.6 \text{ psi.}$$

From Equation 3-7(a) the concrete lining stress, with $r_i = 10.5$ ft and $r_o = 12.0$ ft, can be calculated. First, assuming $p_o = p$ and thus $\bar{p} = o$:

$$\sigma = 9.6 (12^2 + 10.5^2)/(12^2 - 10.5^2) = 71.2 \text{ psi,}$$

which is below the measured values. If it is assumed that $p_o = p/2 = \bar{p}$, then, still using Equation 3-7(a):

$$\sigma = 72.2/2 + \frac{5 \times 4.8 \times 12^2(3 \times 12^2 + 10.5^2)}{(12^2 - 10.5^2)^2} = 1631 \text{ psi.}$$

From these calculations it can be seen that varying degrees of rock pressure concentration probably occurred. Also, a decrease in stress may have resulted from the crushing of some of the hollow blocks behind the concrete, which then redistributed local areas of load concentration.

<u>Pressure from Broken Ground on Rectangular Shafts</u>: Another, possibly more common, action that can result in rock pressure on linings is from local failures in the rock around a shaft (23). With a rectangular shaft, arching action in the ground is less effective in assisting in the support of the broken ground. Fig. 3-8(c) shows the pattern of local failures that have occurred in shafts.

The resultant rock pressure from such local failures is equivalent to the earth pressure resulting from the wedge of ground that tends to slip out. Hence, Equation 1-11(b) would be used for analysing the rock pressure. The height of these wedges would be dependent on the thickness of the weak rock in which they occurred.

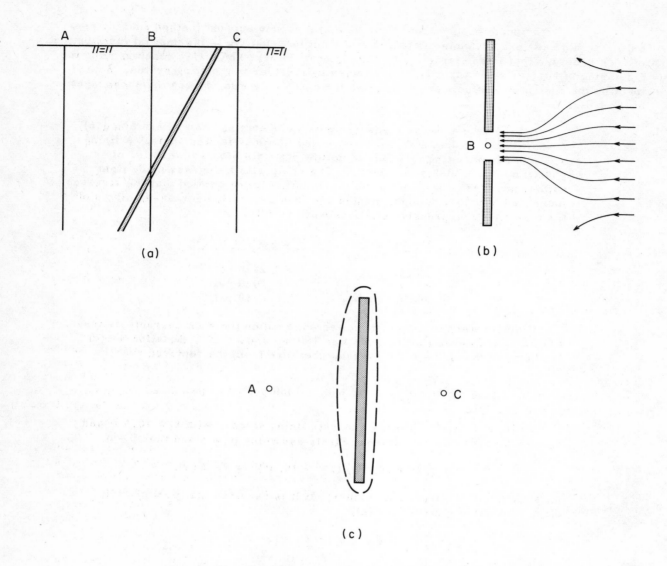

Fig. 3-9 Alternate Locations of a Shaft

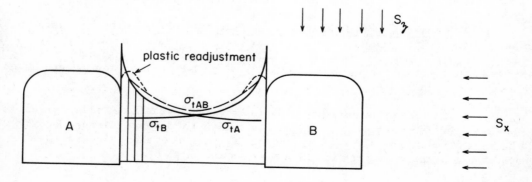

Fig. 3-10 Stress Distribution Between Two Openings

One survey of shaft rock failures showed that about 90 per cent of the shaft walls that failed were, not surprisingly, in rocks such as shale, coal and marls (23). The largest failures were usually associated with faults, folds, and altered zones. Furthermore, the number of failures, although not the size, increased with the volume of ground water in evidence.

A final, very significant finding was that about 65 per cent of the failures were found, in the survey of some 127 shafts extending down to 3000 ft, to have occurred at depths less than 600 ft. This observation does much to diminish the importance in the preceeding analyses of the effects of gravity and of stress concentrations.

Shallow shafts through very soft ground act more in accordance with the predictions based on gravitational effects. In these cases, compressed air can be used during sinking to apply a supporting pressure to the surface of the ground to diminish the shear stress and thus to prevent failure. However, the use of compressed air is most common when the soft ground is combined with a ground water condition that would either promote instability of the ground or produce large volumes of water. For both purposes the depths to which compressed air can be used is limited. (When the compressed air pressure goes above about 20 psi, labour costs increase since most areas have regulations that prevent workers from working 8 hours a day at higher pressures.)

Pressure from Viscous Ground: Another source of rock pressure in some formations can be the long-term deformation of the rock caused by the excavation of the shaft. This is a viscous-type reaction with the rate of deformation of the rock generally decreasing with time (see Fig. 1-6(f)). In these cases, it is useful to obtain a measure of the deformation-time characteristics of the in situ ground around a test tunnel. When this has been done, it is then possible to extrapolate the measured deformation to obtain some measure of the additional deformation that might occur during the life of the shaft.

It must be assumed that beyond shallow depths this deformation is practically irresistible. Consequently, it only remains to calculate the stresses in whatever lining is to be used that will be induced by the predicted deformation. If the lining stresses would become excessive within a short period of time, then one alternative would be to use a temporary or deformable lining until such time as the strain rate has decreased to the extent where the induced stresses over the life of the shaft would be within specified limits. This procedure will be amplified below in connection with tunnels.

Effects of Location with Respect to Stopes: The location of a shaft with respect to the ore zone will affect the stress distribution in the rock. A typical ore zone is shown in Fig. 3-9(a). To avoid the effects on the adjacent ground of the stoping, the shaft could be located some distance on strike away from the mining zone. However, it is common to consider locations adjacent to the mining zone either in the hangingwall, through the vein, or in the footwall.

A hangingwall shaft, as shown by A in Fig. 3-9(a) and Fig. 3-9(c), would have to be located a considerable distance from the mining zone so that the stresses around the shaft would not be affected by the stoping. In addition, unless a pillar is left in the ore zone, the hangingwall shaft must be a sufficient distance from the mining zone so that it will not be affected by subsidence (treated in Chapter 5) or by any zone of failed rock (as indicated by the dashed line in Fig. 3-9(c)) that develops around the stopes.

For a vertical shaft the same factors would apply to a footwall shaft, as shown by Location C in Fig. 3-9(a); however, the distance from the ore zone at the surface can be less to avoid the subsidence effects.

A shaft that goes through the vein (as shown by Location B in Fig. 3-9(a)) would be subjected to increased stresses. It is usually necessary for a shaft located at B to be enclosed in a pillar as shown in Fig. 3-9(b). It can be seen here that there will be a concentration of the field stress in the shaft pillar, which in turn will be concentrated around the shaft itself. This can result in some very high tangential stresses in the surface rock of the shaft. In addition, it might be difficult to avoid the effects of subsidence at the upper levels of the shaft unless a very large pillar is left opposite the shaft throughout the depth of the mining zone.

STRESSES AROUND DRIFTS AND TUNNELS

Elastic Ground: Stresses around drifts in ideally elastic ground can be represented by Equations 3-2, which are applicable to a hole in a biaxial stress field as shown in Fig. 3-10. Normally the horizontal stress, S_x, would be different from the vertical stress, S_z. This case is thus qualitatively similar to that considered in Fig. 3-5 of a shaft in a residual stress field. The stress distribution and possible modes of failure would be similar to this shaft case.

The problem of interaction of openings arises with drifts and raises and with intersecting and parallel tunnels.

In Fig. 3-10 the case is shown of two tunnels, A and B, running parallel to each other and separated by a pillar equal in width to 1.5 times the width of the tunnels in a stress field with the major principal stress being vertical. The curves of the tangential stresses in the sides of each tunnel due simply to a single opening are labelled σ_{tB} and σ_{tA}. The combined effect is shown by the dashed curve σ_{tAB}.

It can be seen for this ideal case that, although the average stress in the pillar is increased somewhat, the maximum stresses at the surfaces of the openings are not affected significantly. Any tendency for plastic readjustment would modify the combined curve, as shown by the dotted line in Fig. 3-10. The variation of the relative significance of these various effects could, of course, produce an infinite number of stress patterns in such pillars.

The interaction of intersecting drifts and crosscuts is probably more common than the interaction of parallel openings. If a crosscut intersects a drift to produce a tee or cross intersection, then it can be visualized that, during driving, the crosscut would be entering ground subjected to stress concentrations arising from the drift. As the effect of the crosscut would be to create its own stress concentrations, a multiple effect would occur.

Although the mathematical problem presented by such stress concentrations has not been solved, it can be imagined that, in a stress field with the major principal stress being vertical, the stress concentrations in the sides at the corners of the intersection would be equal to the square of the stress concentration for one opening. This, of course, would only be valid for ideally elastic ground and would be subjected to all the modifications illustrated in Fig. 3-3. Nevertheless, the sides at these corners will be zones of much higher than usual stress with the consequent probability of failure being greater here than at other locations. (Besides these stress effects, the significant factors affecting strength, e.g., geological weaknesses and deterioration with time would be considered.)

The intersection of a raise and a drift also provides a case of intersecting stress concentrations. Again, the solution of this case has not been rigorously solved. Photoelastic studies indicate that in a gravitation field the stress concentrations may be no more severe than for the drift itself (24). However, from a practical point of view, the brows in the roof of the drift created by a shaft or raise will be zones where any geological weakness can lead more easily to ground failure.

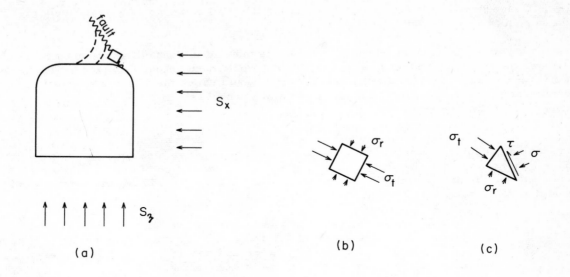

Fig. 3-11 The Consequences of a Fault Intersecting the Roof of an Opening

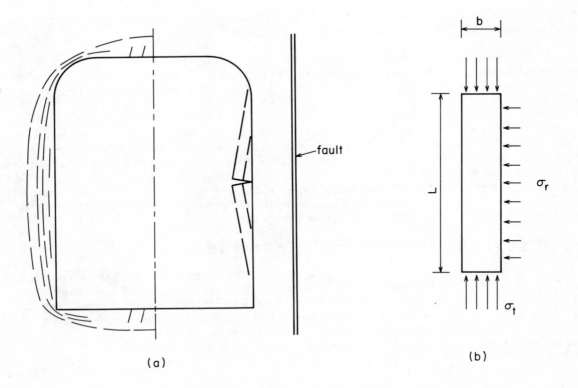

Fig. 3-12 Possible Buckling Failure in the Wall of an Opening

Non-Ideal Ground: For the less ideal case of a fault intersecting a tunnel, as shown in Fig. 3-11(a), the effects on the stress distribution can be examined. An element located at the fault would normally have the stresses as shown in Fig. 3-11(b). At this position the tangential stress, σ_t, would be very high with the radial stress, σ_r, being very low. Consequently, the shear stress, τ, along the fault plane would be relatively high with the normal stress, σ, being relatively low.

Unless the frictional resistance induced in the fault gouge or breccia were greater than the shear stress, slight movement would occur along this place. This movement would induce secondary tensile stresses, which would cause the fall of some rock. The shape of the roof, as shown by experience, would move progressively through such positions as indicated by the dashed lines in Fig. 3-11(a).

In Fig. 3-12 another case is shown of a drift affected by a fault. In this case the fault is running parallel to the side of the drift. Here two mechanisms can be significant. First, there would be a tendency for an increased concentration of stress in the rock between the fault and the drift similar to that described in Fig. 3-4(a). Thus, the tangential stress, σ_t, would have a higher than normal average value.

The second factor that might influence the stability of the ground would arise if the distance of the fault from the side of the drift, b, were small with respect to the height, L, of this element of ground. In this case, the ground could fail as a result of elastic instability at a lower average stress than would be required for crushing.

The analysis of this slab of rock would be similar to that for a column. For a column with ends that are pinned, i.e., no restraint to rotation, the Euler Formula for critical stress is as follows:

$$\sigma_{cr} = \pi E(r/L)^2 \qquad \text{Eq. 3-9}$$

where r is the radius of gyration of the section in the direction in which buckling would occur and L is the length between the equivalent pin connections.

For a column with fixed ends, the points of inflection in the elastic curve, which are the points of the zero moment, can be considered to be the location of the pins of an equivalent pin-connected column. In this case, the effective length, L, is half the actual length of the column.

For columns that are not actually fixed at the ends but have square ends so that rotation is inhibited, the points of inflection in the elastic curve are located somewhere between the two ideal cases of pin connections and fixed connections. For the conditions affecting the stability of the side of the drift in Fig. 3-12(a), it would be necessary to judge where the points of inflection, the location of the equivalent pin connections, are likely to be in the elastic curve of this rock. For a trial calculation, 0.8 of the height of the side from the floor to the start of the roof fillet could be used for the effective length L.

GROUND SUPPORT AROUND DRIFTS AND TUNNELS

Pressure from Broken Ground: If a tunnel lining is placed before rock failure occurs and if the stress field consists of a vertical compressive stress and a smaller horizontal compressive stress, vertical lining pressures greater than horizontal pressures can only arise from secondary causes. The primary action that must occur in such a stress field is the failure of the walls, which would produce side pressure. This action has been demonstrated by model work (19, 20). It has also been established underground (21).

One secondary action that could occur as a result of failure in the walls would be the deformation of the immediate roof, which would give rise to vertical pressure on the lining. A similar action would occur if the modulus of deformation of the material decreased with time, which, together with the high tangential stresses in the walls, would result in the roof being lowered. Another secondary action that can produce vertical pressures is the greater probability of overbreak in the back permitting the development of more loose ground behind the lining, which then produces a dead weight acting in a vertical direction.

The rock pressures for which tunnel linings on civil engineering projects should be designed in blocky and broken ground can be determined by using a simple set of rules. The concept was suggested, as amplified above, that tunnel linings usually are required to support the loose rock that develops around these openings (7). Following this approach and excluding any rocks that would be classified as VISCOUS (see Chapter 1), the following rock loads can be used as vertical pressures for the design of tunnel linings:

Continuity of the Rock	Height of Loose Rock
1. Massive	0
2. Blocky	0 - 0.7 b
3. Broken	2 b

where 'b' is the width of the drift or tunnel and the continuity of the rock can be determined according to the descriptions contained in Chapter 1. These rules are substantiated by some measurements (8).

The amount of safety factor to be included in the design of the support system must be left up to the engineer responsible for the work; however, a safety factor of 2 against rupture of the lining material should be satisfactory in most cases. This safety factor then accounts for the possibility of non-uniform loading conditions. The safety factor should not be excessive as the above loads are the maximum average loads that would normally occur for these rocks.

Typical cases of the development of loose rock around a drift are shown in Fig. 3-13(a). It can be visualized from this figure that the maximum possible horizontal pressure that could develop here would be a function of the height and the width of broken rock adjacent to the legs of the sets. One method of determining the limiting pressure would be to use Equation 1-11(b).

Example - If it was judged that the height of broken rock above the sets that would develop would be 0.7 b in blocky rock, that the angle of internal friction of the broken rock would be 45° and the density 167 pcf, then for a 10 ft high drift the maximum horizontal rock pressure using Equation 1-11(b) would be:

$$p_a = rz/\tan^2(45 + \varphi/2)$$

$$= 167(10 + 0.7 \times 10)/\tan^2 67.5 = 487 \text{ psf.}$$

A more probable analysis of the horizontal rock pressure on the sets would come from using the arching analysis explained in Chapter 5. Using this analysis in the above example, we make the assumption that the width of the loose rock at the sides of the drift is 3 ft and that the ratio of horizontal to vertical stress, k, is 0.5. Hence, the lateral rock pressure is calculated from Equation 5-3(b):

$$p_h = b\gamma(1-e^{-kz \tan\varphi \cdot 2/b})/(2\tan\varphi)$$

$$= 3 \times 167(1-e^{-0.5 \times 17(\tan 45)2/3})/(2 \tan 45) = 250 \text{ psf.}$$

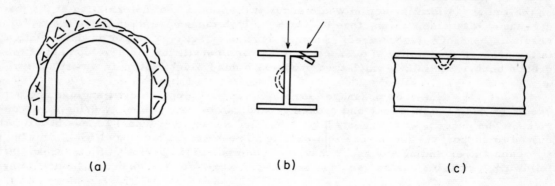

Fig. 3-13 Steel Sets Supporting Loose Rock Developed Around An Opening

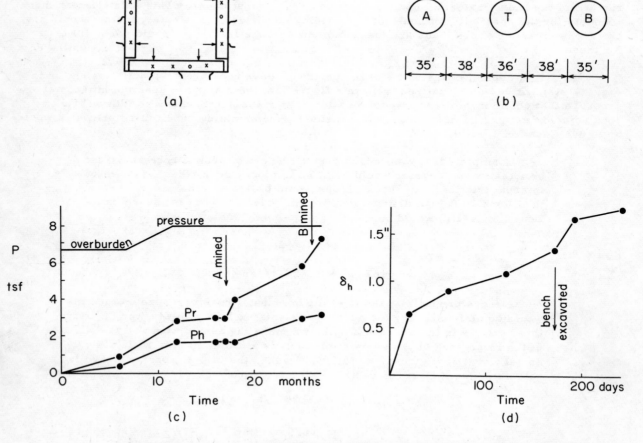

Fig. 3-14 Cases of Tunnels in Yielding, Viscous Shale (9, 10)

The answers from the two analyses are of a similar order.

In both calculations, considering the required condition of the rock to act like a granular mass, the use of a loose rock density of about 100 pcf might be more accurate.

One aspect of rock support that is often ignored, even by those with experience, is that rock pressures are seldom applied uniformly and along the centreline of the supporting member. This can be important, particularly where steel sets are used, as the normal structural shapes, I's and WF's, have been evolved for use where loads can be guaranteed to be applied vertically at the centreline of the section, as shown by the arrow in Fig. 3-13(b).

Where loadings are applied eccentrically and obliquely (as shown by the inclined arrow in Fig. 3-13(b)), these common structural shapes are subjected to stresses they are not designed to resist, i.e., torsional shear stresses, local bending stresses in the flanges, and bending combined with buckling stresses in the webs. As indicated in Figs. 3-13(b) and 3-13(c), the result of rock loads on WF shapes commonly is to cause yielding and buckling of the flanges and buckling of the webs. It can be appreciated after this discussion that timber sets with their square cross-sectional members have the good feature of equal bending and shear resistance around both axes.

In some circumstances a uniform pressure might be exerted on the lining of a tunnel where the ground was broken and in a state of stress as shown in Fig. 3-7(c). However, as the field stress for tunnels is not likely to be equal in the vertical and horizontal directions, this type of analysis cannot be usually applied.

Also, in some cases the lining in circular tunnels may be subjected to concentrated pressures, as shown in Fig. 3-8(b), and analysed as described for a circular shaft using Equation 3-7.

<u>Deformation from Viscous Rocks</u>: In the case of soft, viscous rocks the design of appropriate linings for drifts is quite complicated owing to the unknown reaction of the ground to the excavation before the linings are installed.

An interesting experiment was conducted for a tunnel in soft shale (9). In driving a test drift to examine the rock that would contain the conduits for a high earth dam, it was found that the shale walls of the drift, 6 1/2 ft x 7 1/2 ft, closed by as much as 20 in. in spite of the use of heavy timber sets. The material was soft, jointed and slickened-sided with an average density of only 118 pcf. The drift was 370 ft long and terminated with 64 ft of ground above the measuring station.

To obtain some measurements on the magnitude of rock pressures that would occur on the conduits to be driven later, pre-cast concrete slabs supported with struts and jacks were placed (as shown in Fig. 3-14(a)) around the drift. The load applied to the struts from the jacks initially was about 1.3 times the weight of the overburden, γz. The initial and subsequent loads on the struts were determined using a mechanical strain gauge on the struts. The results of these measurements are shown in the following table:

Time	P_v	P_h
Installed	$1.26\,\gamma z$	$1.32\,\gamma z$
1 year	$1.05\,\gamma z$	$1.34\,\gamma z$

The conclusions that can be drawn from this work are: first, that the vertical field stress at the level of the drift is probably equal to the weight of the overburden and that the horizontal field stress is greater than the vertical stress, possibly being some part of the stress that existed under the several thousand feet of ice and now eroded sediment.

Second, the shale should probably be classified as VISCOUS in view of the vertical pressure on the concrete slabs yielding from the installed pressure of 1.26 γz to 1.05 γz at practically no deformation (just the elastic deformation in the strut resulting from the corresponding change in stress).

Third, in view of the very high pressures that would be applied to a conduit lining that, in effect, permitted no deformation in such rock (i.e., 55 psi vertically and 70 psi horizontally), it can be seen that insufficient information was obtained for the design of a lining that would permit some deformation of the rock. It would have been valuable to have had measurements of the resulting pressures for several magnitudes of deformation between the walls and between the roof and floor.

Another interesting investigation on a similar project was conducted in a formation of shale and lignite of Tertiary origin (10). The ground had an average density of 105 pcf and a modulus of deformation of the order of 3×10^4 psi. The test tunnel, T, as shown in Fig. 3-14(b), was 36 ft in diameter in the rock, 240 ft long and with about 125 ft of cover. Tunnels A and B were excavated after Tunnel T.

Several test sections were established along this tunnel. One section, R, was lined with steel rib WF sections 3 ft apart with timber lagging between ribs. Mechanical strain gauges on the steel were used to determine the stresses induced by the rock pressures. The next section, C, was lined with steel ribs, which were then encased in 3 ft of concrete with steel load transducers across longitudinal slots in the concrete. It was estimated that Section R had an equivalent stiffness between 10 and 15 per cent of the rock that had been excavated. Section C was estimated to have a stiffness 3 1/2 times that of the original ground.

Measurements were taken on these sections for over two years. In Section R the measurements showed that the pressures increased during the first four months after installation and then remained substantially constant during the next sixteen months. The vertical pressure had an average value of 1 tsf and the horizontal pressure had an average value of 0.9 tsf.

In Fig. 3-14(c), the results of the measurements taken in Section C are shown. It can be seen here that, after about twelve months during which the dam fill was placed increasing the overburden pressure from 6.6 tsf to 7.9 tsf, the pressures had become stable at about 3 tsf for the vertical pressure and at 1.7 tsf for the horizontal pressure. Then the adjacent Tunnel A was mined beyond the test station. This resulted in an immediate increase in the vertical pressure followed by a continuous increase in both vertical and horizontal pressures that did not approach stability before the other adjacent tunnel, B, was mined beyond the test station. Tunnel B also caused an immediate increase in the vertical pressure on Test Section C, after which no further readings were available.

The conclusions that can be drawn from this investigation are: first, that the rock pressure varied with the stiffness of the lining. Whereas the relatively flexible Section R with a lining moment of inertia of about 285 in.^{4}/LF was only subjected to a vertical pressure of 1 tsf with an almost equal horizontal pressure of 0.9 tsf, the relatively stiff Section C with a moment of inertia of about 3400 in.^{4}/LF was subjected to a vertical pressure of 3 tsf and a horizontal pressure of 1.7 tsf giving a serious differential pressure condition.

Second, with a lining producing a stiffness greater than the adjacent ground, it would be expected that any added loads to the overlying ground would produce a greater increase in pressure on the lining than on the adjacent ground simply because the adjacent ground would tend to deflect more under equal pressures.

After the fill was placed for the dam, between 6 1/2 months and 11 1/2 months, the overburden pressure was increased by about 1.3 tsf. The vertical pressure on the lining increased, as would be expected, by the disproportionate amount of 1.9 tsf.

The third conclusion is that, whereas for a single tunnel, even with a stiff lining, a vertical pressure of 3 tsf with the horizontal pressure of 1.7 tsf could have been used for design, for the multiple tunnels the appropriate design pressures for a stiff lining would be much greater. It would seem that the increased stresses in the pillar, as shown in Fig. 3-12, were sufficient to exceed the yield strength of the rock. (In soil mechanics terminology, the preconsolidation load was probably exceeded, which would lead to additional consolidation of the pillar shale, but would eventually, if contained, approach stability similar to the reaction of a Kelvin model.)

The fourth conclusion is that, as a result of mining Tunnel A, a continuous increase in pressure resulted between seventeen months and twenty-six months with no indication of the approach of equilibrium. It is possible that the ultimate vertical pressure on the stiff tunnels would be equal to the original overburden pressure plus a disproportionate amount of the fill pressure. On the other hand, the ultimate vertical pressure could be even greater than this amount if the stresses in the pillars had caused some breakdown in the rock so that it could not even support the original overburden pressure without viscous flow. The ultimate horizontal pressure would be a function of the ultimate vertical pressure, the stiffness of the lining, and the flow conditions in the pillar and thus would be difficult to predict.

The fifth conclusion, as for the previous case, is that it would have been interesting for a study of design theory if the basic pressure-deformation-time data had been obtained. With this type of information an explicit analysis could be made for the various lining designs that might be considered.

A third case of tunnelling in VISCOUS rock occurred in a formation containing limestones, sandstones, and shale (11). In this case two 51 ft diameter tunnels were driven with a cover of about 175 ft of rock and 130 ft of soil. The heading and bench method was used for this excavation.

The horizontal change in diameter, d_h, was measured using an invar micrometer tape. The deformation-time measurements are shown in Fig. 3-14(d). Although relatively light steel ribs were used for temporary support, it is improbable that they would have provided a significant amount of back pressure to affect these deformation readings. Consequently, in Fig. 3-14(d) the horizontal deformation is a measure of the ground reaction alone to excavation.

As mentioned in the previous cases, these deformation-time data represent the type of measurements required for general lining design. In this case, an examination of the data showed that, when deformation was plotted against the logarithm of time, a straight line occurred. It was then calculated that, for the concrete lining that would be placed in the tunnel, the extrapolated deformation that would occur during the succeeding thirty years would induce at most a compressive stress of 1500 psi (which would actually be reduced by the creep of concrete)(11).

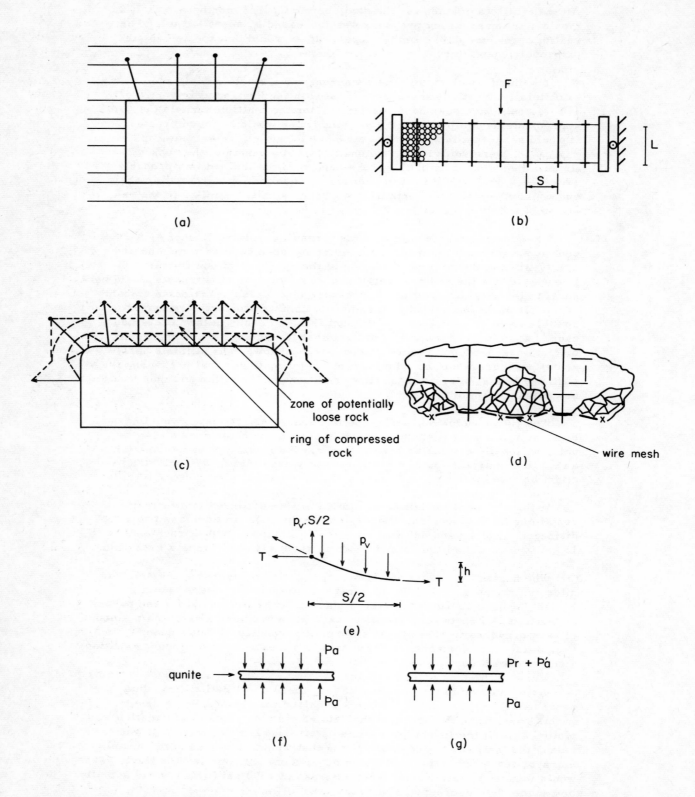

Fig. 3-15 Possible Mechanics of Support of Rockbolts (12), Wire Mesh and Gunite

Rock Bolts: Another major type of rock support around drifts, as well as around other underground openings, is the rock bolt. At the present time a well established theory of rock bolt design together with the specifications for installation has not been evolved. On the basis of research work that has been done, two hypotheses can be identified for explaining rock bolt action and for designing a rock bolt system.

The first hypothesis is that pre-tensioned rock bolts tie together layers of rock in stratified formations (as shown in Fig. 3-15(a)) so that a laminated beam of greater strength than the individual layers is formed over the openings.

The second hypothesis is essentially concerned with reinforcing and keeping intact the natural arch, somewhat as shown in Fig. 2-11(a), that tends to form over any opening in rock. The implication here is that the surface rock around any opening is likely to break down into individual blocks. Originally, these blocks would be tightly interlocked. If this interlocking can be maintained, then a strong boulder arch can be formed in hard rock.

It is postulated here that an ultimate design theory should be used for the design of rock bolt systems. The implication of this type of theory is that the system is designed to resist the ultimate loads that could be applied to it. Also, only a small safety factor is provided against the ultimate loading condition, which is normally a low probability situation. With this orientation it follows that the ultimate condition of a roof in layered rock will be the same as that in a massive rock - both rock types on breaking down will form a series of individual blocks.

Consequently, the second hypothesis for the mechanics of rock bolting (i.e., knitting these ultimate blocks into a self supporting arch that will not fail in any of the modes discussed in Chapter 2) can be seen to be the more general basis for a design theory. In other words, the first hypothesis of forming a laminated beam would be valid only as long as the individual layers maintained their continuity; with the presence of joints or stress cracks normal to the roof the capability of these layers to act as beams is destroyed.

Some experimental work exists to substantiate the reinforced masonry arch hypothesis. First, tests were conducted on cylindrical tubes that were formed into the shape of a beam with the axis of the individual tubes horizontal and transverse to the length of the beam as shown in Fig. 3-15(b)(12). Transverse, vertical bolts were placed in tension through the formed beam. The ends of the beam were free to rotate and contract but not to expand. By varying the distance, s, between the bolts, it was found that, when the ratio s/m (where m was the diameter of the rods) was less than 3, the beam was stable and behaved elastically under load up to the point where failure of the beam occurred. When this ratio was 4, the beam was always unstable.

Similar tests were conducted on crushed rock placed in a box without a bottom (12). Bolts under tension were passed through the crushed rock. The results of the experiments on this box showed that the ratio s/m for stable support could be increased both by supporting the arch of loose rock that tended to fall out between the bolt heads as shown in Fig. 3-15(d) and by increasing the tension in the bolts. Ratios of s/m greater than 6 were made stable in these ways.

To examine the appropriate length for bolts, photoelastic work was conducted with the above experiments (12). This work showed that, when the ratio L/s (where L is the length of the bolt and s is the spacing) was 2 or more, a zone of compression would be formed at right angles to the length of the bolts as shown in Fig. 3-15(c). When this ratio was equal to 2, the width of this zone was small; when the ratio was increased to 3, the width of the zone was equal to about two-thirds of the length of the bolt.

As it is obvious that bolt tension requires good anchorage and that good anchorage is unlikely to be achieved in any ground where tensile stresses exist, the importance of having a compression zone within the bolt length where end anchorage is used in quite clear. From these experiments it could tentatively be concluded that the ratio L/s, to fulfil the reinforced boulder arch hypothesis, should possibly be greater than 2 and certainly greater than 1, and that the ratio s/m (m being the width of blocks between joints) should be less than about 4, unless a membrane is used between the bolt heads. If these conditions are fulfilled, bolting should bind all of the potentially blocky rock into a pre-stressed ring over the opening, as shown in Fig. 3-15(c). This hypothesis has yet to be proven by full scale measurements; furthermore, the length should probably also be governed by the span of the opening.

Besides the above requirements for the reinforced masonry arch, it is obvious that the anchorage capacity of the bolts should be equal to or greater than the strength of the steel; otherwise the steel cannot be used to its capacity. It follows from this and from the above experiments that the bolts should be installed with the maximum tension possible. Normally, it is difficult to obtain an installed tension more than about 50 per cent of the yield point of the bolt. Insufficient work has been done on the reinforcing action of bolts on boulder archs to appraise how important the various levels of tension in a bolt are; consequently, there is no strong reason at the moment in attempting to modify present equipment to achieve the theoretically desirable higher installed tensions.

On the other hand, it is reasonable to postulate that it would be advantageous to have the bolts tightened to a load equal to the weight of the tributary area of rock; in this case, a sudden transfer of the rock load onto the bolts would be prevented, which otherwise could conceivably cause the bolt to fail as a result of a suddenly applied load producing stresses greater than the static stresses under the same load. In addition, if all the bolts could be tightened to this ultimate load, the possibility of the bolts being loaded unequally might be avoided. In this way, the overloading of one bolt, causing it to fail and thus overloading the adjacent bolts in a sequence leading possibly to progressive failure of the entire system of support, might be prevented.

Following the concept of an ultimate design theory, with the possibility of the tributary area of rock below the anchorage of a bolt hanging on the bolt, a bolt would be designed to support this load with possibly some small safety factor applied to the yield point or ultimate capacity of the bolt depending on the nature of the underground opening. Consequently, it would seem to be valuable to be able to place the bolts under tension equal to the yield load; however, with the present torque wrenches and stopers it is not normally possible to achieve this tension. A procedure using a hydraulic jack with steel wedges inside a collar to retain the tension (similar to the post-tensioning technique in prestress concrete work) could be used where the circumstances and expense were warranted. On the basis of experience, these ultimate design requirements might be modified if it is established or judged that the probability of ultimate loads occurring is very remote.

Where it is judged that some membrane is required between the bolt heads to contain the loose rock that would develop (as shown in Fig. 3-15(d)), the strength of this membrane must be analysed. Wood lagging, steel lagging, and wire mesh are probably the most common materials used for this purpose. The vertical pressure on such a membrane can be analysed by using the arching theory described in Chapter 5. Equation 5-3(a) for this case is as follows:

$$p_v = b\gamma(1-e^{-K\tan\varphi \cdot 2z/b})/(2K\tan\varphi)$$

In this equation, b, the width of the arch, can be taken to be equal to s, the spacing of the bolts; z, the height of the caving ground, should be less than s if the bolts maintain their tension; φ should be greater than 45°, in which case K would be less than 0.33 according to the reasoning outlined in Chapter 5 with respect to this equation. If we insert these values into Equation 5-3(a), then the following expression is obtained:

$$p_v \leq 0.727 \gamma s \qquad \text{Eq. 3-10}$$

To simplify the problem we assume that the sag in the mesh, as shown in Fig. 3-15(d), is occurring in only one direction and that the section is representative of a sufficient length to make a two-dimensional analysis valid. Then, to determine the force created in the mesh by the pressure of the loose rock, p_v, the following analysis, considering half the mesh as in Fig. 3-15(e), is often used:

$$\sum M_o = p_v(s/2)^2 - p_v(s/2)^2/2 - Th = 0$$

$$\therefore \quad T = p_v s^2/(8h) \qquad \text{Eq. 3-11}$$

where T is the tensile force per linear foot in the membrane, s is the spacing of the bolts and h is the probable sag of the membrane.

The stress in the proposed membrane can then be determined by dividing the tension, T, by the cross-sectional area per linear foot; or, the required cross-section area in the membrane can be obtained by dividing the force by the permissible stress.

<u>Example</u> - Bolts are spaced 4 ft apart in a hard rock with a density of 170 pcf. Chain-link mesh is to be used to support the loose rock that might develop between the bolts. Determine the maximum probable tensile force that could occur in the mesh.

Using Equation 3-10, the maximum probable p_v can be determined:

$$p_v = 0.727 \times 170 \times 4 = 495 \text{ psf}$$

Over a 4 ft span the maximum probable sag would be about 1 ft; thus, the maximum tension would be from Equation 3-11:

$$T = 495 \times 4^2/(8 \times 1) = 990 \text{ lb/LF}$$

Note that the cross-sectional area of chain-link mesh using No. 11 wire is 0.068 psi/LF; hence, the maximum stress would be:

$$= 990/0.068 = 14,500 \text{ psi}$$

This stress would be much lower than the yield strength of about 100,000 psi and thus should be safe even with local concentrations of load that might occur.

The permissible stress in the membrane is a matter for judgment, recognizing that Equation 3-8 states that the rock pressure will be equal to or less than this quantity; this would mean that the computed T would be conservative, except that local concentrations of load could produce greater than average pressures at some locations. Using the yield stress as the permissible stress in a steel mesh would seem to be conservative in view of the limit conditions implied by Equation 3-8, in view of the ability of mesh to distribute most tendencies for concentrated loads and in view of such a permissible stress having a safety factor against rupture (i.e., a minimum of 1.25 according to normal wire specifications).

<u>Gunite</u>: To increase the quality of support of a rock bolt and mesh system (and in some cases to prevent deterioration of the rock) a thin layer (of the order of 1 in.) of gunite is sprayed onto the mesh and rock surface. From experience it is known that this can produce, even in some cases without rock bolts and mesh, a remarkably effective lining.

In Fig. 3-15(f) an element of such a layer of gunite is shown with equal air pressure, p_a, initially acting on both faces. If the layer of gunite is air-tight, then it can be reasoned that any tendency for loose rock to develop will require an expansion of the rock. This expansion would have to be accompanied by an increase in the volume of voids in the rock mass. Without access to the atmosphere, as the volume of voids increased the air pressure in the voids would decrease. Consequently, the gunite lining might be subjected, as shown in Fig. 3-15(g), to a rock pressure, p_r, plus a reduced air pressure, p_a', on one side and the atmospheric pressure, p_a, on the other side. This unbalanced air pressure would by itself by capable of supporting a moderate rock pressure.

Example - We can postulate an example to illustrate the possible mechanics of the effectiveness of gunite. Assume that initially the porosity of the rock mass is 3 per cent, that the failure strain in developing loose rock is 1 per cent and that the thickness of the loose rock that develops at each stage is 1 foot. The expansion, δ, of this layer is thus:

$$\delta = 1 \times 0.01 = 0.01 \text{ ft}$$

It is then possible to calculate the expansion of the voids, ΔV, that must accompany the deformation of the rock to produce the loose slabs:

$$\Delta V = 0.01 \times 1 \times 1 = 0.01 \text{ cf}$$

From this increase in volume we can calculate the initial volume, V_1, the final volume, V_2, and the decreased air pressure, P_2:

$$V_1 = 1 \times 0.03 = 0.03 \text{ cf}$$

$$\therefore V_2 = 0.01 + 0.03 = 0.04 \text{ cf}$$

$$P_2 = P_1 V_1 / V_2$$

$$\therefore p_a' = 14.7 \times 0.03/0.04 = 11.0 \text{ psi}$$

Or, referring to Fig. 3-15(g):

$$\therefore p_r + p_a' = (167/144) + 11.0 = 12.2 \text{ psi}$$

$$p_a = 14.7 \text{ psi}$$

Hence, the stabilizing pressure is greater than the pressure on the gunite. Alternatively:

$$p_r = 1.2 \text{ psi}$$

$$p_a - p_a' = 14.7 - 12.2 = 2.5 \text{ psi}$$

Hence, the safety factor:

$$F_s = 2.5/1.2 = 2.1$$

From these equations we can see that the air pressure upwards is greater than the air pressure plus rock pressure downwards; or, in an even more significant form, we can see that the unbalance produced in the air pressure caused by the development of the loose rock is more than twice the weight of the loose rock causing this unbalance. In other words, as long as the gunite membrane remains air-tight, the situation seems to be inherently stable.

Ground Water Effect on Weak or Soft Ground

On construction projects soil tunnelling is probably as common as rock tunnelling. The stability problems connected with these projects can arise from two different factors. First, the ground may be too weak to sustain the shear stresses caused by the principal stress differences (as illustrated in Fig. 3-3) so that failure will occur either by yielding or by rupture. If a lining can be placed close enough to the heading (where the full stress concentrations have not developed), the failed ground that ultimately occurs can be supported. However, this usually requires that some part of the heading is not supported by a lining, which might still tend to fail.

Aside from the partial face methods of excavating, one method of supplying ground support is through the use of compressed air. This acts as a back pressure in the tunnel, which decreases the principal stress difference, or shear stress, as shown by the effect of p_i in Fig. 3-2. If the decrease is sufficient, then the ground will be prevented from failing.

Example - A 10 ft diameter circular tunnel is to be driven through soft ground with the centerline 58 ft below the surface. The density of the ground is 105 pcf and Poisson's Ratio is 0.5. Show the reduction in shear stress in the walls that could result from using compressed air in the tunnel at a pressure of 10 psi.

The vertical field stress at the centerline of the tunnel is:

$$S_v = 58 \times 105/144 = 42.3 \text{ psi}$$

From Equation 2-16 the horizontal field stress is:

$$S_h = 42.3/(2-1) = 42.3 \text{ psi}$$

Therefore, from Equation 2-16(b) the maximum surface tangential stress in the ground is:

$$\sigma_t = 2 \times 42.3 = 84.6 \text{ psi}$$

Without any radial pressure or stress on this surface ground, the maximum shear stress is:

$$\tau = 84.6/2 = 42.3 \text{ psi}$$

With the application of compressed air with a pressure of 10 psi, Equation 2-13(b) must be used to calculate the surface tangential stress in the ground:

$$\sigma_t = 2 \times 42.3 - 10 = 74.6 \text{ psi}$$

and

$$\sigma_r = 10 \text{ psi}$$

Therefore, the maximum shear stress becomes:

$$\tau_m = \frac{1}{2}(74.6 - 10) = 32.3 \text{ psi}$$

The second source of ground instability in soil tunnelling can arise from the excavation being below the ground water table. The tunnel will act as a drain, and consequently there will be a flow towards the heading as indicated in Fig. 3-16. The seepage forces associated with such a flow, as analysed in Chapter 6, can add to the shear stresses in the ground at the heading. It should be noted that these seepage pressures can occur regardless of the quantity of flow; in other words, in tight ground, even though there is no apparent flow if the heading is below the ground water table, flow will exist along with the corresponding seepage pressures.

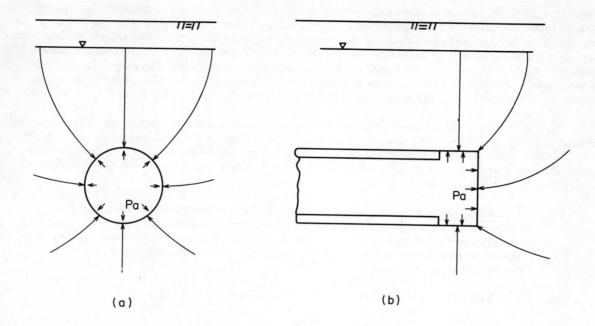

Fig. 3-16 Seepage Forces at a Heading Below the Ground Water Level

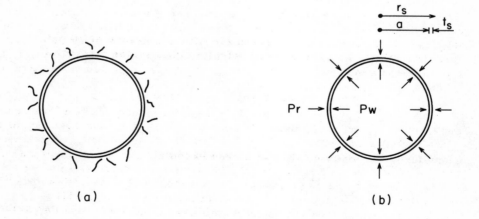

Fig. 3-17 Pressure Tunnel Lining

In the cases where a large quantity of flow occurs, then, at the point of exit, if the hydraulic gradient is high enough, boiling and erosion of the soil particles can occur. Aside from such methods of control as freezing the ground or using some type of grout (chemical or Portland cement), the common way of controlling such ground water is by the use of compressed air in the tunnel. If the pressure in the compressed air is equal to the head of water, then no flow will occur; if it is less than the hydrostatic head, then flow will simply be inhibited. A problem can arise in a large tunnel where the hydrostatic head of water is significantly different between the floor and the crown of the tunnel, whereas the air pressure that can be applied must be equal. The problem is particularly crucial when the magnitude of the air pressure must be limited owing to the danger of lifting the overlying ground and causing a blowout.

Example - A 12 ft diameter tunnel is to be driven through soft ground under a river. It is expected that there will be a minimum cover of 15 ft of ground over the tunnel with a total density of 112.4 pcf and a buoyant density of 50 pcf. The maximum elevation of the water is expected to be 31 ft over the crown of the tunnel. To prevent face instability, compressed air is to be used to prevent the water from flowing into the tunnel. Calculate the required air pressure.

To prevent water flowing in at the crown the air pressure should be:

$$p_a = 31 \times 62.4/144 = 13.4 \text{ psi.}$$

At the floor the required pressure would be:

$$p_a = 43 \times 62.4/144 = 18.6 \text{ psi.}$$

If 18.6 psi pressure is used, there would only be 13.4 psi of water pressure resisting the air pressure plus the weight of the ground, which would be:

$$p = 15 \times 50/144 = 5.2 \text{ psi.}$$

The water plus ground pressure at the crown would just equal 18.6 psi; hence, under an air pressure of 18.6 psi any decrease in the 15 ft of cover, any decrease in density of ground, or any decrease in water level could lead to the air pressure lifting the overlying ground, blowing out to the surface, and permitting the river to enter the tunnel. Under these circumstances consideration would be given to using a lower air pressure, which would theoretically permit some inflow at the floor level; to excavating the tunnel in two benches to reduce the difference in elevation between the top and bottom of the face; and to using some other ground support technique such as freezing.

Pressure Tunnels: With projects requiring the passage of water under high pressure special problems arise. A lining is placed inside a circular rock tunnel as shown in Fig. 3-17. This lining is subjected to an internal water pressure, p_w, which induces the reaction from the rock, p_r. The minimum thickness of lining to resist the water pressure with the help of the rock must be determined. The design must also satisfy the conditions when the tunnel is empty but when water exists behind the lining and applies an external hydrostatic pressure to the lining, which must then resist the tendency to buckle.

The initial design problem is the determination of the pressure of rock reaction, p_r. Whereas the situation as represented by Fig. 3-17(b) is relatively simple, the actual configuration generally includes an inner lining of steel surrounded by an air-gap, resulting from the shrinkage of the concrete or grout which is placed between the steel lining and the rock, with the concrete grout ring being surrounded by a probable ring of fissured rock (i.e., rock unable to sustain tensile stresses), which in turn may be surrounded by the rock mass with some resistance to the tensile stresses. The problem can be analysed as follows:

$$\sigma_s = (p_w - p_r) a/t_s \qquad \text{Eq. 3-12}$$

where σ_s is the stress in the steel liner based on the thin-walled cylinder formula, p_w is the water pressure, a is the internal radius of the steel, and t_s is the thickness of the steel. Ignoring the radial stress in the steel:

$$\delta_r = \sigma_s a/E_s = a^2(p_w - p_r)/E_s t_s \qquad \text{Eq. 3-13}$$

where δ_r is the radial deformation in the lining.

As the ring of grout is assumed to have no tensile strength, it can sustain no tangential stress; hence, its radial deformation is a function of the radial stress, which varies from p_r to p_r', the contact pressure between the grout and rock. It is assumed that the average radial stress is $(p_r + p_r')/2$. Thus:

$$2\pi r_s p_r = 2\pi r_c p_r'$$

where r_s is the outside radius of the steel liner and r_c is the outside radius of the grout ring. Then,

$$\delta'_r = \sigma_r t_c/E_c = (p_r + p_r') t_c/(2E_c) = p_r(1 + r_s/r_c) t_c/(2E_c)$$

where δ'_r is the radial deformation caused by compression of the grout, t_c is the average thickness of the grout, E_c is the modulus of deformation of the grout, and σ_r is the average radial stress assumed equal to $(p_r + p_r')/2$.

If, as a result of blasting or working, there is a ring of fractured rock that cannot sustain tensile stress, then radial deformation, δ''_r, will result, as for the grout, only from compression of the rock. It is assumed that the average radial stress in this rock is $(p_r' + p_r'')/2$, where p_r'' is the contact stress between the fractured rock ring of thickness t_f, and the intact rock mass.

Then,

$$\delta''_r = \sigma_r t_f/E_f = (p_r' + p_r'') t_f/(2E_f) = p_r'(1 + r_c/r_f) t_f/(2E_f)$$

where E_f is the modulus of deformation of the fissured rock and r_f is the outside radius of the fissured rock ring. Also:

$$\delta''_r = p_r(r_s/r_c)(1 + r_c/r_f) t_f/(2E_f)$$

The increment of deformation of the intact rock, δ'''_r, resulting from the hydrostatic pressure, p_w, then is calculated using the thick-walled cylinder formula for deformation:

$$\delta'''_r = (\mu \sigma_r - \sigma_t) r_f/E_r = (\mu p_r'' + p_r'') r_f/E_r$$

where σ_r and σ_t are the stresses in the cylinder at the inside surface (compressive stresses being positive), E_r is the modulus of deformation and μ is Poisson's Ratio of the intact rock. Then:

$$\delta'''_r = p_r(r_s/r_c)(r_c/r_f)(1+\mu) r_f/E_r$$

$$= p_r r_s (1+\mu)/E_r$$

$$\delta_r = g + \delta'_r + \delta''_r + \delta'''_r$$

$$p_w a^2/(E_s t_s) = g + p_r(a^2/(E_s t_s) + (1 + r_s/r_c) t_c/2E_c + (r_s/r_c)(1 + r_c/r_f) t_f/2E_f + r_s(1+\mu)/E_r)$$

$$p_w a^2/(E_s t_s) - g = p_r/A$$

$$\therefore p_r = A p_w a^2/(E_s t_s) - A g \qquad \text{Eq. 3-14}$$

where g is the gap between the steel liner and the grout, and A represents the sum of the coefficients of p_r as indicated.

By using some typical numbers, an appraisal can be made of the relative importance of the various factors. Assume $a = 10$ ft $= 120$ in., $t_s = 0.5$ in., $g = 0.02$ in., $r_c = 126$ in., $r_f = 156$ in., $p_w = 500$ ft $\times 62.4/144 = 216$ psi, $E_f = 4 \times 10^6$ psi, $E_r = 6 \times 10^6$ psi, $E_s = 3 \times 10^7$ psi and $\mu = 0.2$, then

$$1/A = \frac{120^2}{0.5 \times 3 \times 10^7} + \frac{(1 + 120.5/126)6}{2 \times 2 \times 10^6} + \frac{120.5}{126} \frac{(1 + 126/156)36}{2 \times 4 \times 10^6} + \frac{120.5 \times (1 + 0.2)}{6 \times 10^6}$$

$$= 0.960 \times 10^{-3} + 0.003 \times 10^{-3} + 0.008 \times 10^{-3} + 0.024 \times 10^{-3}$$

$$= 0.995 \times 10^{-3}$$

$$\therefore p_r = \frac{216 \times 0.960 \times 10^{-3}}{0.995 \times 10^{-3}} + \frac{0.02}{0.995 \times 10^{-3}}$$

$$= 208 - 20 = 188 \text{ psi.}$$

Using Equation 3-12:

$$\sigma_s = (216 - 188)120/0.5 = 6720 \text{ psi.}$$

From the above it can be seen that: a) the influence of the grout (both t_c and E_c) is small, b) the same applies to the ring of fissured rock, c) in view of the difficulty of determining E_r and μ of the rock mass, a conservative value could be used to include the effects of the grout, of the ring of fissured rock, and of Poisson's Ratio in the factor $(1 + \mu)$. Hence:

$$1/A \simeq \frac{a^2}{E_s t_s} + \frac{r_s}{E_s}$$

Then, as $r_s \simeq a$:

$$\sigma_s = \frac{a}{t_s} \left(p_w - \frac{A a^2 p_w}{E_s t_s} + A g \right)$$

$$\sigma_s = \frac{p_w a}{t_s} \left(1 - \frac{1}{1 + \frac{E_s t_s}{E_r a}} \right) + \frac{E_s g}{a + t_s E_s / E_r}$$

Recognizing that the number used for g will be a guess and hence not very accurate, $t_s E_s / E_r$ can be dropped as being small with respect to a and not affecting the accuracy as much as the second significant figure in g.

Hence,

$$\sigma_s = p_w \left(1 - \frac{1}{1 + \frac{E_s t_s}{E_r a}} \right) \frac{a}{t_s} + E_s g/a \qquad \text{Eq. 3-15}$$

Recalculating the above example and using $E_r = 4 \times 10^6$ psi to account for the grout, fissured rock, and Poisson's Ratio:

$$\sigma_s = 1550 + 5000 = 6550 \text{ psi.}$$

Note that the maximum possible steel stress would occur when there was no rock support:

$$\sigma_s = 216 \times 120/0.5 = 51,800 \text{ psi}$$

which is less than the rupture stress but more than the yield stress. A reasonable precaution would be to keep this ultimate stress below the yield point.

It can be seen in the above example that, if there had been no gap behind the steel liner, the rock would have taken about 97 per cent of the water pressure. Experimental work has shown that in some cases the rock has taken as much as 97 per cent (13), whereas in other cases this figure has been as low as 20 per cent (14) of the water pressure.

It is more difficult to analyse the steel liner action under external pressure, such as arises from grouting and from the hydrostatic pressure in the rock when the internal water is emptied. Failure will be by buckling inwards and, as can be imagined, the critical pressure for this action will be a very strong function of the size of the gap that exists between the steel liner and the surrounding concrete. Difficulty is experienced in making this analysis since not only must a guess be made as to the size of the gap outside the steel liner, but also the hydrostatic pressure that might exist outside the liner is also unknown.

There are several methods that can be used to diminish the possibility of buckling from external pressure. Great care can be taken in grouting. Then, after the liner has been subjected to the maximum internal pressure and has possibly caused some irreversible compression of the surrounding material, a re-grouting program can be followed. In addition, a separate drainage system around the tunnel can be installed to eliminate external hydrostatic pressure. Then, also, steel anchors can be welded to the outside of the liner, which would greatly increase its capacity against buckling. Each of these provisions would, of course, add to the cost of the lining.

REFERENCES

1. Hiramatsu, Y. and Oka, Y., "Stress Around the Shaft or Level Excavated in Ground with 3-dimensional Stress State", Memoirs, Faculty of Engineering, Kyoto University, Vol. 24, Part I (1962).

2. Timoshenko, S. and Goodier, J., "Theory of Elasticity", McGraw-Hill (1951).

3. Fenner, R., "Study of Ground Pressures", Gluckauf, Vol. 74 (1938).

4. Morrison, R. and Coates, D., "Soil Mechanics Applied to Rock Failure in Mines", Trans. CIM Vol. 58 (1955).

5. Terzaghi, K., "Theoretical Soil Mechanics", Wiley (1953).

6. Hiramatsu, Y., Oka, Y. and Ogino, S., "Investigations on the Stress on Circular Shaft Linings", Memoir of Faculty of Engineering, Kyoto University, Vol. 23, Part I (1961).

7. Terzaghi, K., "Introduction to Tunnel Geology", Rock Tunnelling with Steel Supports, Youngstown Printing (1946).

8. Coates, D. and McRorie, K., "Earth Pressure on Multiple Tunnels", 15th Canadian Soil Mech. Conf., NRC Tech. Mem. No. 73, Ottawa (1962).

9. Peterson, R., "Recent Soil Mechanics Studies of the PFRA", 5th Canadian Soil Mech. Conf., NRC Tech. Mem. No. 23, Ottawa (1952).

10. Lane, K., "Effect of Lining Stiffness on Tunnel Loading", 4th Internat. Conf. Soil Mech. and Fdn. Eng., Vol. 2 (1957).

11. Hogg, A.D., "Some Engineering Studies of Rock Movement in the Niagara Area", GSA Eng. Geol. Case Histories, No. 3 (1959).

12. Lang, T., "Theory and Practice of Rock Bolting", Trans. AIME, Vol. 220, p. 333-348 (1961).

13. Leterrier, G., "Le Comportement du Rocher dans les Colleries Blindees: Resultats des Measures Effectuees a Randens, Montpezat, Brevieres Serre-Poncon", La Houille Blanche, Vol. 11, No. A, Mar.-Aug., p. 144-170, 171-172 (1956).

14. Brewer, G., "Dilation Measurements of Steel Sphere and Rock Deformations at Kemano, B.C.", Proc. SESA, Vol. 13, No. 2 (1956).

15. Kujundzic, B., "Anisotropy des Massifs Rocheux", Proc. 4th Inter. Conf. Soil Mech. Fdn. Eng., Vol. 2 (1957).

16. Hiramatsu, Y. and Oka, Y., "A Fracture of Rock Around Underground Openings", Memoir of Faculty of Engineering, Kyoto University, Vol. 21, Part II (1959).

17. York, B. and Reed, J., "An Analysis of Mine Opening Failure by Means of Models", Trans. AIME, Vol. 196, p. 705-710 (1953).

18. Buchanan, J., "The Load Cell Installation in No. 3 Mine, Dominion Wabana Ore Limited, Bell Island, Newfoundland", Mines Branch, PM 194 (1955).

19. Jacobi, O. and Everling, G., "Model Tests Illustrating the Effects of Different Roadway Supports", Proc. Internat. Conf. Strata Control, Paris (1960).

20. Jacobi, O., "Supporting Capacity of Various Types of Supports in the Plasteline Model Experiments", Bergbau-Archiv 13, Helf/4, S 35/44 (1952).

21. Hind, J., "Some Experiments in Roadway Support", Trans. IME, Vol. 119 (1960).

22. Lechnitskii, S., "Theory of Elasticity of Anisotropic Elastic Body", Holden-Day (1963).

23. Onishchev, U., "Ground Stress in Vertical Shafts", Ugol (Dec. 16, 1959).

24. Hiramatsu, Y. and Oka, Y., "The Stress and the Earth Pressure Phenomena in the Rocks Around Shaft Bottom Spaces", Memoir of Faculty of Engineering, Kyoto University, Vol. 25, Part II (1963).

25. Constantino, C., "Stresses in the Vicinity of Deep Underground Shelters", Bull. No. 32, Office of the Director of Defence and Eng., Washington (1963).

CHAPTER 4

PILLARS

INTRODUCTION

Pillars can be defined as the in situ rock between two or more underground openings. The terms height, thickness, and width should be restricted to the dimension normal to the plane of the workings or openings. The length of the pillar is the greatest dimension in this plane and the breadth the lesser dimension.

In examining the mechanics of pillars there are three aspects to be considered. First, the load that is applied to the pillar must be determined. Second, the strength of the pillar, taking into account the various modes of failure, should be appraised. The safety factor can then be determined and judgment exercised on whether it is adequate. The third aspect to examine is the reaction to the pillar stresses of the roof (or back) and the floor.

STRESS DISTRIBUTIONS

The loading of a pillar should take into account both the total load and the stress distribution that results from this load. The second factor of stress distribution is considered first.

In Fig. 4-1(a) two openings, side by side, create a pillar of breadth, B. When the two openings are fairly close together, the stress in the pillar can be considered to be the result of the interaction of the stress concentrations from each opening. Consequently, the maximum tangential stress on the surface of the openings (that is, on the surface of the pillar) is likely to be greater than that which would apply to a single opening. In most cases, the average increase in axial stress in the pillar, as shown in Fig. 4-1(a), is increased more than the maximum stress.

Actually, the stress distribution in most pillars is more likely, as shown in Fig. 4-1(b), to be modified from the theoretical elastic distribution. The relaxing or partial fracturing from blasting or release of confinement of the surface rock produces a zone commonly 2 ft to 6 ft thick that sustains little stress. The peak stress concentration is thus moved into the interior of the pillar. However, purely elastic analyses provide valuable guidance.

In Figs. 4-1(c), 4-1(d) and 4-1(e), the results of photoelastic model tests on gelatine are shown (3). Fig. 4-1(c) shows the axial stress distribution at the mid-height and at 1/8 of the height from the floor. At mid-height the axial stress is a little greater at the centre than at the edge of the pillar. At the 1/8 height the axial stress is greater towards the edge of the pillar.

The transverse stress is plotted in Fig. 4-1(c) for a section 1/8 of the width off the centreline. This curve shows that there is horizontal compression at the ends and horizontal tension in the middle zone. The tension in the middle of these pillars is of interest in view of one of the common modes of failure being by vertical splitting.

In Fig. 4-1(d), the effect of having a soft layer in the pillar at the roof line is shown (3). The main result of this geology is that transverse tension is induced at the top of the pillar. The tendency for the soft layer to be squeezed out can be visualized as causing this additional tension. In addition, the compressive stress concentration at the bottom of the pillar towards the edge is increased when a soft layer occurs at the top.

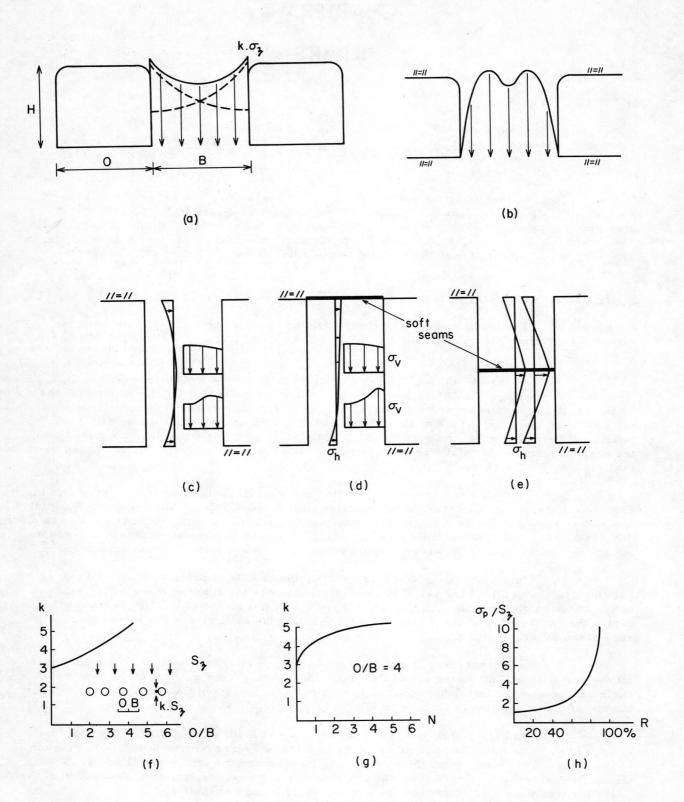

Fig. 4-1 Stress Distributions in Pillars

In Fig. 4-1(e) the effect of having a soft layer in the pillar at mid-height is shown (3). The main result is that relatively large tensions are created at mid-height. Again, this action can be visualized as resulting from the tendency to squeeze out the soft layer. Support is added to this concept by experimental work on brick piers, which has shown that the squeezing action on the mortar causes tension in the bricks and that this tension increases with the thickness of the mortar (13).

To illustrate the effect on the surface stress of the breadth of the pillar, B, in relation to the breadth of the opening, O, Fig. 4-1(f) shows the case of five openings in a uniaxial stress field. The maximum axial stress in the pillar is designated by kS_z, where S_z is the uniaxial field stress acting in the ground in which the openings are made. This graph then shows the variation of the stress concentration factor, k, with the ratio O/B(1). It can be seen that, as the ratio O/B increases (which can be considered as the breadth of the pillar decreasing), the interaction of the openings becomes more significant and the stress concentration factor, k, increases.

Fig. 4-1(g) presents the two-dimensional case where the ratio O/B equals 4 and the number of openings, N, is varied (1). In a uniaxial stress field the maximum stress concentration factor will be equal to 3 for a single opening. As the number of openings increases, it can be seen that the maximum stress concentration factor, k, increases to about 5 for five openings, but tends to be asymptotic to this value.

In Fig. 4-1(h) a simple concept is illustrated for the three-dimensional case (1). Here it is assumed that a square area is being mined. It is assumed that the pillars are also square and that they support all of the overlying ground tributary to their location. The graph in this figure then shows the relationship between σ_p/S_z, where σ_p is the average pillar stress, S_z is the uniaxial field stress, and R is the percentage recovery in the mining area.

The equations for these relations are:

$$R = \frac{(B_x + O_x)(B_y + O_y) - B_x B_y}{(B_x + O_x)(B_y + O_y)} \qquad \text{Eq. 4-1}$$

where the subscripts, x, refer to the dimensions in the x-direction, the subscripts, y, refer to the y-direction, B refers to the pillar, and O refers to the opening; and, it follows that

$$\sigma_p/S_z = 1/(1-R) \qquad \text{Eq. 4-2}$$

From these equations it can be calculated that for 50 per cent recovery the average pillar stress would be twice the original stress due to the overlying ground; for 75 per cent recovery the average pillar stress would be $4S_z$; and as 100 per cent recovery is approached the average pillar stress would approach infinity. This simple concept, of course, ignores the effect of the deformation characteristics of the roof, or back, resulting from the mining operation.

Example - A room and pillar mining plan consists of rooms 28 ft wide and pillars 55 ft x 130 ft. The vertical field stress is 3000 psi. Calculate the percentage recovery and the average pillar stress assuming the vertical field stress is distributed without arching into the pillars.

$$R = \frac{(55 + 28)(130 + 28) - 55 \times 130}{(55 + 28)(130 + 28)} = 0.454 = 45.4 \text{ per cent}$$

$$\sigma_p = 3000/(1-0.454) = 5490 \text{ psi}$$

Mining operations are conducted where the dip of the ore is somewhere between 0 and 90 degrees. Some research work has been done on the stress distribution that occurs in pillars in dipping seams (15). In Fig. 4-2 the results of this work for a pillar with H/B = 1, S_v/S_h = 2, and a vein dipping 60 degrees are shown. The contours in Fig. 4-2(a) connect points of equal maximum shear stress.

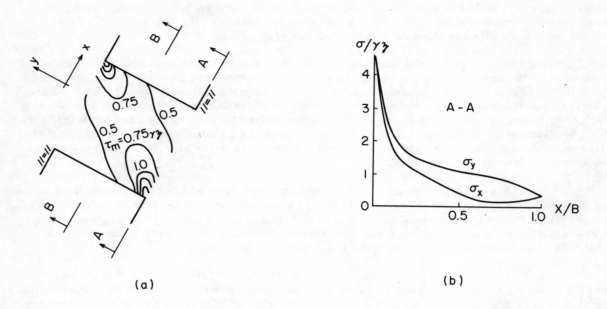

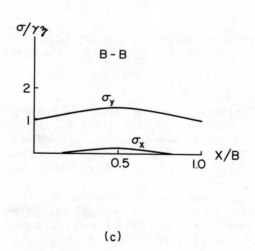

Fig. 4-2 Stress Distributions in Pillars in Inclined Veins (15)

In Fig. 4-2(b), the variation of the principal stresses across the pillar at a section, A-A, through the bottom is shown (15). Very high compressive and shear stresses occur at the floor line on the down-dip side with a similar pattern existing at the roof line on the up-dip side of the pillar.

In Fig. 4-2(c), the principal stresses across the pillar at the centreline section, B-B, are shown. The average axial normal stress is approximately equal to the overburden field stress.

By examining the various cases that were demonstrated in this experimental work, it was possible to draw the curves as shown in Fig. 4-3 (15). Here the average shear stresses on the xy-planes acting at the top of the pillar are expressed as multiples of the vertical field stress due to gravity, γz. The variation of this average shear stress with the dip angle for two cases of pillar geometry is shown.

It was suggested that, with a stress field of $S_v/S_h = 2$ for dip angles between 0 and 30 degrees, it should be feasible for practical purposes to calculate these average pillar shear stresses as if the dip were 0 degrees (15). For the dip angles between 60 and 90 degrees the dip angle could be assumed to be 90 degrees. For dip angles between 30 and 60 degrees, the average shear stresses could then be assumed to vary linearly between these two extreme cases. This suggestion is shown by the dashed lines in Fig. 4-3. This type of guidance is useful for practical purposes.

Mining observations indicate that pillars in inclined stopes often fail in a manner different from those in either horizontal or vertical stopes. Suggestions have been made for changing the shape of these pillars to make them more stable. As a result of these suggestions a photoelastic study was conducted on model pillars in inclined stopes with the down-dip side of the pillar being vertical and the up-dip side being normal to the walls (15). The actual dimensions of the pillars were selected so that the percentage extraction would be the same as in the series of tests run on rectangular shaped pillars. In this way it was possible to have a common basis for comparing the stresses in the specially shaped pillars with those in a conventionally shaped pillar with the sides normal to the walls.

In Fig. 4-4(a), the isochromatics or contours of equal maximum shear stress are shown as multiples of the vertical field stress, γz. It was found that both the shear stresses and the compressive stresses were greatest at the top corners of the pillar.

In Fig. 4-4(b), the variation of the principal stresses across the pillar at Section A-A through the top of the pillar is shown. The maximum compressive stress occurs on the down-dip side.

In Fig. 4-4(c), the variation of the principal stresses across the pillar at mid-height are shown. The compressive stresses are smaller here than at the roof line.

This experimental data indicated that the stress distribution in the specially shaped pillar was somewhat more favourable than in the rectangular shaped pillar. However, owing to the reduced cross-sectional area at the top of the pillar, the stress concentrations here were actually some 50 per cent greater than those in the rectangular pillar.

Tests were then conducted to determine the loads required to cause these pillars to fail in the model material. The results of these tests indicated that the carrying capacity of both types of pillars was about the same. They also indicated that failure always occurred in the specially shaped pillars in the upper part.

From this work it would seem that a specially shaped pillar might have some, but possibly little, advantage over the rectangular shaped pillars. Furthermore, it would probably only be possible to shape pillars, as in the above models, where the ratio of breadth to height is relatively large, so that the upper part of the pillar would not be too small for stability.

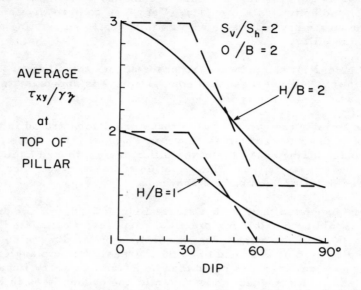

Fig. 4-3 Average Shear Stress in Pillars in Inclined Veins (15)

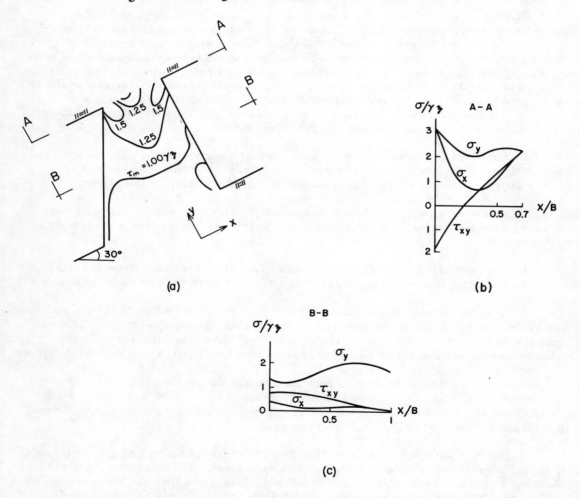

Fig. 4-4 Stress Distribution in Trapezoidal-shaped Pillars in Inclined Workings (15)

PILLAR LOADING IN DEEP, LONG MINING ZONES

The tributary area theory, as expressed by Equations 4-1 and 4-2, can be examined in more detail for the special cases of long mining zones such as shown in Fig. 4-5. The significance of the mining zone being long is that the analysis can be made in plane strain or in two dimensions.

The tributary area theory assumes that each pillar will be loaded by the normal stress, S_o, acting over the area of the wall tributary to that pillar. For a long mining zone this can be expressed:

$$\sigma_p = S_o(B_o + B)/B$$

where B_o is the average breadth of the adjacent openings and B is the breadth of the pillar. For a series of rooms and pillars of equal dimensions, the extraction ratio can be used as follows:

$$R = A_o/A_T = (N+1)B_o/\big((N+1)B_o + NB\big)$$

where A_o is the total wall area adjacent to the openings or rooms, A_T is the total wall area adjacent to the mining zone, and N is the number of pillars.

From these two equations it follows that:

$$\sigma_p/S_o = B_o/B + 1$$

$$= \frac{R}{1-R} \frac{N}{N+1} + 1$$

When $N \to \infty$

$$\sigma_p/S_o \to 1/(1-R)$$

In this way Equation 4-2 is obtained.

However, when $N = 1$, Equation 4-2 gives an answer that is seriously in error with respect to the intention of the tributary area theory. Therefore, it is preferable to express the tributary area theory as follows:

$$\Delta\sigma_p/S_o = \frac{R}{(1-R)(1+1/N)} \qquad \text{Eq. 4-3}$$

and

$$\sigma_p/S_o = \Delta\sigma_p/S_o + 1 \qquad \text{Eq. 4-4}$$

The tributary area theory is applicable under some circumstances; however, it is unsatisfactory as it ignores geometrical properties such as the span or breadth of the mining zone with respect to its depth for horizontal workings, the height of the pillars, the breadth of the pillars, and their location within the mining zone. Also, such geological properties as the nature of the actual field stress (in particular the magnitude of the component parallel to the breadth of the mining zone) and the modulus of deformation of the pillar rock relative to that of the wall rocks are not included.

A more satisfactory solution of this problem is based on solving the statically indeterminant net deflection of the walls resulting from mining. This net deflection at the pillars will be a measure of the increase in pillar stress resulting from mining. The components of wall deflection can be considered to be made up of the inward movement due to the excavation of ground (which is equivalent to applying a stress inwards on the walls), the reverse deflection resulting from the average wall pressure arising from the increase in pillar stress, the relaxation that results from eliminating the side constraint on the pillars, and the local penetration of the pillars resulting from the concentration of the average wall pressure.

To establish an expression for the deflection of the walls resulting from the excavation, we start with the deflection of a point around a circular hole in a plate due to an applied uniaxial stress (16):

v_r = radial displacement

$$= \frac{S_o r}{2E}\left[(1+a^2/r^2)+(1+4a^2/r^2-a^4/r^4)\cos 2\theta - \mu((1-a^2/r^2)-(1-a^4/r^4)\cos 2\theta)\right]$$

v_θ = tangential displacement

$$= \frac{S_o r}{2E}\left[(1+2a^2/r^2+a^4/r^4)\sin 2\theta - \mu(1-2a^2/r^2+a^4/r^4)\sin 2\theta\right]$$

where S_o = field stress

r = radial distance to the point

θ = angle from vertical to point

E = modulus of elasticity of the medium.

It follows from these equations that, on the circumference of the hole where $r = a$, the vertical deflection, δ, can be obtained as follows:

$$\delta = v_r \cos\theta + v_\theta \sin\theta$$

$$= 3 S_o a \cos\theta / E.$$

The following equations apply to the displacement around the circular hole in a plate subjected to a biaxial stress field (16):

$$v_r = \frac{r}{2E}\left\{(S_o+S_t)(1+a^2/r^2)+(S_o-S_t)(1+4a^2/r^2-a^4/r^4)\cos 2\theta\right\}$$

$$- \frac{\mu r}{2E}\left\{(S_o+S_r)(1-a^2/r^2)-(S_o-S_r)(1-a^4/r^4)\cos 2\theta\right\}$$

$$v_\theta = \frac{r}{2E}\left\{(S_o-S_r)(1+2a^2/r^2+a^4/r^4)\sin 2\theta\right\}$$

$$- \frac{\mu r}{2E}\left\{(S_o-S_t)(1-2a^2/r^2+a^4/r^4)\sin 2\theta\right\}$$

where S_t = the field stress acting at right angles to S_o. The vertical deflection of a point on the circumference for $\theta = 0$ and $S_o = 0$ is as follows:

$$\delta = -S_t a/E$$

The deflection of a circular hole due to the excavation of the material from the hole that is under biaxial stress can be determined as follows:

$$\delta = \delta_o + \delta_e$$

$$\delta_e = \delta - \delta_o$$

$$\delta_o = (S_o - \mu S_t) a/E$$

where δ_o = the deflection on the circumference of the hole before the hole is created and due to the pre-existing field stress, and δ_e = the deflection of the circumference of the hole resulting from the excavation. It follows then that the deflection due to excavation is as follows:

$$\delta_e = \left(2S_o a - S_t a(1-\mu)\right)/E$$

It has been shown that the deflection equation for the boundary of an elliptical hole in a plate resulting from excavation of the hole has a strong similarity to that of the circular hole where $\theta = 0$ (17, 18):

$$\delta_e = \left(2S_o a - S_t b(1-\mu)\right)/E$$

where a = the major semi-axis of the ellipse and b = the minor semi-axis of the ellipse. This equation can be used as an approximation for a long, narrow mining zone with elaborations, where necessary, based on the more complete solution for the circular hole.

For a mining zone such as in Fig. 4-6, using equations in plane stress for the moment, and letting $\ell = L/2$, $h = H/L = H/2\ell$ and $S_t/S_o = k$, we can write:

$$\delta_e = S_o \ell\left(2 - kh(1-\mu)\right)/E$$

where L is the span or breadth of the mining opening and H is the height of the pillars.

If a complete excavation of the mining zone is not made, in other words if pillars are left, the internal traction on the walls or the stress equivalent to excavating will have an average value normal to the walls of:

$$S_i = S_o A_o/A_T = RS_o.$$

The deflection resulting from excavation now becomes:

$$\delta_e = S_o \ell\left(2R - kh(1-\mu)\right)/E$$

As a result of the tendency for the walls to deflect (δ_e) due to excavation, the pillars will be subjected to an increased stress. This means that the walls will be subjected to an increased average back pressure, S_p, from the pillars. The deflection resulting from this back pressure is the reverse of the excavation case and can be written:

$$\delta' = 2S_p \ell/E$$

As $S_p = \Delta\sigma_p B/(B_o + B)$ for a series of rooms and pillars of equal dimensions, the equation can be written as follows for a large number of pillars:

$$\delta' = 2\Delta\sigma_p \ell(1-R)/E$$

Where there are only a few pillars the modification following Equation 4-3 can be made:

$$\delta' = 2\Delta\sigma_p \ell(1-R)(1+1/N)/E$$

The relaxation or deflection of the pillar resulting from the release of side constraint due to excavation of the rooms, or stopes, recognizing that the origin for purposes of calculating deflections is the mid-height of the pillars, is as follows:

$$\delta_r = \mu_p S_t H/2E_p$$

The local penetration of the pillars into the walls as a result of the average back pressure, S_p, being concentrated at the pillars causes a local penetration in excess of the general reverse deflection, δ'. The following solution, which exists for the relative penetration of a uniform pressure on the edge of a semi-infinite plate, can be used (19):

$$\delta = \frac{2\sigma}{\pi E}\left\{(B+2x_1)\ln\left(\frac{2x_2-B}{B+2x_1}\right) + (2x_2-B)\ln\left(\frac{2x_2-B}{B-2x_1}\right)\right\} + \frac{\sigma B(1-\mu)}{\pi E}$$

where
- σ = uniform pressure on the edge of the plate,
- B = width of the loaded zone,
- x_1 = distance from centreline of load, within the loaded zone,
- x_2 = distance from centreline of load, beyond the loaded zone to the point to which δ is related.

To use this equation, the coefficient is determined for $x_1 = 0$ and $x_2 = B$. Hence, the maximum local penetration with respect to a point at a distance from the side of the pillar of $B/2$ is:

$$\delta'_p = \frac{\Delta\sigma_p - S_p}{\pi E} B(1-\mu)$$

$$= \left(\Delta\sigma_p - \frac{\Sigma(\Delta\sigma_p B)}{L}\right)\frac{B}{\pi E}(1-\mu)$$

$$= \Delta\sigma_p\left(1 - \frac{\Sigma(B)}{L}\right)\frac{B}{\pi E}(1-\mu)$$

$$= \Delta\sigma_p\left(1-(1-R)\right)\frac{B}{\pi E}(1-\mu)$$

$$\delta'_p = \frac{\Delta\sigma_p B R(1-\mu)}{\pi E}$$

The net pillar deflection, σ_p, will equal the algebraic sum of all the effects produced by the field stresses and by mining. The effects of mining only can be analysed to give the increase in the average stress in the pillar, with the total stress then being equal to the increase plus the original field stress. Thus the increment of pillar deflection, $\Delta\delta_p$, is related to the increase in pillar stress, $\Delta\sigma_p$, as follows:

$$\Delta\delta_p = \frac{\Delta P H}{2A_p E_p} = \frac{\Delta\sigma_p H}{2E_p}$$

where $\Delta\delta_p$ is the increase in pillar deflection due to mining, ΔP is the increase in pillar load, $H/2$ is half the total height of the pillar, A_p is the horizontal, cross-sectional area of the pillar, E_p is the modulus of deformation of the pillar rock, and $\Delta\sigma_p$ is the increase in the average pillar stress.

The increase in pillar deflection is related to the various components of deflection as follows:

$$\Delta\delta_p = \delta - \delta_r - \delta' - \delta'_p$$

From this it follows:

$$\frac{\Delta\sigma_p H}{2E_p} = \frac{S_o \ell}{E}\left(2R - kh(1-\mu)\right) - \mu_p S_t H/2E_p$$

$$- \frac{2\Delta\sigma_p \ell}{E}(1-R)(1+1/N) - \frac{\Delta\sigma_p RB}{\pi E}(1-\mu)$$

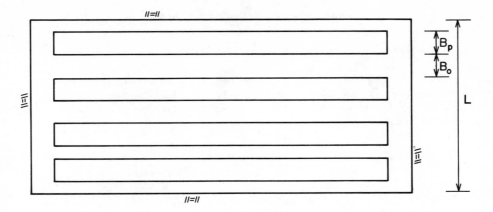

Fig. 4-5 A Long Mining Zone

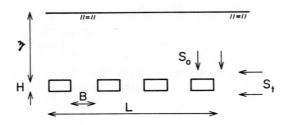

Fig. 4-6 Vertical Section of a Long, Horizontal Mining Zone

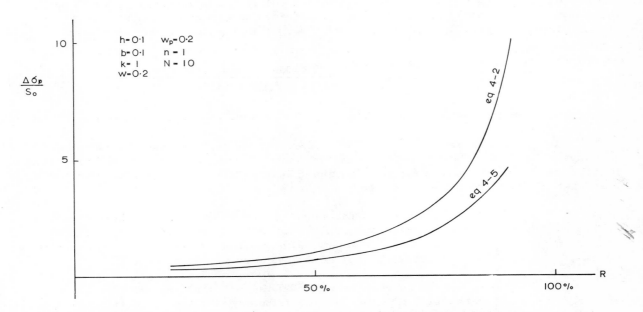

Fig. 4-7 Prediction of Pillar Loading by Tributary Area Theory and by Theory of Deep Mining Zone

It then follows that:

$$\frac{\Delta \sigma_p}{S_o} = \frac{(2R - kh(1-\mu)) - \mu p khn}{hn + 2(1-R)(1+1/N) + 2Rb(1-\mu)/\pi}$$

where $n = E/E_p$, $h = H/L = H/(2\ell)$ and $b = B/L$.

Then this equation can be converted to plane strain from plane stress by changing all E's to $E/(1-\mu^2)$ and all μ's to $\mu/(1-\mu)$. To simplify the resultant equation, let $M = E/(1-\mu^2)$, $w = \mu/(1-\mu)$, and $n = M/M_p$. The final equation for determining the pillar loading is then as follows:

$$\frac{\Delta \sigma_p}{S_o} = \frac{(2R - kh(1-w)) - w_p khn}{hn + 2(1-R)(1+1/N) + 2Rb(1-w)/\pi} \quad \text{Eq. 4-5(a)}$$

This equation ignores the variation in average pillar stress with location within a mining zone as all of the deflection equations were solved for the case of $\theta = 0$, which is at the centreline of the mining zone. Some recent work has shown that the variation of this loading is much less than that which would be indicated by the variation of vertical deflection of the boundary of a circular hole (18).

In addition, the complicating effects of abutment compression at the edges of the mining zone have been ignored. This is justified to some extent by recognizing, as it affects both δ_e and δ', its relative insignificance unless h is very large (18).

The equation, having been derived for the case of plane strain, can only be used for sections that are at least a distance away from the end of the mining zone equal to the breadth or span.

It is interesting to see that, when $k = 0$, $h \rightarrow 0$ and $b \rightarrow 0$, the above equation reduces to that obtained from the tributary area theory.

Equation 4-5(a) is valid where the pillar widths are all approximately equal and equally spaced apart. Where the widths of the pillars and their spacing vary, it has been found that a better form of the equation is to replace the factors in the denominator $(1-R)(1+1/N)$ with $(1+b_o/b)^{-1}$, where b_o equals $B_o/(L/2)$. The equation is then as follows:

$$\frac{\Delta \sigma_p}{S_o} = \frac{(2R - kh(1-w)) - w_p khn}{hn + 2(1+b_o/b)^{-1} + 2Rb(1-w)/\pi} \quad \text{Eq. 4-5(b)}$$

This equation might provide some guidance in planning a pillar extraction operation.

It is judged from empirical studies that the depth of cover for horizontal workings required for Equation 4-5(a) is greater than half the span of the mining zone (18). In these conditions, as well as for steeply dipping seams, this hypothesis will predict pillar loads that are less than those calculated using the tributary area theory.

In Fig. 4-7 a comparison is shown for one case of the variation of pillar loading with extraction ratio as predicted by Equation 4-2 and Equation 4-5(a). At high extraction ratios the differences are substantial with Equation 4-5(a), as shown by empirical studies, being close to the actual loading if all the assumed conditions prevail.

In Fig. 4-8 the variation of pillar loading with the parameter k, the ratio of S_t/S_o, is shown. In this case Equation 4-2, the tributary area theory, does not recognize any influence of the field stress tangential to the mining zone. Equation 4-5(a) shows that this parameter is significant, and with large values of S_t the pillar loadings will be reduced considerably.

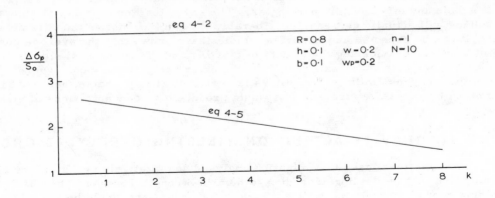

Fig. 4-8 Variation of Pillar Loading with Transverse Field Stress, S_t, or k.

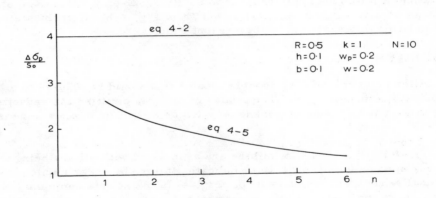

Fig. 4-9 Variation of Pillar Loading with Compressibility of Pillar, E_p, or n.

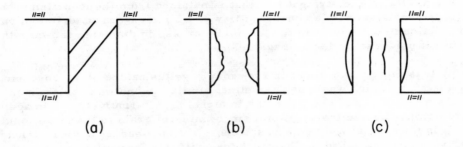

Fig. 4-10 Modes of Pillar Failure

In Fig. 4-9 the variation of pillar loading with the parameter n, the ratio of the compressibility of the wall rock to that of the pillar rock, is shown. Again Equation 4-2 does not include any effect of this parameter. The curve for Equation 4-5(a) shows that the pillar loadings will actually decrease considerably with an increase in the parameter n. Furthermore, it can be expected and empirical studies have shown that the average compressibility of a pillar, considering the effects of blasting and the removal of side constraint, is likely to be less than that of the wall rock. On the other hand, in the cases where $n \geq 1$, in other words the compressibility of the pillars is very low, it is probable that rockbursts could be promoted by excessive stresses that would produce high concentrations of elastic strain energy.

LOADS DUE TO GRAVITY ACTING ON YIELDING OVERLYING GROUND

An additional case that might be analysed quantitatively is that of horizontal workings under rock that is yielding according to Mohr's Strength Theory. The results of having a yielding roof rock would be that local stress concentrations would be diminished and that the vertical shear stresses would extend farther upwards and produce the case of arching described in Chapter 5. Thus the average vertical stress in the roof rock could be calculated using Equation 5-3.

Assuming sufficient deflection to develop the full shear resistance in the overlying ground, the increase in pillar loads would then be obtained by multiplying the tributary area to any pillar by this average roof stress. The answer, however, would not take into account the variation of pillar load with respect to position. Even for this case of yielding ground the variation would, qualitatively, be as shown in Fig. 4-8. Unfortunately, there is no method of calculating this variation at the present time.

PILLAR STRENGTH

After determining the probable loading that would be applied to a pillar, the next step is to determine its strength. The modes of failure, omitting the reduction in pillar strength arising from a major zone of weakness such as a fault or dyke, are generally as shown in Fig. 4-10.

The following modes of failure are associated with pillars being subjected to a uniaxial compressive stress. Fig. 4-10(a) indicates failure by shearing along oblique planes, which results from a crushing failure. Fig. 4-10(b) shows a commonly observed pattern of failure where the central sections slab off producing an hour-glass shape. Fig. 4-10(c) shows a mode of failure where the pillar expands laterally and splits vertically. This lateral expansion has been found in some cases to be a better indication of approaching failure than any measure of the vertical deformation or stress (9, 14).

Another factor that may significantly affect pillar strength is the influence of the size of the pillar. It has been found from uniaxial compression testing that the average strength of a rock usually decreases with an increase in the volume of the sample. A typical relation between the uniaxial compression strength, Q_u, and the diameter of the sample, D, is shown in Fig. 4-11(a). The straight line that is obtained from the logarithmic plot shown in Fig. 4-11(b) is consistent with a concept of flaw, or a micro-crack, distribution. The 50 per cent reduction in strength with the ten-fold increase in diameter shown in this figure is a typical but not universal order of magnitude.

It is thought that flaws in rock may have the nature of microscopic, weak grains within the rock, which would have an effect similar to one weak brick in an otherwise competent mass of bricks or of one weak link in a chain. Alternatively, or in addition, flaws may be cracks varying in size from that of a single mineral grain to the large joints seen in a rock face. It is then visualized that stress concentrations leading to local failure occur around these cracks as described in Chapter 1 for Griffith's Strength Theory.

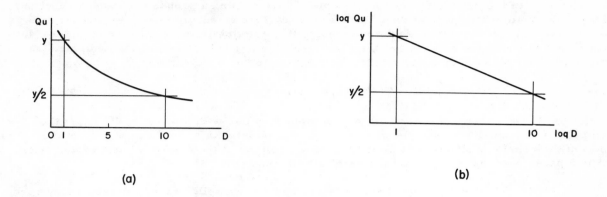

Fig. 4-11 Variation of Uniaxial Compressive Strength of Rock with Size of Sample

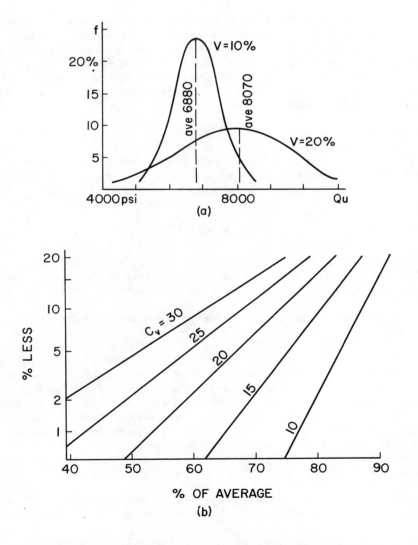

Fig. 4-12 Dispersion of Strength about a Mean

Whatever their nature, it is generally assumed that the maximum size of flaw varies with the volume of the rock sample. By assuming a probability density function of the strength of the imperfections, the average fracture strength of samples and the dispersion of the results can be calculated by using the statistical theory of extreme values (4, 5). It can then be shown, assuming that the number of flaws is proportional to the volume, that the relation between the uniaxial strength of the sample and its volume is:

$$Q_B = Q_o V^{-a} \qquad \text{Eq. 4-7}$$

where Q_B is the compressive strength of a sample or pillar with a breadth, B, Q_o is the compressive strength of a sample with the same shape but with a unit breadth, V, is the ratio of the volume of the sample of width B to the volume of the sample of unit width, and 'a' is a parameter that depends on the particular rock.

This equation has been shown to apply very well, not only to laboratory rock samples, but also to concrete samples (up to 3 ft in diameter) and to brickwork piers (up to 13-1/2 in. square) (6, 7). Consequently, it is probable that the same type of relationship applies to full size pillars.

As mentioned in Chapter 1 in connection with uniaxial compression testing, the ratio of height to width, H/B, might have an effect, independent of volume, on the strength of the pillar. For values of H/B less than one and probably less than two, the preferential plane for the oblique shear type failure shown in Fig. 4-10(a) could not exist from one side of the pillar to the other. Actually, this only affects the strength of samples slightly. Also, the possibility of H/B being large enough (e.g., greater than 4) to introduce the mechanism of elastic instability for pillars is very remote.

Another factor that can affect the ultimate stability of pillars is the ratio of external surface, A_s, to the cross-sectional area of the pillar, A_p. This arises from the total amount of slough, development of loose rock, and other manifestations of deterioration due to exposure being proportional to the area exposed. In some ground, such as potash, this deterioration can seriously reduce the effective area of the pillar that is sustaining any significant stress. Hence, there would be some relation such as

$$Q_B = Q_o (A_p/A_s)^b \qquad \text{Eq. 4-8}$$

where Q_B is the compressive strength based on the nominal pillar size, Q_o is the compressive strength for a pillar with $A_p/A_s = 1$, A_p is the nominal area of the pillar on which the wall loads are applied, and A_s is the surface area of the perimeter of the pillar.

If the length of the pillars was kept constant, the ratio A_p/A_s in Equation 4-8 would reduce to H/B.

The other important consideration in determining the strength characteristics of pillars is the dispersion of strength values that is likely to exist about some mean. In Fig. 4-12(a) two cases are shown. In one case the average strength is 6880 psi with a coefficient of variation of 10 per cent. In the other case the average strength is 8070 psi with a coefficient of variation of 20 per cent. It can be seen by examining the curves that for stresses of the order of 4000 psi there will be a greater probability of failure in the second material, even though its average strength is much higher than the first material. By calculation it can be shown that the probability of failure at a stress of 4800 psi for the first material is of the order of 0.1 per cent, whereas for the second material it could be about 2 per cent.

Using the same statistical theory as for Equation 4-6, it has been established that the standard deviation, $\overline{SD}$, and the number of flaws, n, should be related as follows (5):

$$(\log n_B/\log n_o)^{1/2} \overline{SD}_B / \overline{SD}_o = k \qquad \text{Eq. 4-9}$$

where the subscript o refers to the sample of the reference volume, the subscript B to a sample of some other volume and k is a constant.

Using this equation, the average values of the constant 'k' can be determined from the test data. Then a prediction can be made of the value of the coefficient of variation of a pillar's strength, and the minimum strength can be determined using Fig. 4-12(b).

Thus, to determine the safety factor that is incorporated in pillars, the procedure can be as follows. The rock strength (until field techniques are established) would be determined by laboratory testing. The laboratory program would be arranged to test samples of different sizes so that the exponents 'a' of Equation 4-7 and 'b' of Equation 4-8 could be determined. The data would also be analysed to determine the constant 'k' in Equation 4-9.

The probable average compressive strength of the pillar, Q_B, would then be determined by extrapolating the test data to the order of magnitude of the size of the actual pillars. In addition, the coefficient of variation, C_V, would be predicted using Equation 4-9.

Then the safety factor that would exist in the pillar could be determined. The maximum expected load and the minimum rock strength, not affected by major weaknesses, would be used. (When major weaknesses are discovered during the underground operations, then the analysis of the standard case would be modified to account for their effects.) It is assumed that the effects of the actual stress distributions in pillars (as shown in Fig. 4-1(a)) would be satisfactorily accounted for by the effects of similar non-uniform stress distributions in the laboratory test samples. Thus, this factor would not be included in the analytical procedure.

The minimum strength could be defined as the strength for which 95 per cent of the pillars would have greater values, or in other words, 5 per cent of the cases would have lower strengths. Thus, by following this procedure if a safety factor of 1 existed for the minimum strength, it would be expected that one in twenty pillars would fail. If this were unacceptable, some alternate minimum strength could be used, such as that which would give only a 1 per cent probability of failure.

Example of Pillar Analysis - Determine the probability of pillar failure in a coal mine to be operated in a horizontal seam at a depth of 500 ft. The breadth of the panel is 360 ft and the length of the ultimate mining zone will be about 4000 ft. The room and pillar method is to be used. The height, H, of the coal seam is 10 ft. A 50 per cent recovery is planned during the advance with pillars 40 ft in length by 20 ft in breadth and with entries 18 ft in breadth. The overlying ground consists of rock with a density of 160 pcf, $E = 4 E_p$, $G = 0.42 E$, and $\mu = 0.2 = \mu_p$.

The following laboratory tests were conducted on square samples with H/B = 0.5 to be similar to the planned pillars.

No. of Tests	H in.	Volume in.^{3}	Q_u psi	Standard Deviation psi
44	0.5	0.5	4180	592
29	1	4	3260	332
14	2	32	3120	264
4	5	500	2160	344

Solution:- It is assumed that the flaw density is proportional to the volume of the sample and thus Equations 4-7 and 4-9 are valid for the coal. Using Equation 4-7, $Q_B = Q_o V^{-a}$, log Q_B is plotted against log V as shown in Fig. 4-13. From this curve at the point where V = 1, the value of Q_o is 3790 psi. By solving for the slope of the curve, 'a' is found to be about 0.070, thus:

$$Q_B = 3790 \, V^{-0.070}$$

Therefore, if the volume of the pillar is

$$V = 20 \times 40 \times 10 = 8000 \text{ cf} = 1.38 \times 10^7 \text{ ci}$$

$$Q_B = 3790 (1.38 \times 10^7)^{-0.070} = 1200 \text{ psi}$$

To use Equation 4-9, assume there are 20 flaws per cubic inch (any number could be used):

V	n	$\overline{SD}$ psi	$k = (\log n_B/\log n_o)^{1/2} \overline{SD}_B/\overline{SD}_o$
0.5 ci	10	592	1
4	80	332	0.775
32	640	264	0.748
500	10,000	344	1.160
		avg k =	0.921

Therefore, for the pillar the standard deviation, $\overline{SD}$, can be calculated from Equation 4-9

$$(\log 1.38 \times 10^7 \times 20/\log 10)^{1/2} \overline{SD}/592 = 0.921$$

$$\overline{SD} = 592 \times 0.921/2.90 = 188$$

$$C_v = \frac{188}{1200} \times 100 = 15.7\%$$

From Fig. 4-12(b) the stress as a percentage of the mean strength for a 5 per cent probability of failure and a coefficient of variation of 15.7 per cent is 74 per cent; hence for a safety factor of 1 with a 5 per cent probability of failure the pillar stress must be:

$$\sigma_p = 0.74 \times 1200 = 888 \text{ psi}$$

For the determination of the pillar stress from its loading, Equation 4-5(b) can be used for a first approximation

$R = 0.5$, $h = H/L = 0.0278$, $b = B/L = 0.0556$, $k = S_t/S_o = 1/(5-1) = 0.25$, $\omega = 0.25 = \mu_p$
$n = 4$, $N = 9$.

$$\frac{\Delta \sigma_p}{S_o} = \frac{(2 \times 0.5 - 0.25 \times 0.0278 (1-0.25)) - 0.25 \times 0.25 \times 0.0278 \times 4}{0.0278 \times 4 + 2(1-0.5)(1 + 1/9) + 2 \times 0.5 \times 0.0556 (1-0.25)/\pi}$$

∴ $\sigma_p = 0.800 \, S_o + S_o = 1.800 \times 500 \times 160/144 = 1000 \text{ psi}$

$F_s = 888/1000 = 0.888$

With a safety factor less than 1 the probability of failure will be greater than 5 per cent.

The safety factor on the mean strength is

$$F_s = 1200/1000 = 1.2$$

The average stress being 1000/1200 = 0.833 of the mean strength, from Fig. 4-12(b) with a coefficient of variation of 15.7 per cent it is found that about 13 per cent of the pillars would have strengths less than the stress on the central pillar. Hence, about one in seven of the central pillars (and the stress on the other pillars would not be much different) might fail.

This example indicates a procedure that might be used for checking pillars; however, much research work is required to substantiate the various elements of the analysis.

If in the above example an attempt were made to increase the safety factor by increasing the breadth of the pillars while keeping the extraction ratio constant, it would be found that this would produce very little increase in safety factor. The pillar stress would be reduced only slightly, although the pillar strength taking into account Equation 4-8 could be increased somewhat. The most significant variables that could be modified would be the extraction ratio, R, and the breadth of the mining zone, L.

ROOF REACTIONS

The reaction of the floor or back to the pillar load may affect the stability of the rocks in these zones. Fig. 4-14(a) shows a probable distribution of vertical stress in the floor at the contact between the pillar and floor. This distribution would be applicable where the pillar and floor rocks had approximately the same moduli of deformation. The distribution would vary with the ratio of the floor and pillar moduli of deformation. Actual measurements have indicated that stress concentrations at the edge of pillars can be greater than three times the average pillar stress (12).

Any rounding of the corner at the contact, as discussed in Chapter 2, would decrease the maximum stress concentrations in the floor. Also, with some geological discontinuity between the pillar and the floor rock so that some relative displacement would occur at the contact, the peak corner stresses would be reduced (3).

These stresses may or may not be significant with respect to the stability of the floor or the back, where the same pattern of stress would exist. For weak floors, the possibility of a bearing capacity failure could exist. This situation is not uncommon in coal mines. The method of predicting such a possibility using plate load tests has been described in Chapter 1.

If a bearing capacity failure of the immediate floor rock does not occur, it is possible that weak layers at some greater depth would be stressed to the point where some plastic flow could occur causing excessive deformation and rupture of the overlying more competent rocks. Cases have occurred where an insufficient thickness of harder rocks resulted in the underlying soft layers being pushed up to the extent that the openings were filled (14).

The stresses in such a soft layer could be analysed using Boussinesq's theory for the stresses in a semi-infinite elastic body below a surface load, as described in Chapter 2. Thus, the stresses under a pillar in some layer in the floor could be analysed for a first approximation using the Newmark chart shown in Fig. 2-15. Alternatively, a quick approximation could be calculated based on the relations shown in Fig. 4-14(b) (2).

In Fig. 4-14(b) the rock in the floor is assumed to be perfectly elastic and isotropic. At depth z below the bottom of the pillar and on the plane ab, the vertical stress distribution is as shown. This stress distribution could be approximated by drawing lines at 45 degrees out from the base of the pillar to the plane ab. The assumption is then made that the pillar load is distributed entirely within these 45 degree lines and that the variation of vertical stress is parabolic between these limits.

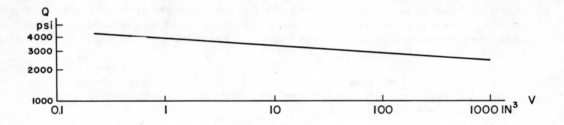

Fig. 4-13 Test Results of Variation of Strength with Volume of Sample

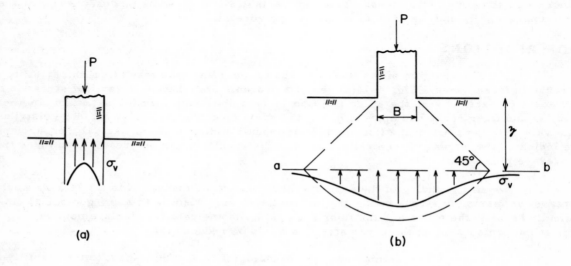

Fig. 4-14 Stresses in Walls Adjacent to Pillars

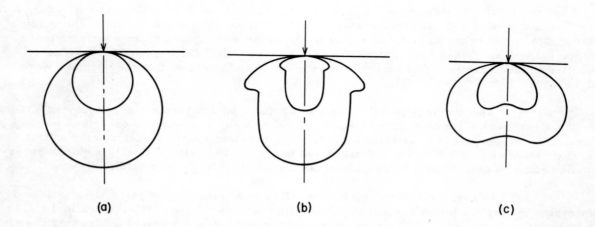

Fig. 4-15 Some Effects of Anisotropy of Stresses in a Semi-infinite Body Under a Surface Load (8)

Thus, the average stress between these limits on the plane ab would be equal to $P/(B+2z)$. This calculation would be suitable where the length of the pillar is very much greater than the width, B. The maximum vertical stress could be obtained by equating the pillar load, P, to the area under the parabolic curve, which would be equal to $0.667(B+2z)\sigma_{v-max}$. From this it follows that

$$\sigma_{v-max} = 1.5\,P/(B+2z) \qquad \text{Eq. 4-10}$$

Thus, the maximum stress from this approximation would be 50 per cent greater than the average stress.

The use of such an approximation would not be unreasonable for most problems as the effect of anisotropy is usually ignored even in the more refined calculations, and the ignoring of such anisotropy can introduce a more significant error than the above approximation.

In Figs. 4-15(b) and 4-15(c), two patterns of equal radial stress contours for cases of anisotropic ground in plane stress (8) can be compared with Fig. 4-15(a) representing isotropic ground. Whereas it is now becoming theoretically possible to analyse cases of anisotropic ground, the limitations on the determination of the different elastic properties is still sufficiently serious to make such complex analyses at the present time of only marginal value.

If a parting exists above the roof or back line, a much larger tangential tensile stress could occur in the roof at the surface than would exist in homogeneous ground. Also, the radial stress in the back would be tensile, and the horizontal stress at the mid-height of the pillar would be greater than in the homogeneous case (3).

The stability of the back, as mentioned above, is influenced by the reaction to the pillar stresses. It is also influenced, of course, by the geometry of the opening, by the rock properties, and by the depth below the surface. In Chapter 5 it is shown that an opening with a length equal to the width has stresses of approximately one-half of those for the opening that is long with respect to the width. The influence of the shape of the back and the depth below the surface are also examined in Chapter 5 (see Figs. 5-2 and 5-3).

REFERENCES

1. Duvall, W., "Stress Analysis Applied to Underground Mining Problems", Part II USBM RI 4387 (1948).

2. Terzaghi, K., "Theoretical Soil Mechanics", Wiley, p. 382, 398, Fifth Printing (1943).

3. Trumbachev, V. and Melnikov, E., "Distribution of Stress in Inter-Room Pillars and Immediate Roofs", Gosgortekhizdat, Moscow (1961).

4. Weibull, W., "A Statistical Theory of the Strength of Materials", Proc. No. 151, Royal Swedish Inst. Eng. Res. (1939).

5. Skinner, W., "Experiments on the Compressive Strength of Anhydrite", The Engineer (Feb. 13 and 20, 1959).

6. United States Bureau of Reclamation "Boulder Canyon Project Report, Part 7", Bull. 4 (1949).

7. Davey, N. and Thomas, F., "The Structural Uses of Brickwork", Structural and Building Engineering Division Meeting of ICI, Paper No. 24 (1950).

8. Lechnitskii, S., "Theory of Elasticity of Anisotropic Elastic Body", Holden-Day (1963).

9. Greenwald, H., Howarth, H. and Hartmann, I., "Experiments on Strength of Small Pillars of Coal in the Pittsburgh Bed", USBM Tech. Paper 605 (1939).

10. Obert, L., "Measurement of Pressures on Rock Pillars in Underground Mines", Part II, USBM RI 3521 (1940).

11. Tincelin, E. and Sinou, P., "Collapse of Areas Worked by the Small Pillar Method", Proc. 3rd. Internat. Cfce. Strata Control, Cherchar, Paris (1960).

12. Mohr, H., "Measurement of Rock Pressure", Mine and Quarry Engineering (May, 1956).

13. Hast, N., "Measuring Stresses and Deformations in Solid Materials", Centraltryckeriet, Esselte AB, Stockholm (1943).

14. Obert, L. and Long, A., "Underground Borate Mining, Kern County, Calif.", USBM RI 6110 (1962).

15. Trumbachev, C. and Melnikov, E., "The Effect of the Dip Angle of a Deposit on the Distribution of Stresses in Interchamber Pillars", Tekhnologia i Eknomika Ugledobychi, No. 3, Moscow (1962).

16. Merrill, R. and Peterson, J., "Deformation of a Borehole in Rock", US Bureau of Mines RI 5881 (1961).

17. Toews, N., personal notes (1964).

18. Coates, D., "A New Hypothesis for the Determination of Pillar Loads", unpublished report (1965).

19. Roark, R., "Formulas for Stress and Strain", McGraw-Hill (1954).

CHAPTER 5

STOPES, CAVING AND SUBSIDENCE

INTRODUCTION

All underground mining methods may be divided into two groups. One group consists of those methods that support and thus attempt to prevent caving of the walls. The second group of methods requires either immediately or ultimately the caving of the roofs(1).

Within the second group are found the mining methods that require caving for their success, e.g., block caving and top slicing. In these cases, instead of trying to prevent rock failure, the design of the layout attempts to ensure rock failure or caving under controlled conditions.

The drawing properties of the broken ore, for those mining methods that are based on ore caving as opposed to cover caving, then become important after successful caving has been achieved. Other mining methods, which do not use caving, e.g., rigid pillar methods, shrinkage stoping and others, are also concerned in some cases with the drawing characteristics of broken ore.

Subsidence effects sooner or later, where the support provided is not as competent as the ore, will follow the excavation of the ore. In this context, subsidence is defined as the failure of ground above the workings resulting from the removal of ore (or any other ground) by underground operations. This concept includes the common surface effects of cracking and slumping, but, in addition, includes some cases of working ground adjacent to the mining area as seen in drifts and crosscuts.

CAVING

Gravitational Stress Field: Certain theoretical concepts are useful where caving is required and also where it is necessary to prevent it. First, the shape in plan of the undercut is significant with respect to the stresses that can be created to cause caving. As an analogue, the effect of the shape of a plate uniformly loaded and simply supported on four sides is shown in Fig. 5-1 (2). The variation of the maximum bending moment, M, in the plate, expressed as a multiple of the maximum moment in a square plate, M_s, is shown as a function of the ratio of the length, L, to the breadth, B. It can be seen from this graph that, as the length of the plate increases, M also increases. When the ratio L/B is 3, M/M_s reaches its maximum value of 2.5. The stresses due to bending will, of course, vary directly with the bending moments.

Whereas the usual geometry of a mining operation is such that the depth in relation to the span of the stope is generally too great for a beam or plate calculation to be valid, the variation of stress with the geometry of the undercut will be similar to that which occurs in a plate. The conclusion here is obvious. If caving is desired and is not being obtained with a square stope, one alternative would be to increase the length of the stope. With the width or span of the stope being held constant, this can increase the caving stresses by more than two-fold. Mining experience confirms this prediction.

The second alternative that may be selected for increasing caving stresses is to increase the absolute dimensions of the stope. Fig. 5-2(a) shows the effect of increasing the span of the opening in elastic ground where the length is large with respect to the breadth, i.e., L/B is greater than about 3 (3). The curves in this figure are shown for the case where the horizontal ground stress, S_x, is one-third of the vertical stress, S_z. The fillets at the top corners of the opening, or undercut, are assumed to have a radius of one-sixth of the height, H. The magnitude of the stresses is expressed as a multiple of the vertical stress in the ground, S_z. The width of the opening, B, is expressed as a multiple of the height, H. The variations of the stresses at the top corner and at the centre of the back are shown in this graph.

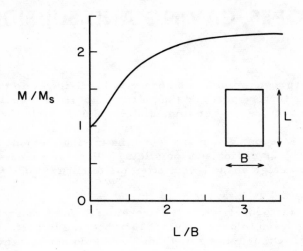

Fig. 5-1 Maximum Bending Moment in a Rectangular Plate Supported on Four Sides Compared to a Square Plate

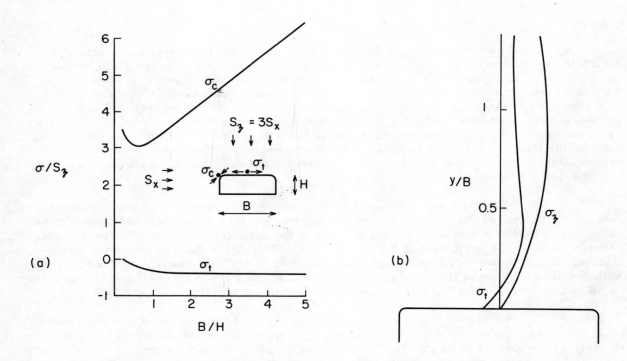

Fig. 5-2 Variation of Roof Stresses with Span of Opening

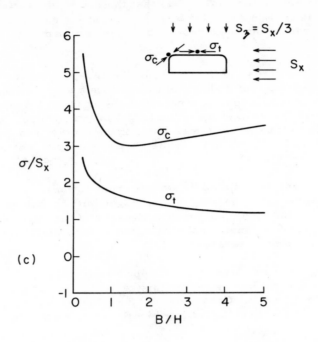

Fig. 5-2 Variation of Roof Stresses with Span of Opening

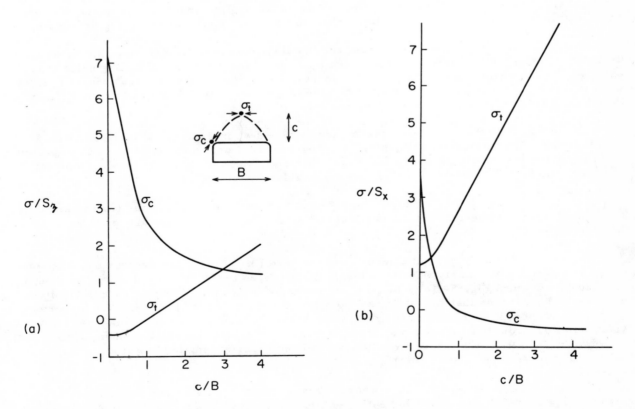

Fig. 5-3 Variation of Roof Stresses with Height of Caving or Arching

It can be seen from Fig. 5-2(a) that the maximum surface stress, σ_c, at the corner of the undercut, is of the order of $3 S_z$ for small spans and increases almost linearly with an increase in the span so that it is about $6 S_z$ when the span is a little more than four times the height of the opening.

The surface stress in the centre of the back, σ_t, varies from close to 0 for very narrow openings to $-0.4 S_z$ for B/H equal to 4. It thus becomes tensile for wide openings. The variation in magnitude of this tensile stress, as shown in Fig. 5-2(a), is insensitive to the magnitude of the span once the span exceeds the height of the opening. (It should be noted that this is contrary to the deduction that would be made if the stresses in the back were conceived in terms of beam theory.) The depth in from the rock surface for which tension exists is small, as shown in Fig. 5-2(b).

Caving might be initiated by two different actions. The tensile stresses in the centre of the back, although of low magnitude, might initiate failure of the rock and start the caving. Alternatively, the high compressive stresses at the upper corners of the undercut might exceed the compressive strength of the rock and cause failure here. It is also possible that both actions could occur simultaneously.

If caving were initiated by the compressive stresses at the corners, it is probable that large blocks would be created by this caving action as a result of the tendency to create an underhang in the back. Consequently, if sufficient control could be exercised, it would seem preferable to have caving initiated by the tensile stresses in the centre of the back.

A third theoretical aspect, the shape of the caving back, can be of significance in a caving operation. Fig. 5-3(a) shows the effect on the caving stresses of the height of the cave, C (4). These curves have been compiled using an original undercut, as described for Fig. 5-2(a), with a ratio B/H of 6 with the shape of the back during caving being an ellipse.

As caving works upwards, the tensile stresses in the centre of the back decrease in magnitude to zero and then become compressive for ratios of C/B greater than one. At the same time, the compressive stresses at the sides of the undercut decrease from their original high magnitude to values, when the height of the cave is more than four times the width, that are little greater than the vertical field stress.

With the surface stresses varying as shown in Fig. 5-3(a), the back would tend to become increasingly stable as caving progressed. Consequently, it could be expected that the caving rate would decrease and might ultimately consist of simply the development of surface loose rock. Field research has shown this to happen (21).

It should be noted that the above deductions are made for the case where the horizontal stress is one-third of the vertical stress and the height of ore above the cave line is always great.

Where the height of ore above the cave line is not much greater than the span of the stope (and particularly when it is less), the stresses in the crown pillar will be modified from those shown in Fig. 5-3(a). Bending action in the crown pillar might produce significant stresses. This means that with negative moments at the top of the beam near the supports, cracking could occur as shown on the right side of Fig. 5-4(a). In addition, diagonal tension could exist out from the supports and produce diagonal cracking through the beam, also as shown in Fig. 5-4(a). This action would, as has been experienced, tend to produce large blocks or poor fragmentation (9, 11).

For this special case where a beam condition exists, two deductions can be made. First, when waste or fill is resting on top of the crown pillar, the bending stresses and the caving action through the pillar should occur more readily. The second point is that, as the caving progresses through the crown pillar and the depth of the pillar is reduced, stresses will be increasing very rapidly. Consequently, with the rate of stress change varying with the cube of the depth of the crown pillar, a situation could exist favouring as mining proceeds the sudden failure of the entire pillar.

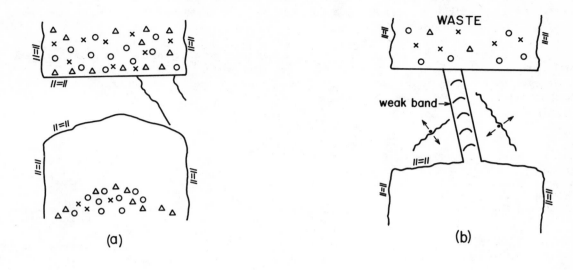

Fig. 5-4 Some Patterns of Caving

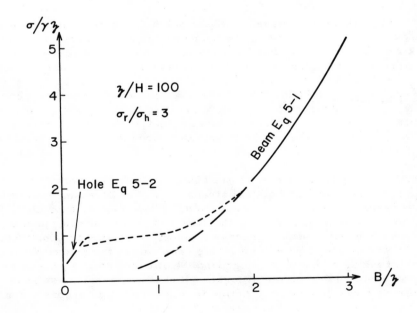

Fig. 5-5 Comparison of Hole Analysis with Beam Analysis for Roof Stresses

It is instructive to examine these two methods (stress concentration and beam analysis) of determining the tensile stress in the centre of a roof or back. The case of a long chamber or undercut can be used for a two-dimensional analysis. A gravitational field stress with lateral confinement giving a horizontal stress of 1/3 (i.e., assuming m = 4) of the vertical stress will be used. Also, a constant ratio of depth of cover to height of room, i.e., z/H = 100, will be used.

It is known that, if the depth of a beam is greater than about half the span, the normal beam Equation 2-4 is not a good representation of the stress distribution. Thus, assuming some restraint at the support (hence the moment coefficient of 1/10), the maximum bending stress is:

$$\sigma = Mc/I = M/S$$

$$= \gamma z/B^2/10 \div 1 \times z^2/6$$

$$= 0.6 \gamma z (B/z)^2$$

Then, adding the axial stress of $0.33 \gamma z$ compression gives:

$$= 0.6 \gamma z (B/z)^2 - 0.33 \gamma z$$

$$= (0.6(B/z)^2 - 0.33) \gamma z \qquad \text{Eq. 5-1}$$

The alternate method of calculating roof stress is to use stress concentration factors. From photoelastic studies on rectangular openings (3) the resultant tension in the roof is found to be approximately $0.3 \gamma z$ for B/H = 1, $0.5 \gamma z$ for B/H = 8 and $0.6 \gamma z$ for B/H = 12. These figures are for $\sigma_v/\sigma_h = 3$ and corner fillet radii of 1/6 H. By assuming z/H = 100, it follows that $B/z = \frac{1}{100} B/L$. From these figures the equation can be established relating σ and B/z:

$$\sigma = (0.27 + 2.7 \, B/z) \, z \qquad \text{Eq. 5-2}$$

Fig. 5-5 shows the curves of Equations 5-1 and 5-2 indicating the geometry for which each is valid. It is possible that the beam analysis would also be valid for rooms with $B/z < 2$ if the rock was distinctly stratified and the layer thickness was less than half the span. However, in this case the loading on the individual layers of rock is usually indeterminant.

Residual Stress Field: The above analyses are for vertical caving in gravitational stress fields with the horizontal stresses being less than the vertical stresses. In an area where the rock is subjected to a residual stress field so that the horizontal stress is greater than the vertical stress, the stress distribution around the stopes would be different from those described above.

In Fig. 5-2(c), the stresses are shown around an undercut where the horizontal stress, S_x, is three times the vertical stress, S_z. The stresses around the undercut vary as the span to height ratio of the undercut, B/H. These curves show that, for the pertinent range of span ratios (i.e., 2 to 5), the stresses in the back, σ_t, and those at the corners, σ_c, do not vary greatly with the change in span. Furthermore, both stresses are compressive, which would tend to produce a stable back unless the compressive stresses at the corners exceed the crushing strength of the ground. However, as it seems to be common for secondary tensile stresses to be induced at right angles to compressive stresses where there is little or no confining stress, the caving action in this case would probably consist of slabbing off the back due to these vertical secondary tensile stresses.

As caving progresses in this type of residual stress field, the variation of stresses around the back also would be quite different from those occurring in the gravitational field. Fig. 5-3(b) shows the variation of the stresses with the height of caved zone. These curves are for an undercut with a span six times its height. It can be seen that as caving progresses the compressive stress in the centre of the back increases very rapidly with the height of the cave line. Under these circumstances one would expect an accelerated caving action.

Fig. 5-3(b) also shows the variation that occurs to the corner stress, σ_c, as the caving progresses in the residual stress field. In this case, the stress drops very quickly from a fairly high value of $3.8 \, S_z$ down to a value of zero when the cave height is equal to the span of the undercut. σ_c then becomes negative or tensile as the caving proceeds from this point. These tensile stresses are not likely to be significant with respect to the caving action; they could produce some loose rock from the walls of the stopes.

As the crown pillar becomes developed during caving in a residual stress field, very high compressive stresses, as shown in Fig. 5-4(a), will become concentrated in this pillar. These stresses would increase as the thickness of the pillar decreased, and (not considering the nature of the material) if caving were proceeding actively this stress situation might produce a sudden failure of the entire pillar. With active caving there would be a tendency for insufficient time to be provided for a possibly moderate crushing failure. The situation would be similar to the rapid application of a load on a compression sample. Normally, this produces a higher compressive strength and a larger strain energy content in the rock at failure.

It has been suggested in the past that in some block caving operations so-called block failures have occurred. In other words, it is assumed that the entire block has failed at the boundaries of the stope and moved downwards like a rigid body. It is difficult to be certain that this action has occurred as conclusive measurements have not been taken to substantiate this action. Furthermore, it does not seem kinematically possible. The horizontal expansion of the peripheral material for such a complete block failure would not be permitted by the solid central section. In addition, as has been shown by experience at some mines, even boundary drifts and corner raises are inadequate to provide for this expansion.

One possibility that might be described as a block failure could occur with the presence of near vertical seams of weak material that could combine to permit large segments of the block to drop.

Induced Caving: If caving is to be assisted or accelerated, it is useful to know the strength properties of the ore. In the rare cases where the ground is strong in tension, it would help, if economically feasible, to destroy this strength by pre-blasting or by the formation of slots in the central part of the stope.

However, most rock masses are very weak in tension. In these cases it would seem to be preferable, if any assistance is to be provided, to attack the haunches of the caving arch where the compressive strength might be inhibiting caving. This would reduce the compressive strength where the compressive stresses are highest thus accelerating the caving. It would also tend to create a flatter back with the probability of increasing the magnitude of the tensile stresses in the central area.

Induced caving may also be desired to avoid preferential caving through weak bands. This piping action, as shown in Fig. 5-4(b), could result in two undesirable consequences. Waste could be drawn down through the pipe before all the ore has caved, and poor fragmentation of the adjacent harder ore could result from the induced diagonal tension. Incidentally, this same cantilever action can also occur where a weak abutment exists adjacent to a caving block. The block itself is then cantilevered from the opposite competent abutment.

Corner raises with drilling stations and horizontal drill-rings into the abutment ground of the caving arch have been used. In this operation it is important to blast when the crown of the caving arch just passes the drilling station. If the blasting is done too sone, a weakness is formed through which preferential caving can occur. If the blasting is done too late, of course, the holes might be lost.

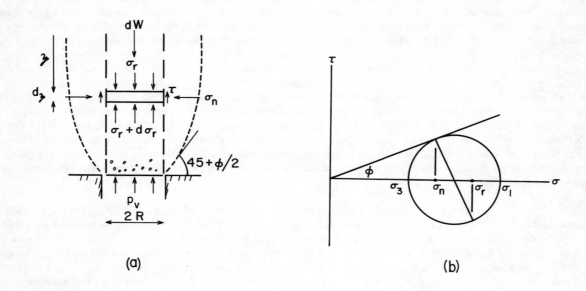

Fig. 5-6 Analysis of Caving or Arching Stresses (14)

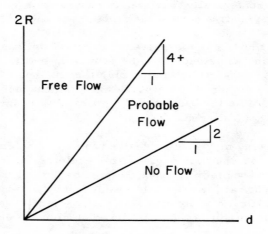

Fig. 5-7 Effect of Particle Size, d, in Relation to Radius of Opening, R, on Flow Condition of Caved Ore (7)

DRAWING

Little theoretical information exists on the subject of drawing caved ore or other granular materials. This is a serious deficiency in the science of rock mechanics. Consequently, any information that is to be used for the rational solution of drawing problems must be obtained from models or from experience in full-scale drawing operations.

A theory that has been derived for the determination of pressures in bins gives some information on the stresses that exist in caved material immediately before drawing occurs (5). In Fig. 5-6(a) the stresses for an idealized case of ground above a circular trap-door of radius R are shown (this figure can also represent the stresses in a bin during incipient flow). To support the ground over the trap-door, a pressure p_v is applied.

Before the trap-door moves down, the major principal stress is vertical and the minor principal stress is horizontal. Consequently, any incipient failure plane would be oriented at an angle of $(45 + \varphi/2)$ as indicated in Fig. 5-6(a). Thus, as the trap-door was lowered for the start of drawing, there would be a tendency for the ground to shear on these incipient failure planes. However, it can be visualized that kinematically this tendency becomes impossible owing to the requirement for horizontal movement from the sides towards the centre-line of the draw-point.

The ground above the edge of the trap-door is a zone of rapidly changing principal stress direction. Over the abutment at the sides of the trap-door the major principal stress would still be vertical, whereas over the trap-door itself the major principal stress would become horizontal. Thus, over a narrow zone at the edge of the trap-door the principal stress trajectories must turn through 90 degrees.

It has been suggested that the actual boundaries of the flow zone above the trap-door are somewhat as indicated by the dotted lines in Fig. 5-6(a). The locations of the boundaries are not as yet theoretically determinable. Hence, it is assumed that flow occurs along vertical boundaries. The column of ground above the trap-door is isolated in a free body diagram and analysed for static conditions (14).

If a horizontal slice of thickness dz is taken out of this column of ground, a vertical stress, σ_v, will act on the top surface and the vertical stress, $\sigma_v + d\sigma_v$, will act upwards on the bottom surface. In addition, the gravitational body force, dW, will be acting downwards on the element. On the sides of the element the shear stress, τ, will be acting upwards and a horizontal normal stress, σ_h, would also be present. During the period of no motion or, strictly speaking, no acceleration, the equilibrium of this element can be analysed.

Let $\sigma_h = k\sigma_v$, where k is assumed to be a constant. By examining the equilibrium of the slice shown in Fig. 5-5(a), an equation can be derived for σ_v:

$$\sum F_v = dW + \sigma_v \cdot A - (\sigma_v + d\sigma_v)A - \tau \cdot P \cdot dz = 0$$

where A is the horizontal area of the slice, P is the perimeter, and dW is equal to $\gamma \cdot A \cdot dz$. Thus:

$$d\sigma_v / dz = \gamma - \tau \cdot P/A$$

With incipient failure or flow along the vertical planes on which τ is acting, the maximum value of τ is:

$$\tau_f = c + \sigma_h \tan\varphi$$
$$= c + k\sigma_v \tan\varphi$$

Thus:

$$\partial \sigma_v / dz = \gamma - c \cdot P/A - k\sigma_v \tan\varphi \cdot P/A$$

The solution for this differential equation gives:

$$\sigma_v = \frac{A/P\,(\gamma - c \cdot P/A)(1 - e^{-kz\tan\varphi \cdot P/A})}{k\tan\varphi} \quad \text{Eq. 5-3(a)}$$

$$\sigma_h = \frac{A/P\,(\gamma - c \cdot P/A)(1 - e^{-kz\tan\varphi \cdot P/A})}{\tan\varphi} \quad \text{Eq. 5-3(b)}$$

Thus, on the trap-door $\sigma_v = p_v$,

if

$$c = 0 \quad p_v = \frac{A/P \cdot \gamma\,(1 - e^{-kz\tan\varphi \cdot P/A})}{k\tan\varphi} \quad \text{Eq. 5-4}$$

if

$$z = \infty \text{ and } c = 0 \quad p_v = \frac{A/P \cdot \gamma}{k\tan\varphi} \quad \text{Eq. 5-5}$$

If A is a circle of radius R, for $z = \infty$ and $c = 0$:

$$p_v = \frac{\gamma R}{2k\tan\varphi} \quad \text{Eq. 5-6}$$

If A is a rectangle, b x L, for $z = \infty$ and $c = 0$:

$$p_v = \frac{\gamma b}{2(b/L + 1)k\tan\varphi} \quad \text{Eq. 5-7}$$

and if $L \gg b$:

$$p_v = \frac{\gamma b}{2k\tan\varphi} \quad \text{Eq. 5-8}$$

$$p_h = \frac{\gamma b}{2\tan\varphi} \quad \text{Eq. 5-9}$$

The preceding analysis is based on the assumption that the boundaries of the flow zone will be vertically above the edges of the trap-door. In this case, Fig. 5-6(b) shows the Mohr diagram of stresses that must exist on these planes of incipient failure or flow. By solving the geometry of this diagram for cohesionless material:

$$k = \sigma_h/\sigma_v = \frac{1}{1 + 2\tan^2\varphi} \quad \text{Eq. 5-10}$$

Measurements have shown that the actual values of 'k' are close to those predicted by Equation 5-10 (19, 20). For a material with cohesion, the ratio of the horizontal to vertical stress is:

$$k = \frac{\sigma_v - 2c\tan\varphi}{\sigma_v(1 + 2\tan^2\varphi)} \quad \text{Eq. 5-11}$$

Equation 5-3 might be used to determine the minimum size of boxhole required to draw a sticky or cohesive ore. One would solve for b where $p_v = 0$.

Example - The cohesion of a broken, sticky ore is found to be 500 psf and the density is 100 pcf. Find the minimum radius of drawpoint to obtain flow conditions.

Using Equation 5-3(a) and placing $p_v = 0$, we have

$$0 = \gamma - c \cdot P/A$$

Therefore

$$P/A = \gamma/c$$

or

$$R = 2c/\gamma$$

$$= 2 \times 500/100 = 10 \text{ ft.}$$

Equation 5-3(a) can also be used for determining the average vertical pressure on the trap-door or plate controlling the draw either from a stope or a bin. However, in this case it must be recognized that a certain minimum amount of vertical deformation of the trap-door is required to produce the incipient shear failure in the broken material so that arching, in effect, occurs and the major part of the superincumbent load is thrown out into the abutments. Some experiments indicate that vertical movement of about 3.5 per cent of the width of the trap-door is required to produce the minimum vertical pressure, p_v (6). This figure is likely to vary with the type of material that is being drawn.

Even for cohesionless material some minimum size of draw point is required to permit free flow. In this case, the required size of opening is not a function of the shear strength of the broken material, but rather it is dependent on the actual size of particles. It can be visualized, as shown in Fig. 2-11(a), that, if the size of opening is not much larger than the size of particles, natural arches can build up over the draw point and prevent free flow.

Experiments have shown relationships between the diameter of the opening, 2R, and the size of particle, d, as shown in Fig. 5-7 (7). The Free Flow zone is generally bounded by the line representing an opening somewhat greater in size than four times the size of particle being drawn. The Probable Flow zone occurs for smaller size openings down to about two times the size of the particle. In this zone, both flow and plugging can occur but neither can be predicted. No flow will occur for openings less than about twice the size of the particles. These boundary curves can be expected to vary with the nature of the materials being drawn.

Several series of model experiments have shown that the ground that is passed through a draw point is that which would be contained before drawing occurred by an ellipsoid above the draw point such as partly indicated by the dotted lines in Fig. 5-6(a). However, detailed examination of flowing material indicated that drawing through a boxhole initially results in vertical flow of the material immediately above the boxhole, as shown in Fig. 5-8(a). Then, with continued drawing, there is an erosion action that pulls some of the adjacent material down and inwards to the boxhole (in fluid mechanics terms, Zone I would have irrotational flow where Zone R would have rotational flow).

It has also been observed, and the concept of erosion would support it, that the material drawn into the boxhole from Zone R, or the width of the boundaries of the original ellipsoid, will increase with the amount of material drawn down through Zone I. Consequently, the use of an ellipsoid, possibly based on model experiments, to predict the amount of material that will be drawn from one boxhole, and particularly the lateral extent of drawing, could be erroneous. It would thus be necessary to analyse any model data in such a way as to have the shape of the ellipsoid a function of the volume of draw.

An example of this procedure is given in the following empirical equation based on model work in one particular material (11):

$$1200 x^2 + (10 y - 14 V^{1/3})^2 = 200 V^{2/3} \qquad \text{Eq. 5-12}$$

where x is the radius of the circular boundary of the draw zone in a horizontal plane or the

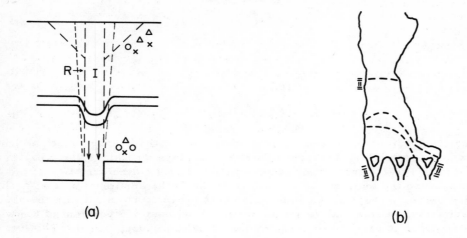

Fig. 5-8 Some Drawing Patterns of Caved Ore (12)

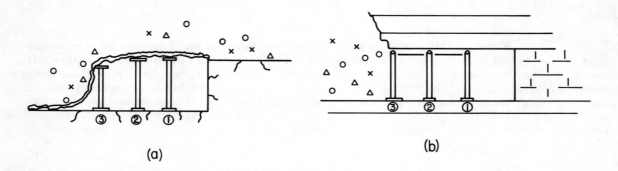

Fig. 5-9 Prop Support for Longwalls
   (a) Under Caved Ground and
   (b) Under Intact Strata

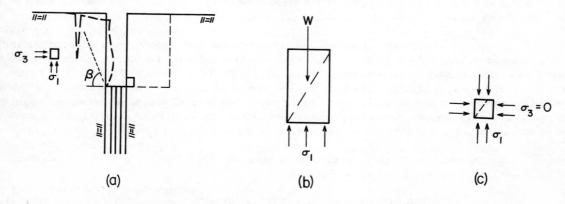

Fig. 5-10 Analysis of Subsidence Adjacent to a Mined-out Vertical Vein

horizontal distance from the centreline of the draw point to the points on the edge of the flow zone, y is the vertical distance from the draw point to points on the edge of the flow zone, and V is the volume of the draw. Other model work has shown, by using particles of pure minerals, that draws are almost identical for a wide range of minerals (16).

The material drawn into a boxhole will only be from an ellipsoidal volume when the top of the draw is not near a free surface. When the ground surface is approached by the draw zone, a cone of draw will be formed. The mechanics describing this surface ground are those that are applicable to slope stability and are described in Chapter 6.

As an indication of how little lateral extent the flow zone can have, in one mine using block caving with 300 ft high blocks it was found that, with 15 ft pillars between blocks, waste was drawn from the previously mined adjacent blocks. On the other hand, with 30 ft pillars the flow zone did not extend far enough to include this waste (9). Also, measurements have shown draw channels in one case 5-6 ft wide for a height of 200 ft (11) and in another case 4-15 ft wide (12).

Another example of the limitation of horizontal movement occurred where the horizontal extent of drawing was to be increased by the mining before caving of a slot at 70 degrees up from the draw point. Vertical drawing occurred in spite of this artificially created weakness (10).

This essentially vertical movement of ore both in caving and drawing makes it difficult to decrease the development costs of driving scram drifts and boxhole raises very close together and makes it expensive to recover the wedge of ore left on the footwall between levels in a dipping vein.

As flow requires, in effect, shear failure in the material, any cohesion in caved material, or any decrease in volume of voids that would increase the shear resistance of broken rock, could be expected to decrease the lateral extent of the draw. It can be visualized from Fig. 5-8(a) that the maximum shear stress is likely to occur vertically above the edges of the draw point. The shear stress should drop off fairly rapidly with distance away from the flow zone. Consequently, the width of Zone R is likely to be a function of the shear strength of the material and thus be reduced by any action tending to increase the shear strength.

At the same time, this concept of maximum shear stress suggests that if there is a zone to the side of the draw point that contains material of lower shear strength, then drawing might occur preferentially towards this material. This phenomenon has been observed during block caving where drawing occurred preferentially along inclined waste bands and also, where a pillar was not left between blocks, into the waste of the previously mined block (11). Also, drawing has occurred obliquely into the zone over previously pulled boxholes (see Fig. 5-8(b) (12)). In this figure the dashed lines indicate the successive volumes drawn.

Returning to the analysis leading to Equation 5-3, if $\varphi = 45°$ and $k = 0.5$, then $p_v = \gamma R$. In other words, the pressure on the trap-door is equal to a column of ground equal in height to the radius of the trap-door regardless of the depth of the ground over the trap-door. This means that the pressure of the ground not carried by the trap-door, $(z - R)\gamma$ is carried by arching action out into the abutments.

Increased pressure is thus applied to the pillars around the boxhole, and in some types of ground where boxhole pillars over the scrams cannot be maintained the pressure is applied to the sets in the scram drifts. From this action it can be concluded that pillar and set loads are likely to be greatest during the actual drawing. Similar results have been shown quite clearly by tests in bins and silos (12, 13).

Furthermore, if a line of boxholes were being pulled simultaneously, the increase in the pillar and set loads would be greater than during the pulling of a single boxhole owing to the superincumbent loads being supported on two sides of the line of draw rather than around four sides of a single draw. A further step in this direction would be a complete perimeter pull around a block that would tend to float the block on the interior pillars or on the sets.

The same type of mechanics that has been applied to the ground being pulled through a boxhole, or to material being drawn from a bin, can be applied to the caved ground within the boundaries of the entire stope. In other words, pressure at the bottom of the caved ground during drawing, if the stope is full from wall to wall, will have an average value less than the weight of the caved ground if the height of the stope is greater than its width. This pressure does not, as has been confirmed, increase with the height of the block (12). However, the ground adjacent to the stope will experience an increase in vertical stress. In some cases this increased stress has been sufficient to cause distress in the adjacent service drifts and in other cases to cause bearing failure of weaker underlying ground with upheaval in the drifts (11).

The same action occurs with fractional panel caving. This is a situation where an entire block is developed for mining purposes, but then only part of the block is undercut and caved at one time. One result of this procedure is to place one of the abutments of the caving zone over the scram drifts. If the ground does not fail under the increased abutment stresses, no serious consequence occurs from this procedure. However, if the ground does fail, the supports in the scram drifts outside the caving zone can experience greater pressures than those under the caved material.

Another aspect of stope pressures resulting from caving action comes from the recognition that the average pressure, p_v, represents a variation from some low value near the boundaries of the stope to a maximum value at the centre. Observations have shown that, in incompetent ground where pillars over the boxholes could not be maintained, the sets were pushed outwards towards both walls and towards the adjacent mining blocks by this larger central pressure (11).

If the compressibilities of the abutments for the arch in the block are unequal, then the distribution of vertical pressure will not be symmetrical. With a longwall operation as shown in Fig. 5-9(a), the caving can be considered to be occurring between the face and some point in the waste behind the mining area. As the rear abutment of the arch will be compressible, the vertical pressures on the props will be lowest in the face row ① and greatest in the rear row ③.

In coal mining, where it is common to have a stratified relatively strong roof (see Fig. 5-9(b)), prop loads have been measured with a typical distribution of load in ① of P, in ② of 2P and in ③ of 3P (17). In one case where props were used in iron ore mining with broken waste rock comprising the back, the relative loads were for ① P, for ② 1.7P and for ③ 1.7P (18).

SUBSIDENCE

As ore is removed by underground operations, the adjacent ground will be subjected to increased stresses. The increased stresses may produce conspicuous deformation or failure in the rock. All these various effects can be described as subsidence. As ore bodies can have any dip, the two extreme cases, with dips of 90 and 0 degrees, will be considered.

The geometry of the ground resulting from mining a vertical seam or vein from the surface downwards is shown in Fig. 5-10(a). The effect of mining in this case is to remove the horizontal stress that previously existed in the immediate wall rock. In other words, whereas previously the wall rock was in a state of triaxial compression, the surface rock is now in a state of biaxial compression, which, assuming the intermediate principal stress is irrelevent with respect to failure, is equivalent to uniaxial compression.

The block of wall rock delineated by the dashed line in Fig. 5-10(a) is shown in Fig. 5-10(b). In this case the simplifying assumption is made that there is no horizontal shear stress on the bottom of this block. Consequently, no horizontal stresses exist on the vertical boundaries. The gravitational body force, W, acts downwards and the reaction upwards from the underlying ground is shown as the major principal stress, σ_1. Superficially, this block can be considered to be equivalent to a uniaxial compression sample. In this case,

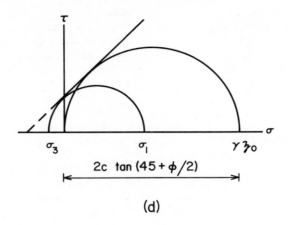

(d)

Fig. 5-10 Analysis of Subsidence Adjacent to a Mined-out Vertical Vein

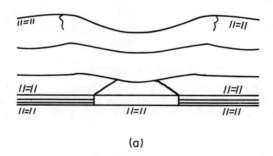

(a)

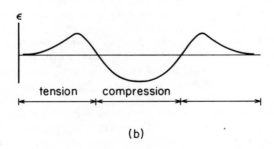

(b)

Fig. 5-11 Subsidence Effects over a Mined-out Horizontal Vein

when the major principal stress, σ_1, exceeds the uniaxial compression strength of the rock, failure will occur by shear on planes at $(45 + \frac{\varphi}{2})$ to the major principal plane as shown by the dashed line in Fig. 5-10(b).

This analysis can be refined slightly by recognizing that the stress σ_1 has its maximum value at the bottom of the block. Consequently, if this stress is just sufficient at the bottom to cause failure, it will not be sufficient to cause failure at some higher elevation. However, once failure is initiated on the element at the lower left hand corner of the block, as shown in Fig. 5-10(a) and Fig. 5-10(c), then the weight of the upper part of the block would be thrown on a progressively smaller shear plane. Thus, through progressive failure the entire block could be sheared.

On the left side of Fig. 5-10(a), the deformation of the wall rock for homogeneous, elastic ground is shown. It can be seen intuitively that tensile stresses must exist parallel to and near the horizontal surface. These stresses produce the commonly observed cracks in the crests of rock cuts and slopes.

If it is assumed that the walls are in a state of plastic equilibrium (i.e., incipient sliding exists and hence Mohr's Strength Theory governs the principal stress ratio), as opposed to assuming that the walls are an elastic mass, then the major and minor principal stresses will be related as shown by the Mohr diagram in Fig. 5-10(d) (14). For ground near the surface, the stresses on an element as shown on the left side of Fig. 5-10(a) will be represented by circles similar to those drawn in Fig. 5-10(d). Although the failure envelope is shown as linear beyond the origin, the resultant equations would only be modified qualitatively by a curved envelope.

The principal conclusion from this type of analysis is that, above a certain elevation with the ground in this state of plastic equilibrium, the minor principal stress, σ_3, will be tensile. Consequently, the horizontal stresses will be tension for all of the ground down to some critical depth, z_o. At this point the major principal stress, γz_o, will be large enough to eliminate the tensile stresses. From the geometry of Fig. 5-10(d) this critical depth is (14):

$$z_o = (2cH) \tan\left(45 + \frac{\varphi}{2}\right) \qquad \text{Eq. 5-13}$$

Where the ground is not in a state of plastic equilibrium, this analysis does not give a measure of the tensile stresses, and unfortunately the solution of the problem in elastic ground has not yet been established.

If tensile stresses occur, then fracture is likely to follow as rock masses are usually very weak in tension. Tension cracks, as shown in Fig. 5-10(a), actually are very common, and the angle between these cracks and the bottom of the mining zone, β, has been the subject of much discussion. 65 to 75 degrees is often erroneously considered to be the range of values to envelope the zone of surface cracking. However, tension cracking will depend on the ratio of strength to stress, which in turn varies with geometry as well as time (with its influence on strength). Measurements in one case showed that this so-called subsidence or draw angle varied between a minimum of 80 and a maximum of 94 degrees (i.e., an overhang)(21). In another case the angle decreased from 73 to 42 degrees with the passage of time (9).

It could be disastrous if some constant fairly steep angle were assumed to govern the ground reaction, and this angle were used to protect surface structures or to prevent tapping a river or lake.

Unfortunately, it is difficult to predict the extent of subsidence effects. The procedure at this point ought to be the taking of surface movement measurements to obtain empirically some quantitative measure of the ground reaction with respect to the prevailing geometry. Subsequent planning could then be based on this data modified with respect to the geometry of the mining plan and the presence or absence of supporting waste. This procedure has been followed with some degree of success (15).

It might be questioned whether, instead of the walls fracturing above the ore, failure would occur in the ground below the elevation of the top of the ore. This question could then be resolved into whether the wall would fail in bearing or in uniaxial compression. The answer to this question has been given in Chapter 1 where it was shown that the stress required to cause failure in bearing, (Equation 1-17) is many times greater in homogeneous ground than that required to cause failure in uniaxial compression. Consequently, it could be expected that subsidence effects would be manifest by surface tension cracking and ultimately by shear failure along oblique planes above the elevation of the mining zone. Observations substantiate these theoretical deductions.

In addition, it could be expected that the openings in the walls adjacent to the ore might be affected by the more severe stress regime. Actually, it has happened that service drifts and crosscuts have started to work, and in some cases collapsed, as the bottom of the mining operation passed the elevation of these openings.

The second typical case that has given rise to commonly observed subsidence effects is that of the horizontal seam or vein. In Fig. 5-11, the consequence of mining in such a horizontal seam is shown. Here the overlying ground tends to sag into the mining area with bending stresses being induced in this ground. As for a fixed beam, tension would tend to occur on the top fibres over the supports as well as on the bottom fibres at the centre of the span. Surface displacement measurements have shown that tensile strains do occur at these locations (15).

In Fig. 5-11(b) the typical pattern of surface horizontal strain is shown. Compression strains occur over the centre of the mined out area with tension strains on either side. With this beam action, it would be expected that tension cracks would occur close to the zones of maximum tensile strain on the surface and that diagonal tension cracks would occur in the ground immediately over the mining area, both possibilities shown by wiggly lines in Fig. 5-11(a). Between these two sets of fractures, shear failure might occur; however, this depends to a large extent on what happens to the ground immediately over the mining area. If this ground breaks up and is drawn down into the mining area, then the horizontal support for the adjacent ground will be removed and shear failure will be induced similar to the case described in Fig. 5-10(a).

Fig. 5-11(a) could, by ignoring the ground surface, also represent the plan view of a vertical vein. The same shear and tensile stresses would tend to occur in the walls adjacent to the stoping area. Consequently, these manifestations of distress in the walls of vertical veins should also be considered as subsidence effects. As an example of such distress, the collapse of an opening 100 ft into the wall and only 75 ft above the undercut of a block caving operation can be cited (9). This collapse would not be included within the traditional draw or subsidence angles. Many of these cases of deformation and working of ground have been noted in the walls at the bottom level of a stoping zone. These effects grade into those that are commonly described as subsidence at higher elevations, which illustrates how a constant angle of draw or subsidence as an envelope of ground failure is a concept that has no basis in the mechanics of the system.

REFERENCES

1. Morrison, R.G.K., "Mining Philosophy, Ground Control, Mining Methods and Rock Mechanics", unpub. notes, McGill University (1963).

2. Timoshenko, S. and Goodier, J., "Theory of Elasticity", McGraw-Hill (1951).

3. Panek, L., "Stresses about Mining Openings in a Homogeneous Rock Body", Edwards Brothers (1953).

4. Obert, L. et al., "Design of Underground Openings in Competent Rock", USBM, Bull. 587 (1960).

5. Jansen, H., "Versuche Uber Getreidedruck in Silozellen", Z. Ver. Deut. Ing., Vol. 39 D, 1045 (1895).

6. Jahns, H. and Brauner, G., "Stowage Pressure in Steep Measures", Inter. Conf. on Strata Control, Cerchar (1960).

7. Aytaman, V., "Causes of Hanging in Ore Chutes and Its Solution", Can. Min. Jour. (1960).

8. Bergau, W., "Measurements in Grain Silos during Filling and Emptying", Proc. Swed. Geotech. Inst. No. 17 (1959).

9. Fletcher, J., "Ground Movement and Subsidence from Block Caving at Miami Mine", Trans. AIME, Vol. 214 (1959).

10. MacLennan, F., "Miami Copper Company Method of Mining Low Grade Ore Body", Trans. AIME, Vol. 91 (1930).

11. Coates, D. et al., "Ground Mechanics at Steep Rock", unpub. report, (1961).

12. Hardwick, W., "Mining Methods and Costs, Inspiration Consolidated Copper Co. Open Pit Mine, Gila County, Ariz.", USBM IC 8154 (1963).

13. Jenike, A. et al., "Flow of Bulk Solids Progress Report", Bull. Univ. of Utah, Vol. 49, No. 24 (1958).

14. Terzaghi, K., "Theoretical Soil Mechanics", Wiley (1943).

15. Grond, G., "A Critical Analysis of Early and Modern Theories of Mining Subsidence and Ground Control", Univ. of Leeds (1953).

16. Young, H., "The Fundamentals of Drawing Broken Ore", unpub. M. Ap. Sc. Thesis, Univ. Toronto (1954).

17. Schwartz, B., "Measurements of Ground Pressure and of Movements on Longwall Faces in French Coal Mines", Trans. CIM, Vol. 57 (1954).

18. Coates, D. and Gyenge, M., "Mechanics of Support and Caving in Longwall Top Slicing", Proc. 4th Inter. Cfce. Strata Control and Rock Mech., Columbia Univ. (1964).

19. Jakobson, B., "On Pressure in Silos", Belgian Group ISSMFE, Proc. Brussels Cfce. on Earth Pressure Problems, Vol. 1 (1958).

20. Moss, E., "The Design of a Raw Sugar Silo", Inst. Civil Eng'rs Cfce. Correlation between Calculated and Observed Stresses and Displacements in Structures, Prelim. Vol. Paper No. 11, Group 2, P 177-196 (1955).

21. Obert, L. and Long, A., "Underground Borate Mining, Kern County, Calif.", USBM RI 6110 (1962).

CHAPTER 6

ROCK SLOPES

INTRODUCTION

The majority of slope problems in rock arise from open pit mining operations and from cuts for roads and railroads. The cuts made for roads and railroads can be troublesome; however, the cost of solving the slope problems connected with mining is of a much greater order of magnitude.

Five million extra tons of waste would have to be mined as a result of an average slope being reduced by 5 degrees in an open pit mine 1000 ft x 1000 ft x 400 ft deep. Furthermore, this would not be a particularly large pit when it is considered that there are many open pits 5000 ft long and many are being planned to go down 1000 ft deep.

The determination of optimum slopes is particularly important for narrow, steeply dipping ore bodies. In these cases, a lower slope means more waste excavation without any compensation such as is obtained in flat ore bodies or in ore bodies with assay boundaries where the wall material still contains metal units that contribute something to the extra cost of excavation. However, safe slopes and maximum profits are almost always in competition.

The state of technology in dealing with slope problems in rock is not very satisfactory at the present time. Whereas many aspects of the problem can be analysed, and in the case of incompetent rocks the critical slope angles can be determined, in hard rocks it is still not possible to predict slope failures or even to follow a rational, established procedure in designing the slopes.

INFINITE SLOPES

It is possible to analyse theoretically certain slope problems if the complications of having a crest and toe are eliminated; in other words, an infinite slope is assumed for these purposes.

In Fig. 6-1(a) an infinite slope is shown. At equal depths below the surface of this sloping face the stress conditions must be identical at all points along the slope (as the slope goes to infinity in both directions, the gravitational force acting on a vertical column of ground must produce the same stress conditions for equal depths since no slice is different from all other similar slices). Consequently, the vertical column of ground isolated in Fig. 6-1(b) must be subjected to forces, P, on the sides of the column that are equal and opposite. For this reason these forces are ignored in the subsequent analyses.

In Fig. 6-1(b), W is the weight of the vertical column, N is the normal reaction on the bottom of the column, T is the tangential reaction on the bottom of the column, 'b' is the width of the column measured parallel to the slope, 'z' is the depth of the column measured vertically and 'i' is the slope angle. It follows, then, that for a material with no cohesion the stability of the ground can be analysed as follows:

$$N = W \cos i = \gamma z (b \cos i) \cos i$$

$$T = W \sin i = \gamma z (b \cos i) \sin i$$

$$\therefore \sigma = \gamma z \cos^2 i$$

$$\tau = \gamma z \sin i \cos i$$

where σ is the normal stress on the bottom of the column, τ is the shear stress on the bottom of

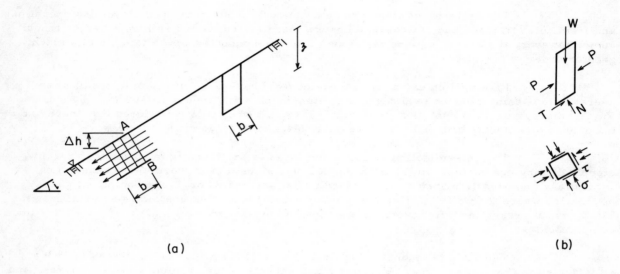

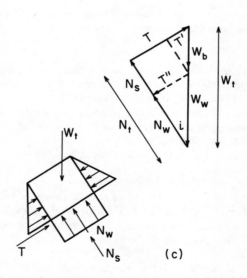

Fig. 6-1 Analysis of Seepage Stresses in a Slope

the column and γ is the density of the ground. From Mohr's Strength Theory, the maximum shear stress, τ_f, is equal to the normal stress, σ, multiplied by $\tan\varphi$:

$$\tau_f = \gamma z \cos^2 i \tan\varphi$$

where φ is the angle of internal friction.

At failure,

$$\tau_f = \tau$$

$$\therefore \tan i_c = \tan\varphi$$

$$i_c = \varphi$$

where i_c is the critical slope angle.

For ground that has an effective cohesion, c, the critical slope can be analysed as follows:

$$\tau = \tau_f$$

$$\gamma z \sin i_c \cos i_c = c + \gamma z \cos^2 i_c \tan\varphi$$

$$\therefore \tan i_c = \frac{c}{\gamma z \cos^2 i_c} + \tan\varphi \qquad \text{Eq. 6-1(a)}$$

$$\text{or} \quad \frac{c}{\gamma z} = \cos^2 i_c (\tan i_c - \tan\varphi) \qquad \text{Eq. 6-1(b)}$$

If
$$z = \infty,$$
$$i_c = \varphi$$

If
$$z < \infty,$$
$$i_c > \varphi$$

If
$$\varphi = 0,$$
$$\tan i_c = (c/\gamma z)/\cos^2 i_c \qquad \text{Eq. 6-2(a)}$$

$$\text{or} \quad c/\gamma z = 1/2 \sin 2 i_c \qquad \text{Eq. 6-2(b)}$$

It can be seen from Equations 6-1(a) and 6-2(a) that the critical slope angle, i_c is a function of cohesion, but, more significantly, a function of the dimensionless parameter $c/(\gamma z)$. This is a fundamental parameter governing the stability of slopes in cohesive ground that will be referred to later.

The effect on stability of seepage can now be examined for the infinite slope. In the left side of Fig. 6-1(a), flow parallel to the surface of the slope is shown with equipotential lines normal to the flow lines or normal to the slope surface. The assumption is made that the phreatic line coincides with the surface of the slope and that the ground is isotropic. The symbol γ_t represents the total density, or the combined weight of the solid rock and water contained within one cubic foot of the ground, γ_w is the density of water, and i_c is the critical slope angle or the slope angle that would just produce failure. The stresses on planes parallel to the surface of the ground are then as follows:

$$\sigma = \gamma_t z \cos^2 i$$

$$\tau = \gamma_t z \sin i \cos i$$

In Fig. 6-1(a) AB is an equipotential line. Consequently, the hydrostatic head at B is equal to $\overline{AB} \cos i$. In other words, the neutral or pore water pressure, u_B, at B is:

$$u_B = \gamma_w \overline{AB} \cos i = \gamma_w (z \cos i) \cos i$$

Therefore, in cohesionless ground the maximum shear resistance would be:

$$\therefore \tau_f = (\gamma_t z \cos^2 i - \gamma_w z \cos^2 i) \tan \varphi$$

At failure:

$$\tau_f = \tau = \gamma_t z \sin i_c \cos i_c$$

$$\therefore \tan i_c = \frac{\gamma_t - \gamma_w}{\gamma_t} \tan \varphi \qquad \text{Eq. 6-3}$$

From Equation 6-3 it can be seen that the critical slope angle is reduced by seepage. To illustrate the nature of this reduction, assume $\varphi = 45°$ and $\gamma_t = 170$ pcf. Then, using Equation 6-3, the reduction in the stable slope angle can be calculated:

$$\tan i_c = \frac{170 - 62.4}{170} \tan 45$$

$$i_c = 32.4°$$

Whereas the slope in cohesionless material would normally be stable up to an angle of 45 degrees, with seepage the critical angle is reduced to 32.4 degrees. Therefore, seepage clearly is a very important factor in slope stability.

The same type of analysis can be made for a slope in cohesive material and the following equations derived:

$$\tan i_c = \frac{c}{\gamma z \cos^2 i_c} + \frac{\gamma_t - \gamma_w}{\gamma_t} \tan \varphi \qquad \text{Eq. 6-4(a)}$$

$$\frac{c}{\gamma z} = \cos^2 i_c \left(\tan i_c - \frac{\gamma_t - \gamma_w}{\gamma_t} \tan \varphi \right) \qquad \text{Eq. 6-4(b)}$$

The nature of the seepage forces can be examined in more detail. In Fig. 6-1(c) the forces and hydrostatic pressures acting on an element adjacent to the surface are shown. As indicated, a force diagram can then be drawn from which the following relationships can be established:

$$N_t = W_t \cos i$$

$$= N_s + N_w$$

$$\therefore N_s = W_t \cos i - N_w = W_t \cos i - W_w \cos i$$

$$\therefore T_f = N_s \tan \varphi$$

At failure

$$T_f = T$$

$$= W_t \sin i_c$$

$$\therefore \tan i_c = \frac{W_t - W_w}{W_t} \tan \varphi$$

The result of the above analysis is essentially the same as that represented by Equation 6-3. The force T in this analysis can be thought of as the reaction in the tangential direction required on the bottom of the element to maintain equilibrium, i.e., it is equal in magnitude and opposite in direction to the component of the total weight W_t acting parallel to the slope.

The buoyant force on this element of ground is equal to the volume of the element multiplied by the density of water and is represented by W_w. The difference between the total weight and the buoyant force can be called the buoyant weight of the element, W_b, as indicated in the force diagram of Fig. 6-1(c).

The component of W_b acting parallel to the slope can be considered to induce the reaction T' as shown in the force diagram of Fig. 6-1(c). The remaining part of T must then be the reaction at the bottom of the element required as a result of seepage. It can now be seen that a force diagram could have been drawn using only the buoyant weight and seepage force; e.g., the two forces tending to produce instability would be, referring to Fig. 6-1(c), W_b and T''. Then the reaction on the bottom of the element would be composed of the interparticle forces N_s and T. The same reaction, T, at the point of failure would be found through this type of analysis as with the total force analysis previously examined.

Letting J be the seepage force, the following relationship can be derived:

$$J = T''$$
$$= W_w \sin i$$
$$= \gamma_w V_t \Delta h/b$$
$$= \gamma_w i_w \text{ per unit volume}$$

where i_w is the hydraulic gradient. From this it can be concluded that the seepage force on a unit element is the product of the density of water and the hydraulic gradient at that point.

It should be noted that the magnitude of the seepage force is in no way dependent on the permeability of the ground. In other words, the volume of flow of ground water does not affect the magnitude of the seepage forces connected with this flow. Consequently, in tight ground where seepage is occurring but may not be apparent to the eye, seepage forces exist and the stability of the slopes will still be affected according to the relationships established above.

The more general case can be examined where the phreatic line does not coincide with the surface of the slope. The depth to the phreatic line is z_w and the dry density of the ground is γ_d. The critical slope angle can then be analysed as follows:

$$u_B = \gamma_w (z - z_w) \cos^2 i$$

$$\therefore \tau_t = c + (\gamma_t z \cos^2 i - \gamma_w (z - z_w) \cos^2 i) \tan \varphi$$

At failure
$$= (\gamma_d z_w + \gamma_t (z - z_w)) \sin i_c \cos i_c$$

$$\therefore \tan i_c = \frac{c/\cos^2 i_c + \left((\gamma_t z - \gamma_w (z - z_w))\right) \tan \varphi}{\gamma_d z_w + \gamma_t (z - z_w)}$$

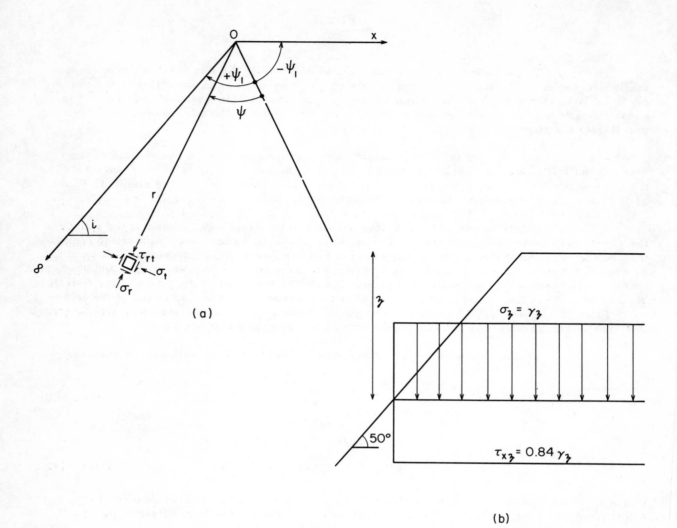

Fig. 6-2 Analysis of Stress Distribution within an Infinite Slope (2)

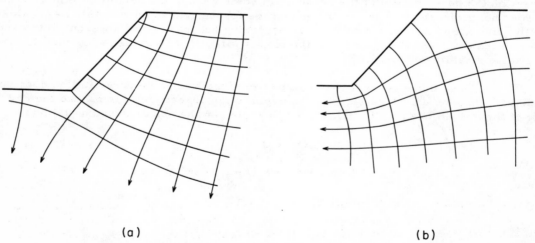

Fig. 6-3 Principal Stress Trajectories
(a) Major Principal Field Stress Vertical and
(b) Horizontal

In rocks

$$\gamma_t = \gamma_d$$

$$\therefore \tan i_c = \frac{c/\cos^2 i_c + ((\gamma_t z - \gamma_w(z-z_w)))}{\gamma_t z} \tan\varphi \qquad \text{Eq. 6-5(a)}$$

or $\quad \dfrac{c}{\gamma_t z} = \cos^2 i_c \left(\tan i_c - (1-(1-z_w/z)\gamma_w/\gamma_t)\tan\varphi\right) \qquad \text{Eq. 6-5(b)}$

ELASTIC STRESS DISTRIBUTION

The elastic stress distribution in a simple slope with crest and toe has not yet been solved mathematically. Solutions have been worked out for semi-infinite wedges, which provide some information on the variation of stresses in the upper parts of slopes. However, they do not give information on the stress concentrations around the toes, which is probably one of the zones of critical stress. Furthermore, these solutions do not indicate the presence of tensile stresses in the crest zone, which we know exist from both experience and gelatine models.

The following equations represent the theoretical solution for the stresses in an infinite slope with a horizontal crest surface as shown in Fig. 6-2(a) (1, 2):

$$\sigma_r = \gamma r \left[\left(\frac{\cos 3i/2}{8 \sin^2 i/2} + \cos i/2 \right) \cos \psi \right.$$

$$+ \left(\sin i/2 - \frac{\sin 3i/2}{8 \cos^2 i/2} \right) \sin \psi$$

$$\left. - \frac{\cos i/2 \cos 3\psi}{8 \sin^2 i/2} + \frac{\sin i/2}{8 \cos^2 i/2} \sin 3\psi \right] \qquad \text{Eq. 6-6(a)}$$

$$\sigma_t = \gamma r \left[\left(\frac{3 \cos 3i/2}{8 \sin i/2} + \cos i/2 \right) \cos \psi \right.$$

$$+ \left(\sin i/2 - \frac{3 \sin 3i/2}{8 \cos^2 i/2} \right) \sin \psi$$

$$\left. + \frac{\cos i/2 \cos 3\psi}{8 \sin^2 i/2} - \frac{\sin i/2 \sin 3\psi}{8 \cos^2 i/2} \right] \qquad \text{Eq. 6-6(b)}$$

$$\tau_{rt} = \gamma r \left[\frac{\cos 3i/2 \sin\psi + \cos i/2 \sin 3\psi}{8 \sin^2 i/2} \right.$$

$$\left. + \frac{\sin 3i/2 \cos\psi + \sin i/2 \cos 3\psi}{8 \cos^2 i/2} \right] \qquad \text{Eq. 6-6(c)}$$

where σ_r is the radial stress from the origin 0 at the crest, γ is the density of the ground, r is the radial distance from the origin, i is the slope angle, ψ is the angle measured clockwise from the bisector of the interior angle at the crest, σ_t is the tangential stress and τ_{rt} the shear stress on radial and tangential planes.

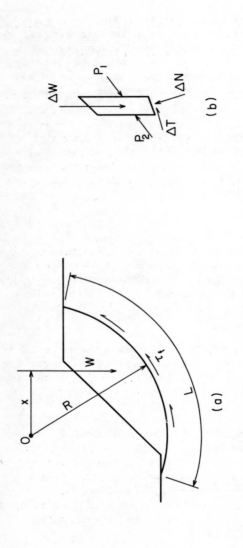

Fig. 6-4 Slip Circle Analysis for Yielding Ground

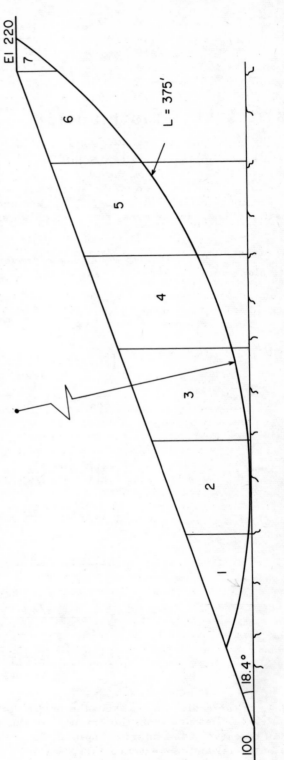

Fig. 6-5 Slip Circle Analysis by Method of Slices

The basic assumption for these equations is that σ_r, σ_t and τ_{rt} all vary linearly with the radial distance, r. This may or may not be true; the result that the horizontal stress, σ_x, increases indefinitely with x indicates that this solution is not a perfect representation of this idealized case.

Using these equations σ_r and σ_t can be calculated for points sufficiently above the toe not to be influenced by the stress concentrations there. Then, if of interest, the horizontal and vertical stresses can be determined by methods as explained in Chapter 1. For example, with a 50 degree slope, it is found, as shown in Fig. 6-2(b), that the shear stress on horizontal planes, τ_{zy}, is $0.84\gamma z$, where z is the depth to the plane, and that this value is constant for various distances x. Fig. 6-2(b) also shows that the vertical stress, σ_z, is equal to γz.

Another way of obtaining some information on the elastic stress distribution in slopes is to draw the principal stress trajectories as shown in Fig. 6-3. Fig. 6-3(a) is an approximate representation of the principal stress trajectories where the major principal field stress is due to gravity and hence vertical.

Fig. 6-3(b) shows the principal stress trajectories where the major principal field stress is horizontal. In this case, these trajectories can be sketched by simply following a trial and error procedure having all trajectories crossing at right angles and all areas with equal sides. This is the procedure that has been established for determining flow nets of fluid flow, which also must produce orthogonal trajectories, and that has been shown to give an accurate representation of the actual flow conditions for various cases of irregular geometry. The maximum normal stress is indicated to be at the toe in this case. In addition, the figure shows a concentration of horizontal stress under the excavation.

SLOPES IN ROCKS THAT YIELD

Research in the field of soil mechanics has established that the failure of slopes in soils usually occurs on cylindrical surfaces. Furthermore, the stability of these slopes can be analysed by determining on a cross-section the most critical circular surface of failure. This procedure is based entirely on empirical observations without having a rigorous solution for the stress distribution in the slopes; hence, the procedure that is generally followed to determine the most critical circle is by trial and error.

The following method of slope analysis can only be used where the material is sufficiently plastic to yield without rupture at points of stress concentrations so that the average shear stress is the important factor with respect to failure. It has been shown that slopes with a safety factor less than about 1.8 will have local stresses greater than the strength of the ground (9).

Centres of circles and radii are selected and the rotational or failure moment about the centre of the circle is compared to the resisting moment. In Fig. 6-4(a) a typical surface of failure is shown. The failure moment, M_F, is the moment of the weight of the failure segment, W, about the centre 0, i.e., (Wx). The resisting moment is the product of the resisting stresses, or the strength of the ground, along the surface of failure, τ_f, multiplied by the length of the arc L, to give the resisting force, multiplied by the radius of the circle, R, to give the moment of resistance. This method has been shown to be applicable not only to slopes in soils but also to slopes in some incompetent rocks (3).

The resistance to failure based on Mohr's Strength Theory is obtained from Equation 1-15: $\tau_f = c + (\sigma - u)\tan\varphi$. Unless the ground is purely cohesive, the shear resistance, τ_f, will vary along the length of the circular arc owing to the variation in normal stresses. For this reason when analysing a trial circle it is common practice to divide the failure segment into a series of vertical slices. One such slice is isolated in Fig. 6-4(b). The forces on this slice are the weight, dW, the component of the reaction normal to the surface of failure, dN, the tangential component of reaction on the surface of failure, dT, and the two lateral forces acting on the sides of the slices, P_1 and P_2.

The weight of the slice can be easily determined and the components of the reaction on the surface of failure follow. However, the lateral forces on the slice are indeterminate. It has been found that, if these forces are considered to be equal and opposite, the error in the resultant analysis is less than 10 per cent and is on the conservative side of the answer given by a more rigorous analysis (9). Consequently, the practice normally followed is to ignore these lateral forces in analysing the stability of the segment.

As mentioned above, the failure moment is equal to Wx; thus, when considering the segment divided into vertical slices, M_F equals $\Sigma(dW x)$. The force, dW, can be divided into two components that are normal and tangential to the surface of sliding. The component normal to the surface of sliding will pass through the centre of the circle of failure and thus have no moment about this point. This then leaves the moment of the tangential component to equal the failure moment, i.e., $\Sigma M_F = (dT R)$.

The resisting moment is $\Sigma(dN \tan\varphi + c\, dL)R$. As the safety factor, F_s, is then the ratio M_R/M_F, the radius of the circle, R, cancels out, and the safety factor then becomes:

$$F_s = \Sigma(dT)/\Sigma(dN \tan\varphi + c\, dL). \qquad \text{Eq. 6-7}$$

In Fig. 6-5 an example is given of a trial circle in a slope. In this case the incompetent ground rests on a layer of hard rock which limits the depth of the circle failure. The total density of the ground is 143 pcf, the angle of internal friction is 14 degrees and the cohesion is 1 ksf. The table below gives the numbers that are determined from this scale diagram of the trial circle.

Slice	Area	Weight	dT	dN
1	1140 sf	160k	-25k	160k
2	2225	318	0	318
3	2875	411	65	400
4	3050	436	140	412
5	2775	397	195	350
6	1800	257	170	195
7	220	31	25	22
			570	1857

$M_R = (1857 \times 0.25 + 1.0 \times 375) R$

$M_F = 570 R$

$F_s = M_R/M_F = 1.48$

The safety factor turns out to be 1.48 for this particular circle; it should be noted that other circles might give a lower and hence more critical safety factor.

For a slope within which seepage is occurring, the effects of seepage forces must be included in the analysis. When seepage forces were examined in infinite slopes, it was found above that one way of accounting for their effects is to use the total density of the ground (i.e., the weight of solids and water in 1 cf) and the effect of inter-granular stresses. Following this alternative it only becomes necessary to determine the phreatic line together with the flow net for the ground included in the various trial segments of failure.

From the phreatic line and the equipotential lines in the flow net it is possible to determine the hydrostatic pressure, u, from which the effective inter-granular normal stress can be determined. In terms of forces produced by this pressure, the hydrostatic force on the bottom of a slice would be $U = u\, dL$; hence, the inter-granular force would be $dN' = dN - U$.

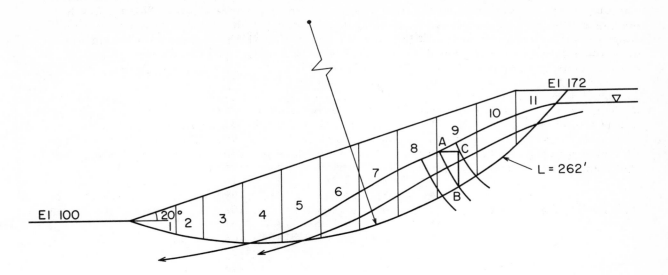

Fig. 6-6 Method of Slices including Pore Water Pressures

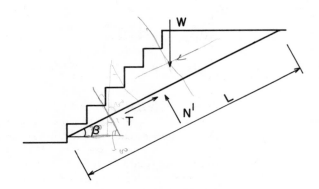

Fig. 6-7 Analysis of Plane Shear Slide

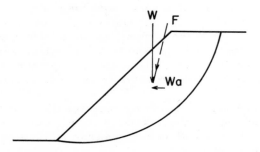

Fig. 6-8 Effect of Horizontal Acceleration due to Earthquake or Blasting

In Fig. 6-6 an example is given of a slope within which seepage is occurring. The equipotential lines defining the flow net are drawn in the area of Slice No. 9. The hydrostatic force at B is represented by the equipotential line AB and is equal to a head of $\overline{CB}$. With this information the hydrostatic force, U, on each slice can be determined as given in the table below. For this ground Y_t = 125 pcf, φ = 22° and C = 300 psf, hence:

Slice	dW	dT	dN	U	N'
1	26k	-7k	25k	0k	25k
2	34	-7	33	0	33
3	67	-7	66	0	66
4	95	-1	95	2	93
5	110	11	110	5	105
6	125	24	123	7	116
7	124	37	118	9	109
8	123	49	113	11	102
9	116	59	100	13	87
10	100	61	79	13	66
11	62	45	44	7	37
		264			839

$$F_s = (839 \times 0.40 + 0.30 \times 262)/264 = 1.57$$

In some circumstances the relatively simple case of plane shear failure, as shown in Fig. 6-7, will govern the stability of a slope. In this case rigid body mechanics can be applied to determine the safety factor:

$$F_s = W \sin \beta / (N' \tan \varphi + cL) \qquad \text{Eq. 6-8}$$

The main practical difficulty in this case is to locate the surface of weakness along which the slide would occur.

The effect of earthquakes or blasting on the stability of slopes is often questioned. One way to analyse these effects is to determine the maximum probable horizontal acceleration (see Chapter 8) that is likely to be applied to the embankment. If the acceleration, a, is expressed as a multiple of the acceleration due to gravity, then this horizontal force will be (W a) as shown in Fig. 6-8. This horizontal force is a body force in the same way that the gravitational force, W, is a body force. The two forces then combine to produce the resultant force, F, acting on the incipient segment of failure.

It can then be argued that the slope is now in an inclined stress field, which requires that the slope angle be reduced by the same angle that the resultant body force has been rotated. For example, if the relatively large horizontal acceleration of 0.2 g occurred, then the resultant force, F, would be inclined to the vertical at about 11 degrees. Thus, to maintain the same safety factor the slope angle should be reduced by 11 degrees. On the other hand, the normal safety factor may exist to account for such rare occurrences of over-loading.

The other aspect of the problem is the effect on the strength parameters of such dynamic effects. The pore water pressure could be increased with a resultant decrease in inter-granular or effective stresses. Also, the appropriate angle of internal friction might be changed. Little empirical information actually exists to provide guidance in these cases.

The analyses above have been made ignoring the third dimension. In other words, the resistance of the ends of the segment of the cylinder of failure has been ignored. This is normally a conservative design procedure; however, it depends on how the strength parameters of the ground are determined.

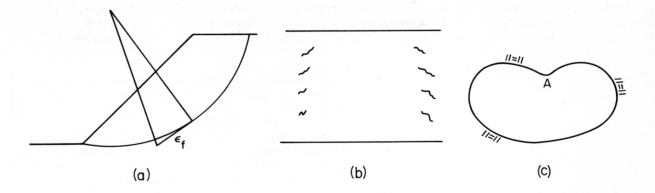

Fig. 6-9 Some Strain and Geometrical Effects

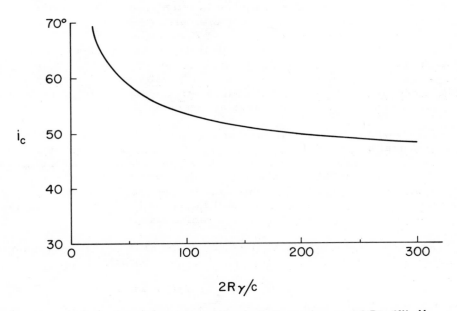

Fig. 6-10 Influence of Radius of Curvature of Pit Walls, R, on Critical Slope Angle, i_c (6)

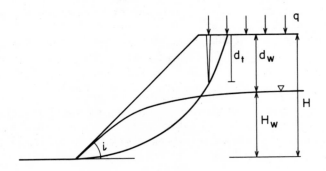

Fig. 6-11 Prototype Slope Condition in Yielding Ground for Slip Circle Analysis

If the strength parameters of the ground are obtained by analysing previous slides, then the ignoring of the end effects can produce calculated strength parameters that will be higher, owing to ignoring the existence of the ends, than are valid for the ground (3). Thus, whereas the slides may have included some cases where the failure segment was quite narrow in plan and the contribution of the ends would be very significant, the design case might be quite long so that no actual assistance of the ends would be effective.

Alternatively, if valid strength parameters are obtained from laboratory testing, then the ignoring of the contribution of the ends would be a conservative design procedure.

The analysis of the resistance of the ends can be made for any specific failure segment. It must be recognized, however, that the strain across the segment, as indicated in Fig. 6-9, will provide an area over which the full potential shear resistance is not mobilized at the time the base of the segment is failing. Whether it ultimately is mobilized or not depends to a large extent on the plastic properties of the ground.

Alternatively, failure at the ends can be through a series of diagonal tension cracks, as shown in Fig. 6-9(b), in which case the analysis of the end shear resistance would not be pertinent. Also, the actual failure segment is often spoon-shaped rather than cylindrical, which makes the analysis of the end effects more difficult.

However, it can be said that configurations that eliminate the assistance of the ends, such as at the point A in Fig. 6-9(c), which represents the plan of an open pit, should be avoided if possible. Furthermore, weak areas in the walls of the pit excavation would be more stable with a concave configuration than with a straight outline in plan. Where the slopes in plan have concave curvature, horizontal arching can come into play.

Fig. 6-10 shows the results of an analysis that has been made on the effectiveness of this horizontal arching for ground with an angle of internal friction of 40 degrees (6). In this figure, i_c represents the average angle of the slope at failure, and the cohesion appears in the dimensionless parameter $2R\gamma/c$, where R is the radius of curvature in plan of the slopes, γ is the density of the ground and c is the cohesion. In this case horizontal arching becomes significant when the radius of curvature of the wall in plan is less than about $150c/(2\gamma)$ or $75c/\gamma$ so that, if c = 3400 psf and γ = 170 pcf, this radius equals 1500 ft.

From the preceding discussions and analyses it is seen that the maximum slope angle is a function of the following dimensionless parameters:

$$i_c = f(c/\gamma H, q/\gamma H, d_w/H, d_t/H, \varphi) \qquad \text{Eq. 6-9}$$

where d_w is the depth to the ground water table behind the slope as shown in Fig. 6-11, d_t is the depth of tension crack at the crest, and q is the surcharge load on the crest of the slope.

With the definition of safety factor being the ratio of the resisting moment to the failure moment, this safety factor also equals the ratio of shear strength to shear stress. Consequently, the safety factor, F_s, where c is a measure of part of the shear resistance and γH is a measure of the major part of the shear stress, can be written in the form:

$$F_s = Nc/\gamma H \qquad \text{Eq. 6-10}$$

In this equation N, described as the stability number, includes the effects of φ and i. The remaining factors shown in Equation 6-9 are then applied as special corrections as explained below. The stability number, N, has been established as a function of another parameter, λ, which is a measure of the relative shear resistances arising from friction and from cohesion (7) as represented by the following equation:

$$\lambda = \tan\varphi \gamma H/c \qquad \text{Eq. 6-11}$$

Thus, by calculating λ the value of N can be determined for a particular i by using stability curves such as those in Fig. 6-12.

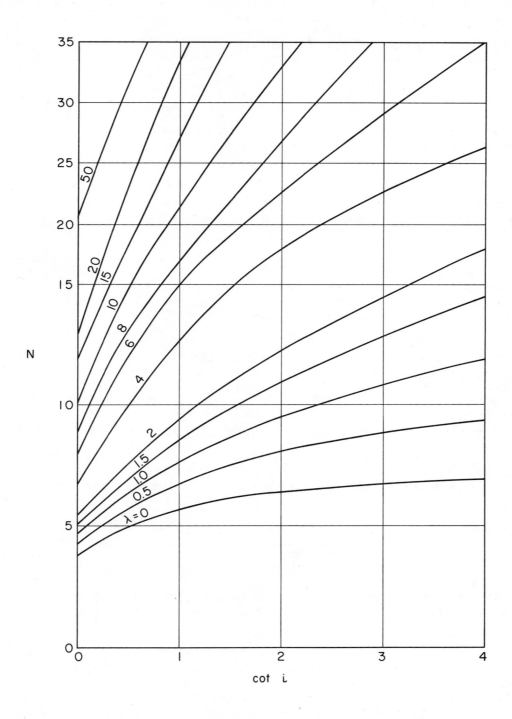

Fig. 6-12 Effective Slope Angle, i, on Stability Number, N (7)

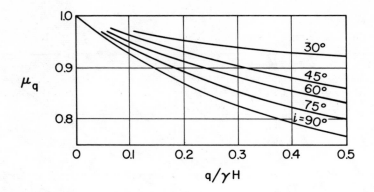

Fig. 6-13 Effect of Surcharge, q, on the Reduction Factor, u_q(7)

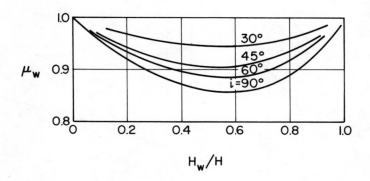

Fig. 6-14 Effect of Height of Ground Water, H_w, on the Reduction Factor, u_w(7)

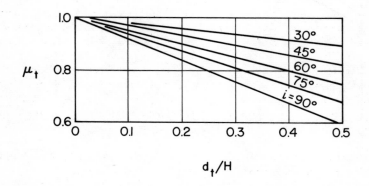

Fig. 6-15 Effect of Depth of Tension Cracks, d_t, on the Reduction Factor u_t(7)

Where a surcharge pressure, q, is applied to the crest of the slope, the safety factor equation can be modified to account for this condition (7):

$$F_s = \frac{\mu_q N c}{\gamma H + q} \qquad \text{Eq. 6-12}$$

where μ_q is a reduction factor that varies, as shown in Fig. 6-13, with $q/\gamma H$ and i (7).

For the cases where a ground water table exists behind the slope, as shown in Fig. 6-11, the appropriate density to use in the equation for the safety factor is the total density (i.e., weight of water plus solids); however, the effective density should be used for the parameter λ as shown in the following equation:

$$\lambda = \tan\varphi \frac{\gamma_t H - \gamma_w H_w}{\mu_w c} \qquad \text{Eq. 6-13}$$

where H_w is the height of the water table behind the slope measured from the elevation of the toe of the slope, as shown in Fig. 6-11, and μ_w is a reduction factor, as shown in Fig. 6-14, that varies with H_w/H and i (7).

Where tension cracking occurs, the equation for the safety factor can be modified by a reduction factor, μ_t, that varies with the depth of the crack, d_t, and with i, as shown in Fig. 6-15 (7). The assumption is made that the crack can be filled with water, which provides a hydrostatic force adding to the failure moment. Hence:

$$F_s = \frac{\mu_t c N}{\gamma H} \qquad \text{Eq. 6-14}$$

Combining all these factors, the equations for determining the safety factor can be summarized:

$$F_s = cN/p_d \qquad \text{Eq. 6-15}$$

$$p_d = \frac{\gamma_t H + q}{\mu_q \mu_t} \qquad \text{Eq. 6-16}$$

$$\lambda = \tan\varphi \, p_e/C \qquad \text{Eq. 6-17}$$

$$p_e = \frac{\gamma_t H + q - \gamma_w H_w}{\mu_q \mu_w} \qquad \text{Eq. 6-18}$$

With the general curves shown in Figs. 6-12, 6-13, 6-14 and 6-15, it is unnecessary to use the time consuming procedure of determining critical circles by trial and error. It should, however, be recognized that the following assumptions have been made: (1) that the surface of failure is cylindrical; (2) that the two-dimensional analysis is valid, i.e., plane stress exists and the assistance of the ends is ignored; (3) that the shear resistance of the ground is governed by Mohr's Strength Theory; (4) that the maximum shear resistance is mobilized at the time of failure at every point on the surface of failure; (5) that the horizontal stresses on the vertical sides of the slices can be ignored; and (6) that the slope has the simple geometry as shown in Fig. 6-11. For cases that depart from these assumptions, in particular where the geometry is complex, an individual trial circle analysis should be made.

It must be recognized, as explained in Chapter 1, that when dealing with failure phenomena, even with supposedly uniform material, the strength demonstrated by individual samples or cases will vary. This variation will generally have some statistical distribution about a mean value. The ultimate objective is to recognize this variability and to deal in probabilities of failure.

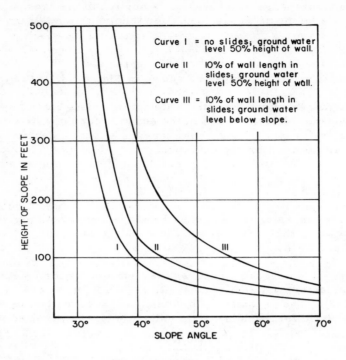

Fig. 6-16 Typical Stability Curves for Incompetent Rock

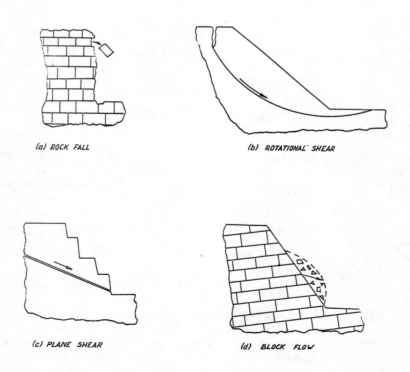

Fig. 6-17 Classification of Types of Slope Failure

For example, Fig. 6-16 shows a series of design curves relating height of slope to maximum slope angle (3). Curve I applies to the requirement that there should be no slides and that the ground water height can be equal to half the height of the slope. In this case a safety factor of 1 is applied to the strength for which 99 per cent of the test results would be higher than the specified strength.

Curve II in Fig. 6-16 represents the same ground water condition as Curve I, but the strength is selected such that there is a 10 per cent probability of failure. This means that much more liberal slope angles would be used for any given height of slope.

Curve III represents the same probability of failure as Curve II, but with no ground water affecting the stability of the slopes. Hence, much steeper slopes can be used.

Besides illustrating the importance of ground water in determining the appropriate slope angle, these curves indicate that, even with a moderate variation of strength properties, the slope angle required to eliminate all failures would generally be unacceptably low.

By recognizing that some slides should be expected for maximum economy of operations (i.e., the cost of a few slides in the life of an open pit would be less than the cost of the extra stripping to eliminate all slides), it becomes essential that inspection and operating procedures be adopted that will safeguard lives. These may consist of having scheduled observations of critical wall crests, of rigorously excavating all the initially loose rock, of scaling all the subsequently loose rock, and of giving some consideration to minimizing the volume of loose rock through modified blasting procedures.

TYPES OF SLOPE FAILURE

Although slope failures are often classified into many different groups with distinguishing characteristics, most of the failures can be divided into one of four groups (see Fig. 6-17). The following four types of failures are distinguished principally by the mechanics of the breakdown of the ground.

(a) _Rock Falls_ occur when loose rock develops on the face of a slope with the slope angle being greater than the angle of repose of this loose material. The blocks of rock fall or roll to the bottom of the slope. The cause of this type of slope failure can be considered to be the result of a breakdown in the tensile strength of the rock mass, e.g., weathering may attack any cementing action that may exist between the joint surfaces.

(b) _Rotational Shear_ failure occurs when a segment of the slope fails by rotation of a mass, as described above, on a more or less circular arc. The cause of failure here can be said to be the inadequacy of the shear resistance for the geometry of slope at the time. This type of failure requires some plastic adjustment of stress concentrations as described in the section on slopes in yielding rocks.

(c) _Plane Shear_ describes the failure, as shown in Fig. 6-17(c), where sliding occurs along a geological plane of weakness. The cause of failure can be said to result from inadequate shear resistance along this plane for the existing slope geometry.

(d) _Block Flow_ failure is imagined to occur in a uniform, hard rock mass that does not contain weaknesses permitting plane shear failure. The method of failure is visualized as being a general breakdown of the rock mass as a consequence of the crushing at the points of highest stress in the brittle rock blocks comprising the mass.

The rock at the points of stress concentrations probably breaks down and loses whatever cohesion had been acting. An increased load is thrown on the adjacent rock whose strength in turn is then exceeded. This progressive action is thought to continue until a general breakdown of the rock mass occurs. This type of failure does not produce a distinct segment of ground that moves, but rather is characterised by the internal deformation associated with most types of flow. The initial failure plane is not likely to be either circular or planar, and at this stage it cannot be predicted.

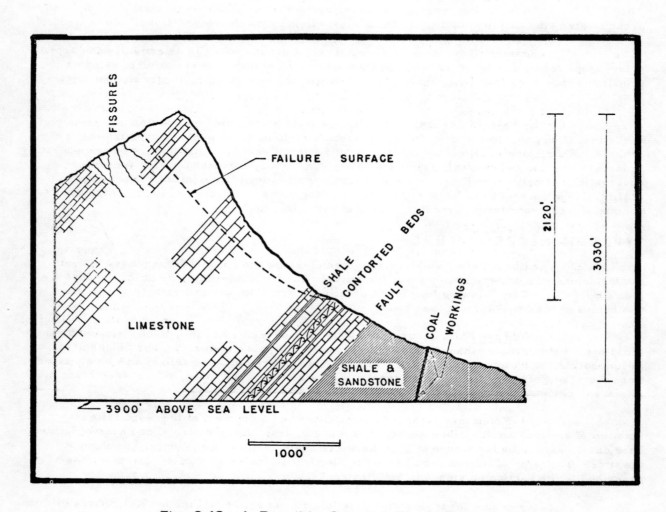

Fig. 6-18 A Possible Case of Block Flow Failure

A good example of a typical block flow is provided by the famous Frank slide, where 80,000,000 tons of rock flowed as a fluidized mass 2 1/2 miles in less than 100 seconds. Eyewitnesses described the movement of the mass as being like a viscous fluid. The blocks varied in size from powder up to 40 ft with the common sizes being between 3 and 20 ft (8).

The geometry of the slope at the Frank slide is shown in Fig. 6-18. The maximum slope angle was 53 degrees with a height of 2100 ft. The limestone dipped 49 degrees into the slope and contained a well-developed family of joints and fracture planes. The outline of the failed mass is also shown in Fig. 6-18.

Several factors combined to either increase the shear stresses or decrease the shear strength of the mass so that failure occurred. The temperatures for the immediately preceding days varied from 70 degrees F during the day to 0 degrees F during the nights causing melting of the surface snow in the day and freezing of the resultant water in the cracks at night. General ground water seepage added to the shear stresses. Compressible layers of shales and sandstones below the limestone at the toe of the slope probably aggravated the stress distribution. Underground workings in the toe area added to this effect. In addition, earthquakes that occurred during the preceding months could have decreased the strength of the rock mass.

PREVENTION AND CONTROL

It can be considered in general that failure occurs as a result of either an increase in stress or a decrease in strength. Prevention and control thus require action that prevents either the increase of stress or the decrease in strength.

Rock falls can be avoided by having slopes at angles less than the angle of repose of the loose rock. This explains why many open pit slopes are not more than 45 degrees. If this solution is adopted for preventing rock falls, then inspection may be required to prevent the slopes from being steepened, particularly where soft rock underlies a hard cap, by loss of toe material either through erosion on river projects or from indiscriminate excavation in open pit mines.

If steeper slopes than the angle of repose are accepted, then the factors causing the decrease in tensile strength in the rock should be recognized. The strength of the rock may have been decreased by the blasting that created the slope. In this case a scaling program may be sufficient to insure that the fall of loose is controlled. Alternatively, the extra cost of using closer blast holes and lighter charges might be repaid by the reduction in the amount of loose rock created.

The deterioration of the surface rock can also be due to the expansion of freezing water in joints, due to the weathering of joint material by air and water action and due to the expansion of joint material such as micas and anhydrite on exposure. In these cases, drainage of water away from the top of the slope might help.

The fall of rock from such faces might be prevented in special, limited areas by rock bolting. Also, the falls might be controlled by covering limited areas with mesh either with or without bolting.

In some types of rock it would be possible to have steep slope angles without any danger of a deep-seated failure. At the same time, the development of loose rock on the face would be of such a quantity to flood the benches and, in effect, establish the angle of repose as the actual slope angle. In these circumstances it would not generally be feasible to attempt to use a steeper slope than the angle of repose of the loose rock.

Rotational shear failure might be caused by an increase in shear stresses. The two most direct actions leading to such increases are the increase in the slope angle and the increase in the total height of the slope. These factors would generally be the subjects of design; however, it is possible for construction or operating crews to excavate toe material thereby effectively increasing the slope angle and to ignore the implication with respect to slope stability. Also, the presence of unknown caverns in the toe zone can also produce, when encountered, unplanned steepening of slopes.

Shear stresses in slopes can be increased by the dumping of waste material at the top of the slope, i.e., this becomes a surcharge pressure, q, as shown in Fig. 6-11. Where this would create a critical condition, the dumped material should be kept sufficiently back from the crest of the slope so that its stability is not affected, i.e., normally about half the height of the slope.

The other major way in which shear stresses can be increased in slopes is by the seepage of water parallel to the slope. It is common for slides to be initiated during the spring runoff or after heavy rain storms. The effect on an otherwise stable slope, as illustrated in the calculations above, can be great.

If the ground water causing the seepage can be directed away from the slope, then the stability conditions would be improved. This might be done by intercepting the water before it entered the slope zone by surface drains, and by underground drainage tunnels. Vertical wells from the crest of the slope or horizontal wells from the toe might also be used. It is, however, very difficult in most rock masses to be able to lower the ground water level effectively and economically. The success of such operations in soils is in no way a guarantee that the same procedure can be used successfully in rock.

Shear stresses can also be increased by the common tension cracks that develop in the crest and reduce the length of the failure surface. These cracks can become filled with water, and the hydrostatic force from this water adds to the failure moment of the rotating segment. Fortunately, the influence of this factor is not particularly large. The more important aspect is that these cracks, if filled with water, would provide a reservoir to supply water for seepage through the slope. Consequently, if feasible, it would be helpful to drain the water from any such tension cracks.

A factor that can cause both an increase in stress and a decrease in strength but, surprisingly, is often considered unimportant, is the effect of underground workings in the zone under an open pit slope. Even a moderate amount of stoping, as treated in Chapter 5, can cause arching in the overlying ground, which increases stresses in the arch abutments. These abutment areas may already be zones of high stress due to the slope. The combination could, as has occurred in several cases, produce failure in this part of the pit wall. At the same time, any caving action could cause a breakdown in the strength of the rock mass.

The prevention of a decrease in shear strength is largely concerned with the control of surface and ground water. The presence of water itself assists in the degradation of rocks by physical and chemical processes. Hence, if water can be kept out of the slope, this will help in preventing a decrease in shear strength.

Another mechanism by which shear strength can be reduced is associated with soft shales. The unloading of shale by an excavation results in a tendency to expand. The ultimate expansion can be impeded by the water in the pores of the shale. The tendency for the shale to expand puts the water in tension or creates a negative piezometric pressure. Eventually this suction in the pore water is dissipated by the inflow of more water. With an increase in moisture content or an increase in volume of voids, a decrease in shear strength results. This mechanism explains many of the apparent cases of loss of strength with time and slope failures in shale.

In such soft shales the decrease in shear strength might be anticipated by the installation of piezometers to measure the pore pressures in the ground. Alternatively, the results of releasing pressure on this type of rock can be studied in the laboratory, and the reduced strength provided for in the design. In addition, the expansion of some shales on release of overburden pressure has been inhibited by rock bolting.

Most slope failures are preceded by tension cracks appearing in the crest material. It is probable that the shear failure surface is initiated at the bottom of these tension cracks or at least connects with them. Consequently, if the tension cracks could be prevented, increased stability might be achieved. Heavy rock anchors might be an appropriate solution in special, limited areas.

Plane shear failures can only be anticipated and controlled if the presence of the critical geological features producing the weak planes are known. It is usually difficult to be certain that such conditions might not apply to a particular slope. However, if the expense is warranted, detailed sub-surface investigations can establish whether this type of failure is pending or not. The comments above regarding the control or prevention of increases in stress apply also to incipient plane shear failure.

Zones of weakness such as faults, dykes or layers of weak rock, which are themselves a source of trouble, are usually also sensitive to decomposition. Consequently, the above comments regarding the degradation effects of flowing water apply particularly to these zones.

Block flow failures are presumably caused by the same factors that increase stresses and decrease shear strengths, as discussed above for rotational shear and plane shear failures. It is possible that in some cases of incipient block flow that the conditions might be improved by grouting. Normally this type of action would not be considered owing to the possibility of increasing the joint spaces as a result of grout pressures lifting the overlying rock. However, under light pressure otherwise open joints have been filled with grout making a more competent and stable mass. The possibility of local stress concentrations leading to progressive failures might be reduced by this grouting action.

REFERENCES

1. Fillunger, P., "Drei Wichtige Ebene Spannungszustande Des Keilformigen Korpers", Z. Math. Physik., Vol. 60, pp. 275-285 (1912).

2. Terzaghi, K., "Theoretical Soil Mechanics", Wiley (1943).

3. Coates, D.F., McRorie, K.L. and Stubbins, J.B., "Analyses of Pit Slides in Some Incompetent Rocks", Trans. SME (Dec. 1963) and Trans. AIME, Vol. 226 (1963).

4. U.S. Corps of Engineers, "Stability of Earth and Rock Fill Dams", Manual EM 1110-2-1902 (1960).

5. U.S. Corps of Engineers, "Soil Mechanics Design", Engineering Manual, Part 119, Chap. 2 (1962).

6. Jenike, A., and Yen, B., "Slope Stability in Axial Symmetry", Proc. 5th Symp. Rock Mech., Univ. Minn., Pergamon (1963).

7. Janbu, N., "Stability Analysis of Slopes with Dimensionless Parameters", Harvard Soil Mechanics Series No. 46 (1954).

8. Coates, D.F., "Pit Slope Stability", Mining World (April, 1962).

9. Bishop, A., "The Use of the Slip Circle in the Stability Analysis of Slopes", Geotechnique, Vol. 5, No. 1 (1955).

CHAPTER 7

FOUNDATIONS

INTRODUCTION

Many problems in foundation design and construction are concerned with rock mechanics. Whereas these problems have been resolved in the past by good engineering judgment, it is becoming possible with the advancement of the subject of rock mechanics to analyse and to test so that quantitative answers can be obtained on some aspects to, at least, provide guidance for the engineer's judgment.

The major rock foundation problems arise in the field of dam construction. The magnitude of these structures, the drastic consequences of failure and the occurrence of sites with relatively poor rock conditions combine to make this field one where clear incentives exist to improve current practices by research and analysis.

The mechanics problems associated with foundations arise principally from the deformation of the foundation strata under the applied loads. In some cases absolute deformation or settlement is significant; in other cases relative deformation between different points on the structure are more significant. In addition, there are cases where actual rock failure is a possibility, although in most rock foundations the imposed stresses are an order of magnitude lower than the strength of the rock.

Another group of mechanics problems arises from the need or from the economy resulting from the use of rock anchors. Many types of structures ranging from cantilever hangars to retaining walls can be constructed more cheaply if competent rock anchors can be designed as opposed to other methods of resisting the forces that these anchors counteract. In other cases, rock anchors are essential, such as in the case of a potentially spalling abutment that must be stabilized while the dam is under construction.

SETTLEMENT

The analysis of deformation or settlement in foundations follows the procedure of selecting an idealization of the problem and then using the most appropriate elastic solution to obtain a first approximation. In Fig. 7-1 some of the problems that have been analysed in the theory of elasticity are shown.

In Fig. 7-1(a) the case of a circular, rigid foundation resting on a semi-infinite elastic body is shown. The settlement, d, of this foundation can be calculated using the following equation (1):

$$d = \frac{Q(1-\mu^2)}{2RE} \qquad \text{Eq. 7-1(a)}$$

where Q is the foundation load, μ is Poisson's Ratio, and R is the radius of the footing.

For points beyond the loaded area the deflection, d_r, can be calculated from the following formula (4):

$$d_r = \frac{Q(1-\mu^2)}{\pi RE} \sin^{-1} R/r \qquad \text{Eq. 7-1(b)}$$

where r is the distance from the centre of the loaded area.

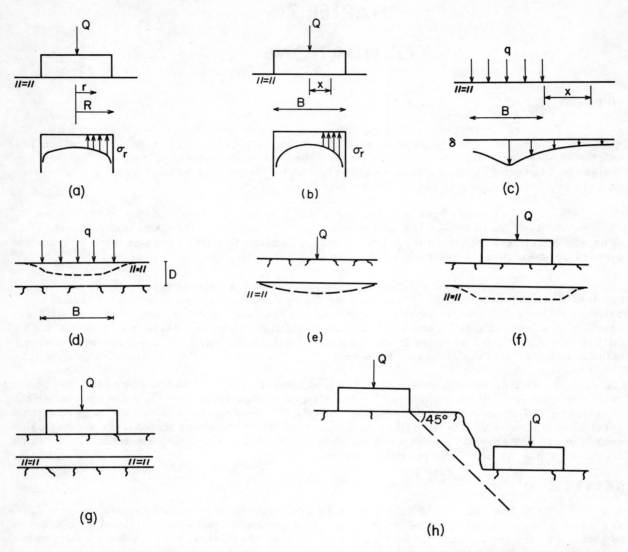

Fig. 7-1 Stress Distribution Effects for Various Types of Foundation Loadings

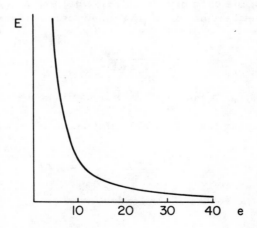

Fig. 7-2 Variation of Modulus of Deformation, E, with Void Ratio, e (3)

Example - A building column carries a load of 1000 kips. Its footing has an equivalent radius of 2.5 ft. The effective modulus of deformation of the subgrade is 4×10^5 psi. Calculate the settlement of the footing assuming it to be rigid and assuming for the rock $\mu = 0.25$. Also, calculate the settlement of the rock surface 4.0 ft from the centre of the load.

From Equation 7-1(a):

$$d = \frac{10^6 (1 - 0.25^2)}{2 \times 12 \times 2.5 \times 4 \times 10^5} = 0.04 \text{ in.}$$

From Equation 7-1(b):

$$d_r = \frac{10^6 (1 - 0.25^2)}{\pi \times 12 \times 2.5 \times 4 \times 10^5} \sin^{-1} 2.5/4.0 = 0.017 \text{ in.}$$

The contact pressure for this rigid circular footing would be (1):

$$\sigma_v = \frac{Q/(\pi R^2)}{2(1-(r/R)^2)^{1/2}} \qquad \text{Eq. 7-2}$$

where r is the distance from the centre of the foundation.

The solution indicates that at the edges of the footing the contact pressure is equal to infinity. This arises from the assumption that the footing is completely rigid, which then results in the tendency for the load to be concentrated at the edges of the footing as a result of the foundation rock tending to deform into a settlement crater with greater deformation at the centre than at the edges. Actually, of course, such infinite pressures do not occur as the foundation structure is not completely rigid; furthermore, strictly elastic relations are not likely to apply at such locations. A moderate amount of plastic yielding either in the footing or in the rock at these points would cause some redistribution of pressure at the edges.

In Fig. 7-1(b) a long, rigid footing of width B is shown. In this case the contact pressure, which is similar to that for the round footing, is as follows (1):

$$\sigma_v = \frac{Q/B}{\pi(0.25-(x/B)^2)^{1/2}} \qquad \text{Eq. 7-3}$$

where Q is the load per unit length and x is the distance from the centre of the footing.

Equations 7-2 and 7-3 could be of some assistance in predicting dangerous stresses for large foundations, such as dams, that are to be placed on soft rock.

In Fig. 7-1(c) the case of the foundation applying a uniformly distributed pressure, q, is shown. By using Boussinesq's Equation 2-14(a) and integrating the effect of point loads over an area, the following general solution has been obtained for the deflection of the corner of the loaded area, d_c, on the surface of a semi-infinite elastic body (2):

$$d_c = qBI'/E \qquad \text{Eq. 7-4(a)}$$

where

$$I' = \left(\ln \frac{1+\sqrt{a^2+1}}{a} + \ln(a+\sqrt{a^2+1})\right)(1-\mu^2)/\pi \qquad \text{Eq. 7-4(b)}$$

with $a = y/B$ and y = the length of the loaded area. For a square area the settlement of the centre point, d_o, is twice that of the corners:

$$d_o = 2 d_c \qquad \text{Eq. 7-4(c)}$$

If in the case of the square footing $\mu = 0.3$, then the settlement of the centre point is:

$$d_o = qB/E \qquad \text{Eq. 7-5}$$

Problems of this type, particularly where irregular loaded areas are involved, can also be solved using the Newmark chart described in Chapter 2. The procedure would be to divide the subgrade rock into several layers and then determine the increase in vertical stress at the centre of each layer under the point on the foundation for which the settlement is to be calculated. This would be done by using, as explained in Chapter 2, a sketch of the loaded area on transparent paper to the scale of 1 in. = (depth to middle of layer). The sketch is placed over a Newmark chart and the number of elemental areas covered by the sketch counted. The increase in vertical stress would then be:

$$\sigma_v = 0.005 \, q \, N \qquad \text{Eq. 2-16}$$

where q is the surface bearing pressure and N is the number of elemental areas covered by the loaded area.

Example - A square footing, 10 ft x 10 ft, is placed on rock with $E = 1 \times 10^5$ psi and $\mu = 0.3$. The bearing pressure is 400 psi. Calculate the settlement of the centre of the footing, assuming it is flexible, using Equations 7-4 and then using the procedure with the Newmark chart.

Equations 7-4 reduce to Equation 7-5, hence:

$$d_o = \frac{400 \times 10 \times 12}{10^5} = 0.48 \text{ in.}$$

Using the Newmark chart, divide the subgrade into layers 2 ft, 4 ft, 8 ft, and 16 ft thick. Then draw sketches of the 10 ft x 10 ft footing to the scales of 1 in. = 1 ft, 1 in. = 4 ft, 1 in. = 8 ft, and 1 in. = 16 ft.

For the first layer:

$$\sigma_v = 0.005 \times 400 \times 200 = 400 \text{ psi}$$

$$\Delta d_o = \frac{400 \times 24}{10^5} = 0.096 \text{ in.}$$

For the second layer:

$$\sigma_v = 0.005 \times 400 \times 150 = 300 \text{ psi}$$

$$\Delta d_o = \frac{300 \times 48}{10^5} = 0.144 \text{ in.}$$

For the third layer:

$$\sigma_v = 0.005 \times 400 \times 68 = 136 \text{ psi}$$

$$\Delta d_o = \frac{136 \times 96}{10^5} = 0.131 \text{ in.}$$

For the fourth layer:

$$\sigma_v = 0.005 \times 400 \times 24 = 48 \text{ psi}$$

$$\Delta d_o = \frac{48 \times 192}{10^5} = 0.092 \text{ in.}$$

$$\Sigma \Delta d_o = 0.096 + 0.144 + 0.131 + 0.092 = 0.463 \text{ in.}$$

In this example, the use of Equation 7-4(a) is more efficient; however, when the loaded area is irregular in plan, the alternate procedure using Newmark's chart is very useful.

Fig. 7-1(c) shows that the settlement crater under the loaded area of width B extends out beyond the actual loaded area. At a distance, B, from the edge of the loaded area the settlement of the surface is about half the amount that occurs at the edge of the loaded area.

To obtain some idea of the magnitude of the settlement crater beyond the loaded area for a strip load, one simple solution that can be used at the present time is that for a loaded area on the edge of a semi-infinite plate. In other words, this solution is only valid for a state of plane stress, whereas normally it would be desirable to have the solution for a state of plane strain. To simplify the equation, the assumption is made that settlements beyond the loaded area that are less than about 10 per cent of the maximum settlement under the loaded area are insignificant; this produces the following equation (1):

$$d_x = \frac{q}{\pi E} \left(2((B+x) \ln \frac{3B}{B+x} - x \ln \frac{3B}{x}) + (1-\mu)B \right) \qquad \text{Eq. 7-6}$$

where x is the distance from the edge of the loaded area.

In Fig. 7-1(d) a foundation that applies a uniform pressure, q, on a soft layer of thickness D overlying a hard layer is shown. The settlement of the foundation is assumed to result from the compression of the soft layer only. The equation of the settlement of a corner of the loaded area is as follows (2):

$$d_c = qB I''/E \qquad \text{Eq. 7-7(a)}$$

and

$$I'' = \left(\ln \frac{(1+\sqrt{a^2+1})\sqrt{a^2+z^2}}{a(1+\sqrt{a^2+z^2+1})} + \ln \frac{(a+\sqrt{a^2+1})\sqrt{1+z^2}}{a+\sqrt{a^2+z^2+1}} \right) \times (1-\mu^2)/\pi$$

$$+ 0.5 \left(\tan^{-1} \frac{a}{z\sqrt{a^2+z^2+1}} \right) (1-\mu-2\mu^2)/\pi \qquad \text{Eq. 7-7(b)}$$

where a = y/B with y being the length of the loaded area and z = d/B.

In Fig. 7-1(e) the reverse case of a hard layer of thickness D over a soft layer is represented. In analysing this case, it has been assumed that the hard layer would act like a beam resting on a series of springs. The spring constant applicable to the soft layer (often called the coefficient of subgrade reaction) is defined as k = q/d, where d is the deflection of the spring resulting from the pressure q. In this case, the deflection of the line load, Q, is (1):

$$d = Q\beta/(2k) \qquad \text{Eq. 7-8(a)}$$

where

$$\beta = (k/4EI)^{1/4}, \qquad \text{Eq. 7-8(b)}$$

E is the modulus of deformation, and I is the moment of inertia of the cross-sectional area of the beam, or hard layer, per unit width.

Example - The wall of a tall building is founded on a layer of hard rock 5 ft thick with $E = 4 \times 10^6$ psi. Under the hard rock there is a deep layer of soft rock with $k = 10^4$ pci.

The wall load is 200^k/LF. Calculate the settlement due to the soft rock. Ignore the spreading action of the footing.

From Equation 7-8(b):

$$\beta = \left\{ \frac{10^4}{4 \times 4 \times 10^6 \times 12 \times 60^3/12} \right\}^{1/4} = 0.00735$$

then

$$d = \frac{200{,}000 \times 0.00735}{2 \times 10^4} = 0.0735 \text{ in.}$$

Where for the above case the foundation load is distributed over an area, more complex solutions have been obtained; however, as the width of the loaded area increases with respect to the thickness of the hard layer, the situation is approached where Equation 7-4(a) can be used with equal accuracy.

In Fig. 7-1(f) a rigid foundation sitting on a hard layer overlying a soft layer is shown. As this is a prototype situation that often occurs in practice, it would be of interest to be able to analyse the settlements and contact pressures. Where the thickness of the rigid layer is large with respect to the width of the foundation, say more than twice, Equations 7-1, 7-2 and 7-3 would provide some guidance. Where the thickness of the rigid layer is small with respect to the width of the footing, say less than one-quarter, the rigid layer could probably be ignored and the settlement based on the modulus of deformation of the underlying soft material. For the intermediate cases the type of approximation used in Chapter 4 (see Fig. 4-14(b)), might be useful; here the equation for a long footing of width B was derived as:

$$\sigma_{v-max} = 1.5 Q / (B + 2D) \qquad \text{Eq. 4-10}$$

In the case of a round footing (both these cases are for a uniformly applied foundation pressure), assuming that the pressure distribution between the hard and the soft layers can be represented by a paraboloid of revolution, the following equation has been derived (2):

$$\sigma_{v-max} = \frac{2Q}{\pi (R + D)^2} \qquad \text{Eq. 7-9(a)}$$

For a square footing Equation 7-9(a) can be modified by assuming that footings of equal area will have a similar pressure distribution and equal maximum pressure, i.e., $\pi R^2 = b^2$, hence:

$$\sigma_{v-max} = \frac{2Q}{\pi (B/\sqrt{\pi} + D)^2} \qquad \text{Eq. 7-9(b)}$$

The calculation of settlement in these cases would be very difficult. For an approximation it would not be unreasonable to assume that the hard layer contributes little to the settlement, and moreover that it is sufficiently jointed to provide little resistance by bending action to the settlement. The further assumption then could be made that the contact pressure between the hard layer and the soft layer is uniform and equal to the average pressure:

$$\sigma_v = Q/(B + 2D)^2 \qquad \text{Eq. 7-10}$$

Then Equation 7-4(a) could be used. Alternatively, the pressure could be assumed to be similar to that for a rigid footing on an elastic body and Equation 7-1 used.

A problem that often arises in rock foundations is shown in Fig. 7-1(g). Here there is a layer of soft material within an otherwise competent rock. If there is no concern about extruding this soft layer through an adjacent cut face, such as shown in Fig. 7-1(h), then the only remaining consideration is that of settlement. Probably the only feasible way of analysing this problem is to use Equations 7-9 to determine the average stress on the weak layer resulting from the structural load. Then, with the determination of the modulus of deformation, it would be possible to calculate the compression or change of thickness of this layer.

Equation 7-10 is based on the approximation, as shown in Fig. 4-14(b), used to derive Equation 7-9(a). One assumption in this derivation was that the increase in pressure at any level below a foundation is effectively included within lines drawn at 45 degrees outward from the edges of the foundation. From this approximation, which is quite reasonable from a theoretical point of view, some guidance is obtained when constructing adjacent foundations at different elevations as shown in Fig. 7-1(h). The safety of the upper foundation should be assured if the cut for the lower foundation does not encroach on the ground that is effectively supporting the upper foundation, i.e., within lines drawn at 45 degrees from the edges of the footing.

Of course, in competent ground that is not subjected to high stresses with respect to its strength, it would be possible to cut a vertical face at the edge of the upper foundation; however, as a general rule it would be good practice to keep lower foundations and their cuts beyond these 45 degree lines. On the other hand, if a soft layer such as shown in Fig. 7-1(g) existed in the rock below the upper foundation but above the subgrade of the lower foundation, the consequence of having a cut face close to the upper foundation would have to be examined with great care. The possibility of extruding the soft material would be the major concern.

The determination of the modulus of deformation for the foundation rock to be used in the above theoretical equations is a practical problem that has not been entirely satisfactorily solved. Laboratory tests can be conducted on diamond drill core, the seismic velocity rock can be measured in the field, or plate load tests can be conducted either on the foundation subgrade or in tunnels in the same formation. On most sites, considering the jointing that must be expected in near-surface bedrock, the only method that can be considered for realistic calculations is the use of plate load testing. Unfortunately, this type of test is expensive, and furthermore a large number of tests should be conducted to obtain a mean value of the modulus of deformation as well as information on the variation of the ground properties around this mean.

Neither the laboratory tests on core nor the field tests on seismic velocity are likely to give information on the deformation properties of the subgrade that takes into account the closing of joints and the compression of infilling material (3). In one extreme case, laboratory compression tests on core samples gave moduli of deformation between 3 and 5×10^6 psi; seismic velocity measurements on the core gave values between 4 and 6×10^6 psi; whereas plate load tests on the shattered rock in situ gave values between 50,000 and 200,000 psi (12).

The results of an interesting study have shown that the modulus of deformation varies with the void ratio in the manner as shown in Fig. 7-2 (3). With a void ratio of less than about 5, the material would be distinctly rock. With a void ratio of more than about 40, the material would be normally classified as a soil. In between these two limits, rock with various degrees of alteration can exist. These materials, as can be seen from Fig. 7-2, generally provide moduli of deformation that are very low.

It is common practice in dam construction to carry out a program of consolidation grouting. This consists of a pattern of shallow drill holes, commonly between 15 ft and 30 ft deep, into which grout is pumped. This helps to fill the voids in the joints and thus produce a foundation with a higher effective modulus of deformation. Plate load tests have actually shown the large improvement that can be effected by consolidation grouting (3).

Aside from the difficulties mentioned above in applying theoretical settlement formulae when the deflection of the subgrade under a dam is considered, an added difficulty arises from a triangular pressure diagram as shown in Fig. 7-3. Without a theoretical solution for this case it is possible to divide the pressure curve into a series of steps as shown in Fig. 7-3. Then the theoretical equations can be used to compute the settlements at several points such as A, B, C and D. It should be remembered in such calculations that the load at one point not only produces deflection at that point, but also contributes to the deflection of the adjacent point, such as the contribution of the force at D to the settlement at C. Equation 7-6 can be used to obtain approximate values for the contribution to settlement of areas adjacent to the loaded area.

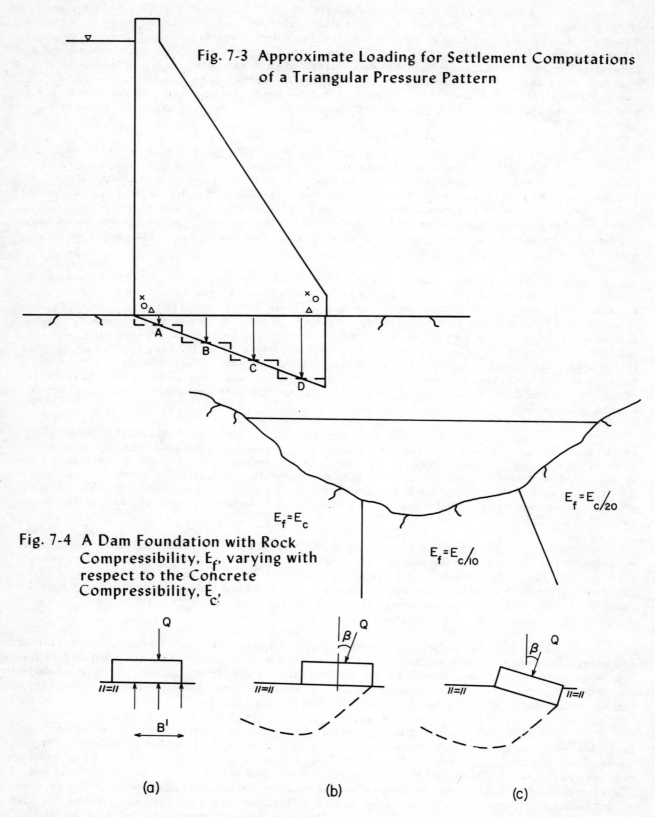

Fig. 7-3 Approximate Loading for Settlement Computations of a Triangular Pressure Pattern

Fig. 7-4 A Dam Foundation with Rock Compressibility, E_f, varying with respect to the Concrete Compressibility, E_c.

Fig. 7-5 Foundations with Eccentric Loading

This type of approximate analysis does not, however, take into account all of the important interactions that model studies can show. For example, it has been found that, as the ratio E_f/E_c (foundation to concrete modulus of deformation ratio) decreases, the compression in the concrete at the toe and the diagonal tension in the heel both increase (11). For some dams it is thus important to measure the modulus of deformation of the foundation and to make an adequate analysis of the effects of settlement, if necessary, by the construction of a model.

A further complication arises in attempting to calculate the probable settlements under a dam foundation when structural or intrusive action has produced a subgrade with different elastic properties. A dam built across a valley with varying foundation rock properties is shown in Fig. 7-4, where E_f is the modulus of deformation of the foundation rock and E_c is the modulus of deformation of the concrete in the dam. In a case such as this there would be a redistribution of contact pressure with the soft subgrade tending to shed load over to the adjacent, stiffer rock. This action would induce shear stresses in the concrete with the possibility, particularly at the heel of the dam, of more serious tensile stresses. Furthermore, where E_f is less than E_c, there is a tendency for the foundation rock to expand laterally under load more than the concrete thus producing an additional component of tensile stress.

BEARING CAPACITY

There are sites, although they are not common, where the bearing capacity of the rock subgrade for a structural foundation might be critical. This situation would usually occur where there is severely altered bedrock. In this case, on rock where some plastic reaction will occur to redistribute stresses, Equation 1-17, established for soils, could be used to determine the safety factor that would exist under the foundation:

$$q_f = 0.5 \gamma B \cdot N_\gamma + c N_c + q N_q \qquad \text{Eq. 1-17}$$

It should be realized that, even on these sites of poor rock, differential and absolute settlements are likely to cause more severe design problems than the limitations on bearing capacity.

For a rock foundation on brittle rock, Equations 1-24, derived in Chapter 1 using Griffith's Strength Theory, might be used:

$$q_f = 24 T_s \qquad \text{Eq. 1-24(a)}$$

$$q_f = 3 Q_u \qquad \text{Eq. 1-24(b)}$$

It is seldom that foundation stresses are of the same order of magnitude as the bearing capacity for rocks that are brittle.

However, there are a few cases where rocks can be brittle and still of low strength; for example, a material such as coal would fall into this category. For these materials under a thick, rigid foundation, the stress concentrations at the edge of the foundation, as shown in Fig. 7-1(b), could be significant. Using Equation 7-3, the average bearing pressure at failure would be as follows:

$$q_f = Q/B = \pi \sigma_v \left(0.25 - \left(\frac{B/2 - y}{B} \right)^2 \right)^{1/2}$$

where y is the distance in from the edge of the foundation, which is used for the simplification that can be obtained as shown below.

From Equation 1-24(b) we can postulate that failure will occur when the stress in the foundation, σ_v, is equal to three times the uniaxial compressive strength. Also, we can assume that towards the edge of the concrete foundation the lack of confinement together with the high stress level would cause some plastic reaction in the concrete so that the theoretically infinite stress will not occur; consequently, the maximum stresses would occur at some distance, y, in from the edge of the foundation. The following equation is then obtained:

$$q_f = 3\pi Q_u (y/B)^{1/2}(1-y/B)^{1/2} \quad \text{Eq. 7-11(a)}$$

This equation is based on the concept that failure will be initiated at a point under the foundation at a distance y in from the edge and, because the foundation rock is brittle, a progressive breakdown of the subgrade would occur.

If we make the further assumption that the distance y is about one foot and that the foundation to which Equation 7-10(a) would be applied will be greater than 10 ft wide, then the following simple expression is obtained:

$$q_f = 9.4 Q_u/B^{1/2} \quad \text{Eq. 7-11(b)}$$

where B must be in feet, although q_f and Q_u can be in any units as long as they are the same. Equation 7-11(b) would give an answer that would be theoretically for the above assumptions 5 per cent too high for a foundation with a width of 10 ft and an answer that would be 1 per cent too low for a foundation of width of 100 ft. These would not be significant deviations.

The above equations for bearing capacity show that the effect of the width of the footing, B, varies with the type of material. For a frictional, plastic rock that approaches the properties of a soil, Equation 1-18 indicates that the bearing capacity will increase with the width of the foundation. If the foundation material can be considered as frictionless, e.g., in a shale with high positive pore water pressures, or where the bearing is occurring on a vertical face, the first term in Equation 1-18 is eliminated, as shown in Equation 1-29, and the bearing capacity is independent of the width of the foundation. Equation 1-24, which might be valid for brittle rock, also is a case where the bearing capacity is independent of the width of the foundation.

Equation 7-11(b) suggests that for a rigid footing on a brittle foundation the bearing capacity should vary inversely with the square root of the width of the foundation. This occurs as a result of the stress concentration at the edge of the footing, such as shown in Fig. 7-1(b), increasing with the width of the footing. With this in mind it is then easy to imagine that, in the case shown in Fig. 7-1(f) with a rigid footing on a hard layer over a soft layer, the stress concentrations under the footing would be even greater than in the case shown in Fig. 7-1(b) and that the stress concentrations would be a stronger function of the width of the footing. Tests have shown this type of variation actually occurs with the index applied to B varying between 0.43 and 0.49 (5). Equation 7-12 then suggests that, for a rigid footing on layered rocks, the dependence of the bearing capacity on the width of the footing will be even greater. Again, tests have shown that this type of variation occurs with the bearing capacity varying inversely with B raised to a power between 0.89 and 0.95 (5).

Although it is not possible at the present time to analyse this situation quantitatively, it can be said that the bearing capacity in this situation would vary inversely with the width of the footing raised to some power:

$$q_f \propto Q_u/B^n \quad \text{Eq. 7-12}$$

Other, more complicated cases of bearing capacity arise on weak foundations. It is not uncommon to have an eccentric load on a foundation as shown in Fig. 7-5(a). In this case, if the centroid of the bearing area is made to coincide with the centroid of the foundation load and a uniform contact pressure assumed, then the width of the bearing area, B', can be used in the bearing capacity formulae that have been derived for soils (7). This rule would then apply to Equations 1-18 and 1-19, which, as mentioned above, are only applicable to foundations that are sufficiently plastic to flow under stress concentrations without fracturing or initiating progressive failure.

Another case that may be of concern is that of the inclined load, as shown in Fig. 7-5(b). In this case, the length of the failure surface in the rock is decreased and consequently the bearing capacity is decreased. For a foundation that does not include adhesion between the structure and the subgrade, the limiting value of the angle of inclination, β, will be the angle of friction between the concrete and rock. A reasonable assumption for this maximum angle is the angle of internal friction either of the rock or concrete.

Some theoretical work together with model verification has been done to determine the reduction in the bearing capacity factors for inclined loads on soils (7). On the basis of this work and limiting ourselves to the most common cases where there is no adhesion between the structure and the subgrade, where the angle of internal friction of the rock is between 40 and 50 degrees and where inclinations vary between zero and 60 degrees, the following equation might be used for estimating the reduction of the bearing capacity for a long inclined load on a subgrade of rock that has not been altered into a residual soil:

$$q_f = (1 - 0.8(\beta/\varphi)^{1/2})(0.5\gamma B N_\gamma + c N_c + q N_q) \qquad \text{Eq. 7-13}$$

with Equations 1-18 and 1-19 being used for the determination of the N-factors.

Where the foundation has been constructed to be perpendicular to the inclined load, as shown in Fig. 7-5(c), the reduction in bearing capacity has been found to be less than for the horizontal foundation (7). Again, using the limitations described for Equation 7-13, the reduced bearing capacity can be estimated using the following equation:

$$q_f = (1 - 0.3\beta/\varphi)(0.5\gamma B N_\gamma + c N_c + q N_q) \qquad \text{Eq. 7-14}$$

Example - A 10 ft wide long foundation is constructed on a soft rock with $\varphi = 40°$, $c = 80$ psi and $\gamma = 155$ pcf. The load is inclined to the vertical at 20°, and the foundation is inclined at 20° to the horizontal to be normal to the load. Calculate the bearing capacity of the foundation.

From Equations 1-18:

$$N_\gamma = \tan^6(45 + 40/2) - 1 = 96$$

$$N_c = 5 \tan^4(45 + 40/2) = 105$$

From Equation 7-14:

$$q_f = \left(1 - 0.3 \times 20/40\right)\left(0.5 \times \frac{155}{1728} \times 10 \times 96 + 80 \times 105\right)$$

$$= 0.85 \times 8443 = 7170 \text{ psi.}$$

For most foundations with inclined loads on rock any concern with failure of the subgrade is likely to arise from the presence of weak layers in the rock (13). A typical situation is shown in Fig. 7-6 of a dam on a foundation with the weak zones in an otherwise hard rock. The main concern of the designer would be the possibility of a horizontal shear failure along the weak zones. The analysis would thus involve the determination of the safety factor against such failure.

The forces involved in such an analysis are shown in Fig. 7-6(b) where H is the horizontal force on the dam due to the head of water, W is the weight of the concrete in the dam, R is the reaction of the concrete in the toe area of the dam (where, if failure occurred, the shear fracture would likely pass through the concrete), N is the normal component of the reaction of the rock at the weak zones, F is the tangential component of the reaction of the rock at the weak zones, and U is the uplift due to the hydrostatic head in the foundation rock. All of these forces can be independently determined except N and F, which can then be calculated by using the equations of equilibrium.

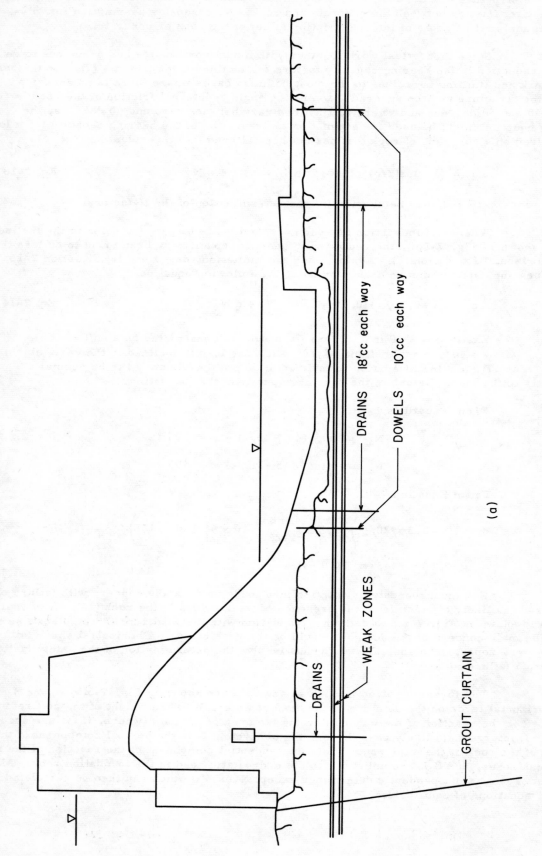

Fig. 7-6 A Dam Foundation with Weak Zones and Some of the Construction Provisions that can be Taken

(a)

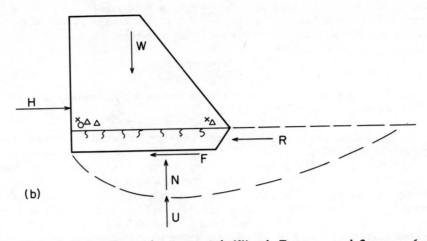

Fig. 7-6 A Dam Foundation with Weak Zones and Some of the Construction Provisions that can be Taken

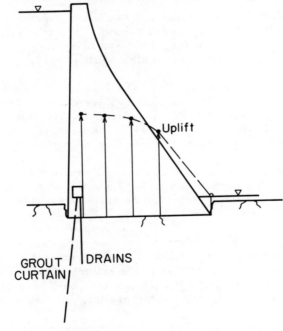

Fig. 7-7 A Typical Pattern of Hydrostatic Uplift on the Base of a Dam

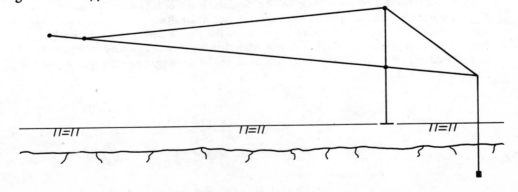

Fig. 7-8 Rock Anchors for a Large Cantilever Hanger

By determining the shear resistance of the weak zones, the value F can be compared with the maximum resistance and in this way a safety factor would be established. There may be some cohesion in such weak zones, or the shear may arise only from friction. Where there is no cohesion the safety factor would then be:

$$F_s = N(\tan \varphi)/F$$

Field tests show that in zones of fractured but unaltered rock the angle φ is usually more than 37 degrees and often as high as 53 degrees (11).

In Fig. 7-6(a) various devices are shown that could be incorporated to increase the safety factor against shear failure. First, the grout curtain at the upstream side would be constructed to cut off the high water pressures from acting on the subgrade of the foundation, in other words, to make U as small as possible. The drains leading into the drainage gallery within the dam would be designed to accomplish the same purpose; if any hydrostatic pressures occurred in the foundation, these drains would reduce them to the head of the drainage gallery. Additional drains in the toe area would be designed for the same purpose as well as to prevent high uplift pressures that would tend to lift the concrete in the stilling basin area. Then, in case these drains were not effective, steel dowels anchored in the rock would hold the concrete down against any tendency for uplift. In this way the maximum value of R would be guaranteed.

As the uplift force, U, is one of the most important factors governing the stability of a dam, great effort is expended to keep it to a minimum. Workmanship is very important in such foundation treatment for, as shown in Fig. 7-7, in spite of the design provisions for a cutoff grout curtain and drains, high uplift pressures can occur.

The analysis of the stability of the rock abutments of an arch dam can be based on the same mechanics as for the subgrade. Excessive deformation could induce or permit stresses in such a dam that would be critical. Alternatively, soft layers parallel to the rock surface might be sheared similar to such layers in the subgrade. The most important aspect of these problems is to obtain sufficient subsurface information to identify the potential problems.

For example, the Malpasset Dam in France was a very thin arch dam, 200 ft high with the thickness varying from 5 ft at the crest to 23 ft at the bottom (15). The radius of curvature at the crest was 344 ft and the chord distance was 590 ft.

The high abutment pressures together with what seems to have been an extensive seam of weak material in one abutment zone, more or less parallel to the rock surface, led to the failure of the dam. It was difficult after the event to determine if such a seam could have been detected or identified before construction by a more intensive subsurface investigation.

In a similar situation a lock failed as a result of the presence of a clay seam about 1/4 in. thick in the foundation rock (14). In this case, the subsurface investigation included 3 in. diameter diamond drill core holes spaced about 30 ft apart together with three 36 in. diameter holes for visual inspection. Careful examination of the holes and cores did not identify the seam. Obviously some risks in engineering work cannot be eliminated.

ROCK ANCHORS

Many uses are being found for rock anchors where reliability can be guaranteed. In Fig. 7-8 a large cantilever hangar is shown. Normally the weight of the roof is counter-balanced with a large mass of concrete. Where bedrock conditions are favourable, a cheaper solution is to use rock anchors to provide the reaction against the upward pull.

In Fig. 7-9 the use of rock anchors in dam design is shown. In this case effective rock anchorage permits the cross-section of the dam to be reduced with considerable savings in concrete costs. The reduced base section, of course, will give rise to tension at the heel of the dam which must then be resisted by the rock anchors. The reliability of these anchors is extremely important; however, methods have been devised for monitoring the tension in the cables so that any corrosion, relaxation or failure of the anchorage can be detected and corrective action taken.

Fig. 7-10 shows the use of rock anchors for holding a retaining wall. In this case, excavation below ground level generally requires an elaborate system of struts or braces to hold the bank of soil until the structure is constructed in the excavation. These struts or braces complicate the pouring of foundations and walls. Consequently, both construction time and costs are reduced where they can be eliminated with rock anchors.

Fig. 7-11 shows a similar use for rock anchors only in this case some poor rock in the abutment area of a dam was held with rock anchors (6). As shown, a concrete abutment was poured against the poor rock, and the anchor holes were drilled through the concrete into the rock. The cables were anchored at the bottom of the holes and pre-stressed. The construction of the dam was then able to proceed without further concern for the fall of rocks from this area.

There are many other uses that can be made of rock anchors. For example, tall buildings owing to wind forces might be subjected to uplift on one side and hence require a reaction similar to the dam in Fig. 7-9. Cantilever bridges have the same force system and anchorage requirements as the cantilever hangar in Fig. 7-8. The hanging wall in a mining stope might be supported by rock bolts or by longer rock anchors thus avoiding internal support similar to the case in Fig. 7-10. Rock slopes for highway cuts or open pit mining operations in special cases might be economically stabilized by rock anchors similar to the case shown in Fig. 7-11. The proviso must be kept in mind that the reliability of the anchorage must, in all these cases, be certain.

Other less common uses of rock anchors include the recompression of foundation soils or soft rock either to minimize the expansion and deterioration that would normally follow the relief of pressure from excavating the overlying ground or to pre-compress such compressible layers so that the settlement under the structural loads will not be excessive.

Dolphins, which provide mooring posts for ships, can be cheaply constructed if good bedrock provides anchorage for both compressive and tensile forces arising from the impact of the ship. Where the force being transmitted to the ground is compressive, the structural element is a pile socketed into bedrock as is used for many other types of structures.

A common type of rock anchor consists of a steel structural shape, rod or cable anchored into rock with Portland cement grout or concrete. The ultimate strength of such a rock anchor is governed by several different modes of failure. First, the shear stress arising at the interface of the steel and grout might fail. It has been found that the shear stress, as shown in Fig. 7-12(a), is fairly evenly distributed along the length of the anchor if the modulus of deformation of the anchor is much greater than that of the ground (2). This distribution would thus not be valid for hard rock. However, where valid, the average shear stress can be calculated according to the equation:

$$\tau = P/(2\pi RL) \qquad \text{Eq. 7-15}$$

where R is the radius of the steel rod and L is the length of the anchor.

Based on experimental work done in the field of concrete construction, the allowable average stress can be determined according to the equation:

$$\tau_a = 0.045 f'_c \quad (158 \text{ psi max}) \qquad \text{Eq. 7-16}$$

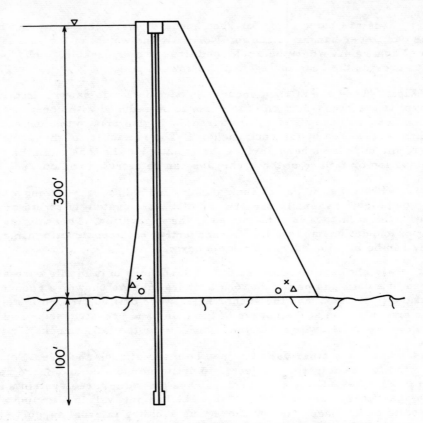

Fig. 7-9 Rock Anchors used in Dams

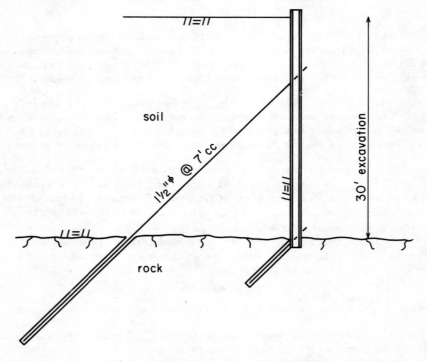

Fig. 7-10 Rock Anchors used with Retaining Walls

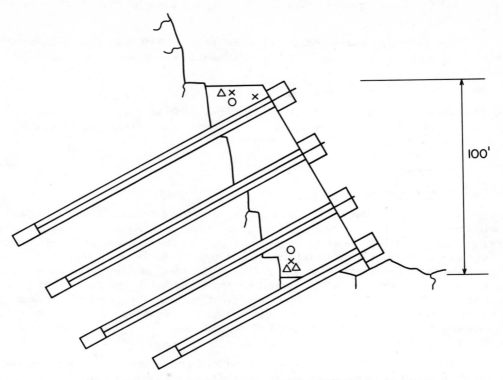

Fig. 7-11 Rock Anchors for Stabilizing Rock Cuts

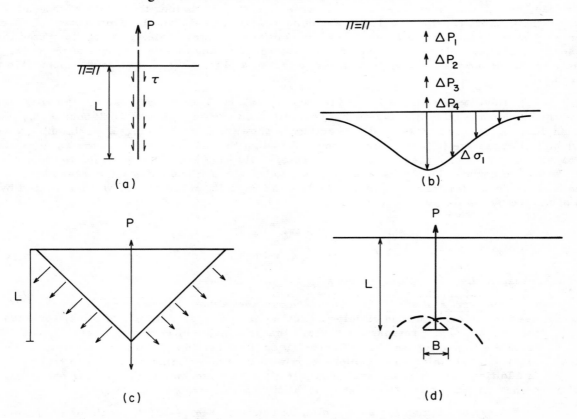

Fig. 7-12 Stress Distribution Patterns for Rock Anchors

where τ_a is the permissible stress in shear, and f'_c is the uniaxial compressive strength of the concrete or grout.

For deformed bars or any structural elements such as lugs that, in effect, throw the failure surface into the concrete, the permissible shear stress can be calculated according to the equation:

$$\tau_a = 0.1 f'_c \text{ (350 psi max)} \qquad \text{Eq. 7-17}$$

These equations incorporate a safety factor against the mean strength of the concrete of the order of 2 to 2.5. If we assume the safety factor to be 2, then the maximum capacity of the rock anchor would be calculated from the following equation:

$$P_f = 4\pi R L \tau_a \qquad \text{Eq. 7-18}$$

where P_f is the anchor load at failure.

In addition to the shear stress created at the interface of the steel and grout, there will be shear stress also at the interface of the grout and rock. Assuming that the strength of the rock is greater than the grout, Equations 7-17 and 7-18 would be used for the determination of the capacity of the anchor with respect to failure at this interface. In this case, as the radius, R, would be that of the hole and thus greater than the radius of the steel element, the capacity of the anchor would not be governed by this mode of failure. On the other hand, the possibility of significant shrinkage in the grout diminishing the intimate contact between the grout and rock might actually provide a weaker surface here than that adjacent to the steel. It is, therefore, important to insure that the grout either includes a small amount of expanding agent or that the mix is such that the shrinkage would be insignificant.

In addition to the shear stresses induced by the load on the anchor, there will be tensile stresses in the rock beyond the end of the anchor as shown in Fig. 7-12(b). In this case, it is possible to calculate the magnitude of the stresses and their distribution by assuming the anchor load, P, is made up of a series of increments distributed along the length of the embedded portion of the anchor. Then, by using Boussinesq's Equation 2-16(a), the contribution of each increment of load to the tensile stresses can be computed. Unfortunately, having computed these stresses it is questionable whether they can be compared to any tensile strength of the rock. The tensile strength of the rock mass, as opposed to that of a piece of core, is almost impossible to determine.

Nevertheless, if failure of the anchor is due to tensile stresses in the rock, failure may be initiated at the bottom of the anchor, but will then create a cone of rock bounded by planes on which diagonal tensile stresses will be acting as shown in Fig. 7-12(c). Experimental work on anchor bolts in concrete (8) can be analysed to show that the maximum load on the anchors at the time of failure, which produced a cone of concrete, could have been closely predicted by assuming a cone of failure bounded by planes at 45 degrees to the surface and using the average tensile strength of the concrete. Thus, we may calculate the capacity of the anchor when governed by this mode of failure by the following equation:

$$P_f = \sqrt{2}\pi L^2 T_s$$

$$P_f = 4.45 L^2 T_s \qquad \text{Eq. 7-19}$$

where T_s is the tensile strength of the rock.

<u>Anchorage of Piles</u>: It was mentioned above that rock anchors have been used in the construction of dolphins for mooring ships. A tripod of piles, socketed into the bedrock, provides a structure that is capable of resisting lateral forces from the impact of the ship. Depending on the direction of the impact, some of the piles will be subjected to compression with the others being subjected to tension. In addition to such use of anchored piles, the same technique is used for obtaining higher capacity piles for other general structural purposes.

A typical design for such a pile, as shown in Fig. 7-15(a), would be to use a 24 in. OD, 1/4 in. steel shell, f'_c = 4000 psi, a permissible stress in compression in the concrete f_c = 900 psi, a permissible stress in the steel f_s = 16,000 psi, and a core consisting of a 14 in. x 14 in. at 170 lb/LF structural member. Using these

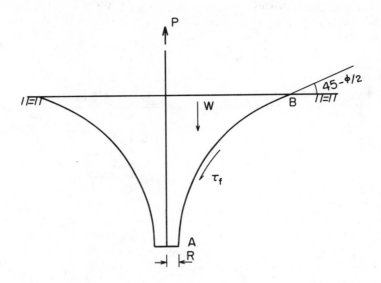

Fig. 7-13 Analysis of Anchor Capacity in Yielding Ground

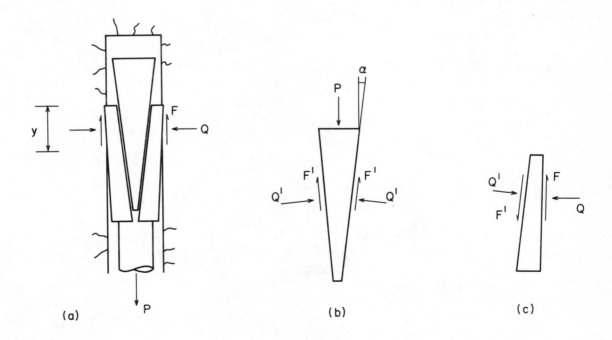

Fig. 7-14 Analysis of Rock Bolt Anchorage Capacity

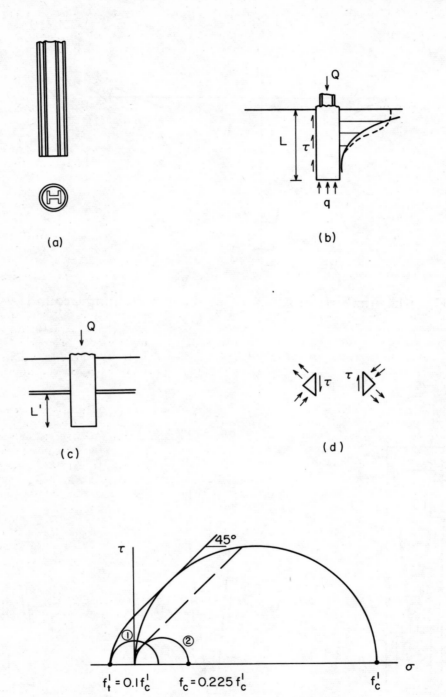

Fig. 7-15 Analysis of a Composite Pile in a Rock Socket

typical allowable stresses and deducting 1/16 in. from the thickness of the shell as not being effective or being subjected to corrosion, the allowable load on the concrete would be 390^k, on the steel shell would be 226^k, and on the steel core would be 551^k making a total allowable load of 1167^k.

The determination of the appropriate length of socket, L, to transfer the load into the rock is usually done by using the following formula:

$$L = \frac{Q - 0.35 f'_c A_c}{0.05 f'_c C_s} \qquad \text{Eq. 7-20}$$

where Q is the design load for the pile, f'_c is the specified uniaxial compressive strength of the concrete, A_c is the cross-sectional area of the concrete and C_s is the circumference of the concrete or the inside circumference of the shell. For the example above, L is 37.6 in. Equation 7-20 implies that $0.35 f'_c$ is the appropriate bearing pressure between the bottom of the pile and the rock. This is not consistent with the normal permissible compressive stress of $0.225 f'_c$ in columns, but is somewhat consistent with the maximum permissible bearing stress, such as occurs on the top of footings, of $0.375 f'_c$ when the area on which the bearing stress occurs is less than 1/3 of the total horizontal area. The formula also implies that $0.05 f'_c$ is the appropriate, allowable shear or bond stress between the concrete and the rock walls of the socket. Furthermore, there is a presumption that the distribution of the load between shear stresses on the sides and bearing stresses on the bottom will be according to these allowable stresses.

The mechanics of the socket can be examined by assuming initially that the moduli of deformation of the pile and of the bedrock are equal and that no shrinkage occurs in the concrete to prevent an intimate bonding of concrete to rock. As shown in Fig. 7-15(b), the pile load, Q, will give rise to shear stresses, τ, on the sides of the socket and bearing stresses, q, on the bottom of the socket.

With the perfect matching of the moduli of deformation, the distribution of the shear stresses can be determined by using Boussinesq's Equation 2-21(c). This equation shows that at a depth z = 0 the shear stresses are equal to infinity. At lower depths the distribution of shear stresses will be as shown by the solid curve on the right of Fig. 7-15(b). If it is assumed that close to the surface where the shear stresses tend to be very high a plastic re-distribution of stresses occurs such as shown by the dashed line in Fig. 7-15(b), then the proportion of the load carried by the shear stresses over the length of the socket can be calculated.

Using Boussinesq's Equation 2-21(a), the load transmitted to the bottom of the socket can also be calculated. For the above pile, even if it is assumed that only 5 ft out of the 6 ft is effectively sustaining shear stress, it is found that only about 4 per cent of the load is transmitted to the bottom of the socket.

If, as a result of this discovery, we make the normal type of calculation for average shear or bond stress (assuming all the load is taken by the sides of the socket) the following value is obtained:

$$\tau = \frac{1,167,000}{\pi \times 24 \times 72} = 215 \text{ psi}$$

This is higher than the normally assumed maximum stress in sockets of 200 psi; however, it is lower than the maximum average bond stress of 350 psi that would be permitted for this concrete.

If, on the other hand, the calculated required length of socket, 37.6 in., were used or if soft, altered or weathered seams occurred down to a depth of 3 ft from the surface, as shown in Fig. 7-15(c), then the length of the socket, L', that would be actively sustaining shear stress would only be about 36 in.

The average shear stress would then be 430 psi, which would exceed even the permitted maximum average bond stress of 350 psi. In this case the assumptions included in the formula for calculating the length of socket would not be fulfilled and excessive shear stresses would be created.

The effect of the shear stresses might be as shown in Fig. 7-15(c). The element on the left side shows that the shear stresses will induce both compressive stresses and tensile stresses in the rock. The element on the right side of Fig. 7-15(d) shows that the corresponding stresses would also be created in the concrete. With the tensile strength of concrete normally being about $0.1 f'_c$, it would seem that, with average shear stresses of the magnitudes calculated above and recognizing that maximum shear stresses as shown in Fig. 7-15(b) would be even higher, tensile failure in the concrete at least, if not in the rock, might occur.

In Fig. 7-15(e) the Mohr circle, ①, shows the stress conditions at failure for a condition of pure shear. With the failure envelope as shown being typical for concrete, Circle ① represents failure in tension. However, together with the shear stresses at the surface of the socket there will also be normal, compressive stresses acting on vertical and horizontal planes. The effect of these stresses would be to move Circle ① to the right.

With the specification that the permissible compressive stresses in the concrete part of the pile be $0.225 f'_c$, the allowable envelope of stresses could be considered to be as shown by the dashed line in Fig. 7-15(e). Circle ② represents the permissible uniaxial compressive stresses. The maximum shear stresses associated with this stress circle would be $0.1125 f'_c$, which for the above concrete would be 450 psi. As the compressive stress on horizontal planes in the concrete below the top of the socket is likely to be less than the value above the socket in the pile and as the shear stress will be greater than the average value, the reason for using a maximum permissible average shear stress of either 200 psi or 350 psi instead of 450 psi can be appreciated.

If the modulus of deformation of the pile is less than that of the rock adjacent to the socket and if no shrinkage occurs in the concrete, the proportion of the load transmitted to the bottom of the pile will be even less than when the moduli are equal. Consequently, the pattern of stress distribution will be much the same.

If the modulus of deformation of the pile is greater than that of the rock surrounding the socket or if some shrinkage occurs in the concrete, the proportion of the load transmitted to the bottom of the pile will be greater than the above two cases. Where the shrinkage in the concrete is greater than the roughness of the socket, the limit will be reached of 100 per cent transmission of load to the bottom of the pile. In these circumstances, the example used above will produce an average bearing pressure of:

$$q = \frac{1167}{\pi \times 12^2} = 2.58 \text{ ksi} = 186 \text{ tsf}$$

This bearing pressure is much higher than the normal maximum allowable pressure cited in various building codes of 50 to 60 tsf. However, as shown by Equation 1-24(b), the ultimate bearing capacity of a brittle rock should be equal to at least three times the uniaxial compressive strength. If a safety factor of three were used against bearing failure, then the allowable bearing pressure would be equal to the uniaxial compressive strength of the rock. As it is a very poor rock that does not have a uniaxial compressive strength of more than 5000 psi, the lack of shear resistance in the socket and the transmission of the full load to the bottom of the pile should not produce a serious condition unless structural features such as open joints or beds of soft material are located immediately under the bottom of the socket. The possible occurrence of these types of features, especially in near-surface bedrock, accounts for the conservatism in the tabulated values for permissible bearing pressures in most building codes.

From the above it would seem that a rational procedure for designing such a high capacity pile would be to drill a socket down from the surface of the bedrock to a depth sufficient to eliminate any open joints, altered rock or soft layers. The construction procedure could include the drilling of a small diameter hole below the bottom of the socket to establish absolutely the absence of such structural features. Then there would seem to be an advantage in not only not counting on the transfer of some load through shear into the sides of the socket, but in preventing the development of such shear stresses, by using a liner of thin steel or cardboard, so that the occurrence of possibly very high, damaging shear stresses near the top of the socket would be avoided.

Another mode of failure that occurs in soil anchors and may or may not apply to some cases in rock is that shown in Fig. 7-12(d) where a plate or projection at the bottom of the anchor exists. Failure in yielding ground if the length, L, is greater than about five times the width of the bottom plate, B, is then by shear in the ground. This type of failure is comparable to a bearing capacity failure. Consequently, Equation 1-17 can be used for the calculation of the capacity of the anchor if governed by this mode of failure. The appropriate modification for this geometry produces the following equation:

$$P_f = B^2 (\gamma L N_\gamma + c N_c) \qquad \text{Eq. 7-21}$$

where N_c is obtained from Equation 1-19. As for all the uses of Equation 1-17, this analysis is only valid if the ground is sufficiently plastic so that stress concentrations will be eliminated by plastic flow without fracture.

Another mode of failure that can occur in soils and may occur in rock is that shown in Fig. 7-13. As for the last case, this anchor requires some plate or projection at the bottom of the anchor that will engage the surrounding ground. Failure is then by shear along the surface that is governed by Mohr's Strength Theory. In other words, the shear stress at failure, τ_f, will be equal to the cohesion plus the normal stress multiplied by the coefficient of friction. In this case the capacity of the anchor will be as follows:

$$P_f = W + T_v \qquad \text{Eq. 7-22}$$

where W is the weight of the ground bound by the failure surface and T_v is the vertical component of the shear stress multiplied by the area of the failure surface. The calculation of these two components of capacity has been simplified by using the following expressions (9):

$$W = \gamma L^3 F_1 \qquad \text{Eq. 7-23}$$

$$T_v = \gamma L^3 (F_2 C/(\gamma L) + F_3) \qquad \text{Eq. 7-24}$$

Table 1 gives the value of the F-factors for the appropriate range of parameters.

TABLE 1

Value of F-factors

L/R	φ =	0°	30°	40°	50°
2	F_1	0.50	0.62	0.66	0.70
	F_2	2.39	2.42	2.12	1.75
	F_3	0	0.48	0.56	0.58
4	F_1	0.25	0.37	0.41	0.45
	F_2	1.60	1.78	1.61	1.47
	F_3	0	0.31	0.37	0.39
∞	F_1	0.05	0.17	0.21	0.25
	F_2	1.40	1.60	1.52	1.32
	F_3	0	0.30	0.36	0.37

In the case of a rock bolt anchored by a wedge or shell, the mode of failure, unless it is in the steel itself, is a case of bearing failure under an inclined load such as shown in either Fig. 7-5(b) or Fig. 7-5(c). Although the outside surface of a rock bolt anchor is likely to bear on the wall of the hole at some angle, this angle in most cases will be small, and in the case of shell anchors they are usually designed with the objective of having the entire outside surface of the shell expand equally and thus be parallel with the side of the hole. Consequently, it will be assumed that the case represented by Fig. 7-5(b) will apply to all rock bolt anchors.

In Fig. 7-14(a) a typical two-leaf bail anchor is shown. The leaves are expanded with the central wedge drawn down by threads engaging the bolt, which is subjected to a torque. Under working conditions the bolt is under a tension, P, that exerts a downward pull on the anchor. The anchor is supported by frictional forces, F, whose maximum value is dependent on the normal force, Q. The maximum value of Q is in turn dependent on the bearing capacity of the rock at that level.

The forces on the wedge, as shown in Fig. 7-14(b), can be analysed. By taking the sum of the vertical forces acting on the wedge, the following relationship can be established:

$$P = 2(Q' \sin \alpha - F' \cos \alpha)$$

where P is the load in the bolt, Q' and F' the normal and tangential reactions on the sides of the wedge, and α the angle to the sides of the hole of the wedge. As slippage will occur between the faces of the wedge and the leaves, the maximum force, F', will be related to the normal force, Q', by a friction coefficient; hence:

$$F' = Q' \tan \psi$$

where ψ is the angle of friction between the two metal surfaces. Combining these equations we obtain the following expression:

$$P = 2Q'(\sin \alpha + \tan \psi \cos \alpha)$$

By taking the sum of the forces in the horizontal and vertical directions on one of the leaves, as shown in Fig. 7-14(c), the following equation can be established:

$$Q' = Q(\tan(\alpha + \psi) + \cot \alpha) \sin \alpha.$$

By replacing Q' with Q in the above equation for **P**, the following equation can be obtained:

$$P = 2Q(\tan(\alpha + \psi) + \cot \alpha \sin \alpha (\sin \alpha + \tan \psi \cos \alpha)$$

By using Equation 7-13 for the bearing capacity under an inclined load on a horizontal foundation, but recognizing that the load in this case is applied to a vertical wall (which eliminates the 0.5 γB-factor), that there is no surcharge on the surface adjacent to the load (which eliminates the q-factor), and that the increase appropriate for a square bearing area would apply to this geometry, the following expression can be obtained that would be applicable to yielding rock:

$$P = 2A' c N_c \left(1-(\alpha+\psi)/\varphi^{1/3}\right)\left(\tan(\alpha+\psi) + \cos \alpha\right) \sin \alpha (\sin \alpha + \tan \psi \cos \alpha) \quad \text{Eq. 7-25(a)}$$

where A' is the area of contact with the rock of one of the two leaves of the shell, c is the cohesion of the rock, N_c is a bearing capacity factor (see Equation 1-19), α is the wedge angle, and ψ is the friction angle between the wedge and shell. This equation can be abbreviated to the following form:

$$P = A c N_c I S \quad \text{Eq. 7-25(b)}$$

where A is the total area of all the leaves in the shell, c is the cohesion of the rock, N_c is the appropriate bearing capacity factor obtained from Equation 1-19, I is the reduction factor for an inclined load (which is included in Equation 7-13 and can be taken equal to $1-((\alpha+\psi)/\varphi)^{1/3}$), and S is the shell factor being equal to $(\tan(\alpha+\psi)+ \cot \alpha) \sin \alpha (\sin \alpha + \tan \psi \cos \alpha)$.

REFERENCES

1. Roark, R., "Formulas for Stress and Strain", McGraw-Hill (1954).

2. Terzaghi, K., "Theoretical Soil Mechanics", p. 382, 423, 398, Wiley (1943).

3. Serafim, J., "Rock Mechanics Considerations in the Design of Concrete Dams", Internat. Conf. on State of Stress in the Earth's Crust - preprint of papers, Rand Corp. RM-3583 (1963).

4. Habib, M., "Determination du Module d'Elasticite des Roches en Place", J. Fdns. et de la Mech. des Sols, p. 35 (Sept. 1950).

5. Jenkins, J., "A Laboratory and Underground Study of the Bearing Capacity of Mine Floors", 3rd Internat. Conf. on Strata Control, Cerchar, Paris (1960).

6. Ischy, E., "Castillon Dam", 3rd Congress on Large Dams, Stockholm (1948).

7. Meyerhof, G., "The Bearing Capacity of Foundations under Eccentric and Inclined Loads", P. 3 Internat. Conf. Soil Mechanics Fdn. Eng., Vol. 1, p. 440 (1953).

8. Adams, R., "Some Factors which Influence the Strength of Bolt Anchors in Concrete", J. Amer. Conc. Inst., Vol. 27, No. 2 (1955).

9. Balla, A., "The Resistance to Breaking Out of Mushroom Foundations for Pylons", P. 5 Internat. Conf. Soil Mechanics Fdn. Eng., Vol. 1 (1961).

10. Luga, A. et al., "Bearing Piles and Open Caissons for Foundations", P. 5 Internat. Conf. Soil Mechanics Fdn. Eng., Vol. 2, p. 85 (1961).

11. Rocha, M. et al., "Experimental Studies of Buttress and Multiple Arch Dams", T. 7 Cong. Large Dams, Ques 26, Rome, p. 641 (1961).

12. Monfore, G., "Laboratory Tests of Rock Cores from the Foundation of Dam BR-9, India, and Analysis of Load-Bearing Tests", USBR, Denver, Lab. Rpt. No. C-731 (1954).

13. Rettie, J. and Patterson, F., "Some Foundation Considerations at the Grand Rapids Hydro-Electric Project", Eng'g J., EIC, Vol. 46, No. 12, p. 32 (1963).

14. Anon, "Clay Seam Wrecked Wheeler Lock", Eng. New. Rec. p. 19 (Jan. 4, 1962).

15. Walters, R., "Dam Geology", Butterworth, p. 51 (1962).

CHAPTER 8
ROCK DYNAMICS

INTRODUCTION

In this chapter only those engineering problems in rock mechanics that can be assisted by analysis are included. Consequently, although ground motion is dealt with, structural dynamics and foundation design are excluded as these do not fall properly in the field of rock mechanics. Also, whereas there are sections on cratering and blasting, no space is devoted to subjects like crushing and grinding as, at the moment, problems in these areas are not being assisted by the techniques being used in the field of rock mechanics.

The chapter starts with a brief review of simple harmonic motion and forced vibrations to reacquaint the student with the basic concepts of harmonic motion and to provide some appreciation for the significant factors in structural response to ground motion.

Wave transmission concepts follow, as they are required for understanding empirical data in the various topics such as blasting, cratering, rockbursting, and the stability of underground openings.

SIMPLE HARMONIC MOTION

When any elastic body is temporarily forced from its position of equilibrium and then allowed suddenly to return to its original position, the individual particles of the body will tend to describe simple harmonic motion, i.e., the particles will describe a periodic motion about the position of equilibrium such that their acceleration towards the position of equilibrium is always proportional to the distance away from this position. Such elastic bodies can be, among others, a mass on a spring, a mass on a beam or column of a structure, a foundation on elastic ground, or a mass of rock subjected to a momentary displacement or impact.

In Fig. 8-1 a vibrating elastic system is represented. The spring in this figure could represent the elastic deformation properties of a beam, a column, elastic ground under a foundation or, in general, an elastic rock mass.

In Fig. 8-1(a) the forces acting on the mass when it is displaced from its position of equilibrium are shown. In this configuration the force of gravity, W, is acting downwards and the reaction of the spring, W + kz, is acting upwards. This reaction includes the reaction of the spring at the position of equilibrium, W, plus the additional reaction due to the compression of the spring through the distance z. The coefficient, k, is the spring constant in typical units of lb/in.

From Newton's Second Law we can write the equation:

$$(W - W - kz)g = Wa = W\, d^2z/dt^2$$

where a is acceleration, t is time and g is the acceleration due to gravity or the conversion factor to be applied to gravitational units of mass to obtain absolute units. The equation of motion is thus a linear, homogeneous, second order differential equation:

$$d^2z/dt^2 + kgz/W = 0$$

The solution of this equation is as follows:

$$z = z_o \cos(t\sqrt{kg/W}) \qquad \text{Eq. 8-1}$$

where z_o is a constant.

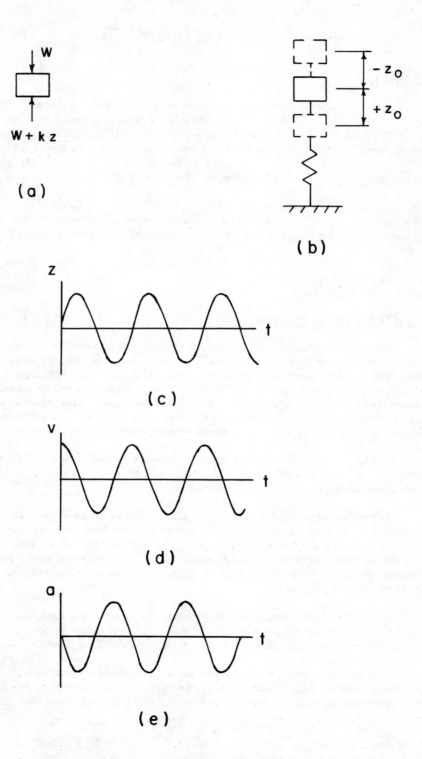

Fig. 8-1 One Degree of Freedom Prototype for Elastic Structures

By taking derivatives of Equation 8-1, it follows that:

$$v = dz/dt = z_0 \sqrt{kg/W} \sin(t\sqrt{kg/W}) \qquad \text{Eq. 8-2}$$

and

$$a = dv/dt = z_0 (kg/W) \cos(t\sqrt{kg/W}) \qquad \text{Eq. 8-3}$$

where v is the velocity and a the acceleration of the body. The starting conditions are that when $t = 0$, $z = z_0$ and $v = 0$.

In Fig. 8-1(c) the graph of displacement versus time is shown for the mass as it oscillates freely describing simple harmonic motion. The velocity of the body at any time is the derivative of displacement with respect to time or the slope of the displacement-time curve. Thus Fig. 8-1(d), the graph of velocity versus time, shows the velocity to be one-quarter of a cycle in advance of the displacement curve. Similarly Fig. 8-1(e) is the acceleration-time curve, which can be considered as representing the slope of the velocity-time curve. The acceleration curve in turn is one-quarter of a cycle ahead of the velocity curve.

Equation 8-1 can be used to determine the period of the oscillations by solving for the increment in time between the passage of a particle through two identical positions, i.e., when $z = z_0$. From this it is found that the period is:

$$T = 2\pi \sqrt{W/kg} \qquad \text{Eq. 8-4}$$

The reciprocal of period is, of course, the frequency:

$$f = \frac{1}{2\pi} \sqrt{kg/W} \qquad \text{Eq. 8-5}$$

In Fig. 8-2(a) a dash-pot is connected to the mass sitting on the spring. This represents diagrammatically the common situation of an elastic system including some damping forces. As these damping forces are often proportional to the velocity of the particles in the system, they are represented by this dash-pot, which mechanically is considered to be a loose piston in a cylinder of fluid. In this case the force equation becomes:

$$(W - W - kz - c\, dz/dt)g = W\, d^2z/dt^2$$

where c is the damping constant.

The solution of this differential equation shows that, when the coefficient of damping is large, the body will not oscillate when released from a disturbed position, but will merely return slowly to its original position as shown by the dashed lines in Fig. 8-2(b). Where the coefficient of damping is small, the vibrations after the initial displacement are represented by the oscillating curve in Fig. 8-2(b). In this case, vibrations occur but they are of decreasing amplitude.

It is well to recognize here that structural dynamic problems can be divided into two types. First, abnormal stresses can result from impact on an elastic structure such as the sudden ground motion resulting from earthquakes or nearby explosions acting on foundations of buildings. In this case, the complex of masses and springs, of which the structure can be considered to be composed, is given an initial, sudden displacement and then allowed to respond in a manner governed by its own properties. A simple system then acts as shown in Fig. 8-2, owing to all structures having some damping. Consequently, the oscillations that occur eventually die out.

The second type of dynamic problem, as opposed to the impact case, results from continuous vibrations. In this case, the external force causing the motion is usually continuous and of an oscillating nature itself. In Fig. 8-3(a) a force, P, is applied to the mass supported by the spring. P varies according to the relationship $P = P_0 \sin wt$ where P_0 is a constant with units of force, w is a constant with units of radians per second, and t is time.

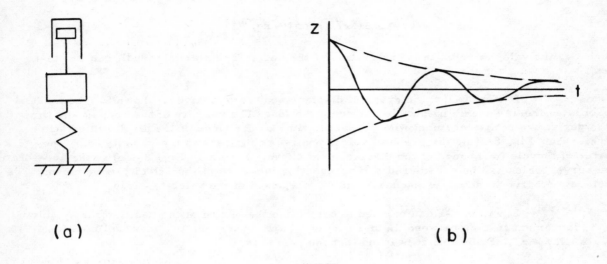

Fig. 8-2 Plastic Prototype with Damping or Viscous Friction

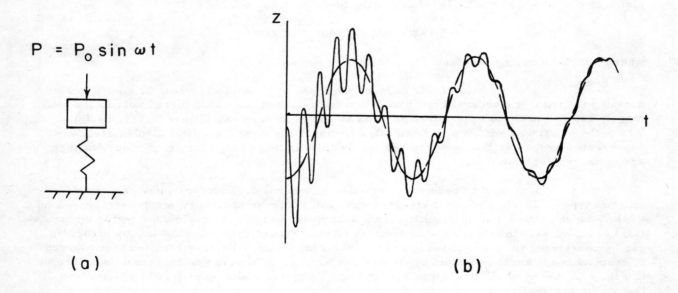

Fig. 8-3 Elastic Prototype under Forced Vibrations

By solving for the times when $P = P_o$, it is found that the period of oscillation of the force is equal to $2\pi/w$ or, in other words, the frequency is equal to

$$f = w/2\pi.$$

The force equation for the system shown in Fig. 8-3(a) is as follows:

$$(W - W - kz + P_o \sin wt)g = W d^2z/dt^2$$

The solution of this differential equation gives the following expression:

$$z = z_o \cos(t\sqrt{kg/W}) + (1 - (kg/(Ww^2)))^{-1} (P_o/k) \cos wt \qquad \text{Eq. 8-6}$$

where, at $t = 0$, $z = z_o$ and $v = 0$.

Equation 8-6 contains two terms. The first term is the same as Equation 8-1. In other words, it is the reaction of the mass-spring system to a disturbance. This type of reaction for most systems includes a damping factor, which as shown in Fig. 8-2, means that the motion will eventually die out.

The second factor in Equation 8-6 represents the continuing motion resulting from the applied force. As the first factor in the equation, when damping is included, eventually becomes insignificant, Equation 8-6 can be rewritten as follows:

$$z = (1 - (kg/(Ww^2)))^{-1} (P_o/k) \cos wt \qquad \text{Eq. 8-7}$$

It can be seen from this equation that the frequency of the resulting steady-state motion is the same as the frequency of the applied force. The graph of displacement versus time for such a system is shown in Fig. 8-3(b). In this case, depending on the damping factor, after about three cycles the contribution to the motion of the first factor in Equation 8-6 is likely to be insignificant, and Equation 8-7 will then describe the subsequent motion.

From Equation 8-7 it follows that the maximum displacement of the mass will be equal to the displacement that would occur under the maximum force applied statically, P_o/k, multiplied by the factor $(1 - (kg/Ww^2))^{-1}$. This latter force is known as the magnification factor and can be rewritten as follows:

$$MF = (1 - (f/f_n)^2)^{-1} \qquad \text{Eq. 8-8}$$

where $f = w/2\pi$ and f_n, the natural frequency of the structure, is $\frac{1}{2\pi}(kg/W)^{1/2}$.

The variation of MF with the ratio f/f_n is shown in Fig. 8-4. As f/f_n approaches 1, i.e., resonance, the MF approaches infinity for a system with zero damping. As the frequency ratio becomes greater than 1, the magnification factor becomes negative, which simply means that the motion of the mass is in the opposite direction to the motion of the force.

As the frequency ratio approaches infinity, the magnification factor approaches zero. That is, as the frequency of the applied force becomes very high with respect to the natural frequency of the system, the displacement it causes is almost zero. In other words, there is insufficient time for the mass to move in one direction before the direction of the force is reversed.

In Fig. 8-4 the reflection of the negative part of the curve for zero damping is shown on the positive side of the x-axis as a dashed line. The readings from this curve then simply indicate the magnitude of the magnification factor ignoring the direction of the motion. It is common to present this type of data simply as scaler values (i.e., magnitudes only).

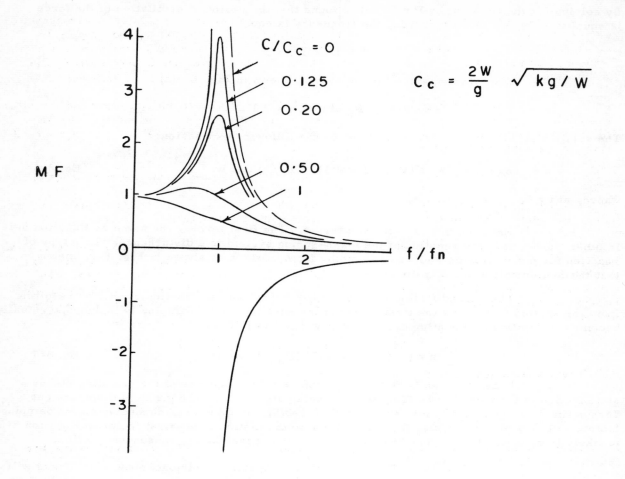

Fig. 8-4 Magnification of Displacement Response of an Elastic Structure

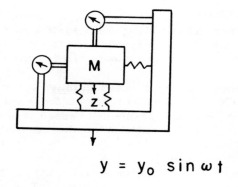

$y = y_o \sin \omega t$

Fig. 8-5 A Seismic Instrument

The curves of the magnification factor for systems with damping are shown by the family of curves in **Fig. 8-4** for different ratios of the coefficient of damping to the critical coefficient. It can be seen from these curves that, with damping even when the frequency ratio is 1, that is at resonance, the magnification factor has some relatively low finite value.

The mechanical systems analysed above, besides being prototypes of structures, machinery, and even in some cases rock masses, are also closely representative of some practical instruments. **Fig. 8-5** is a somewhat simplified sketch of a seismic instrument. A mass, M, is supported vertically on springs and held by a horizontal spring.

When such a device is used as a seismograph, the mass is generally quite heavy and the base is anchored in bedrock. The instrument can also be used for measuring the amplitude of mechanical vibrations, in which case it is called a vibrometer. As explained below, the same instrument might also be used as an accelerometer.

In the case of vertical ground motion, the motion of the base of the instrument in Fig. 8-5, $y = y_o \sin wt$, represents the forced oscillations resulting from such events as earthquakes or explosions. As a result of the displacement of the base, the mass is also displaced by some amount z. The relative motion of the mass with respect to the base, which is equal to the extension of the springs, is (z-y). The force equation for the mass can then be written:

$$(W - W - (z-y)k)g = W d^2z/dt^2.$$

A differential equation is then obtained of the following form:

$$\frac{d^2z}{dt^2} + \frac{kg}{W} z = y = y_o \frac{kg}{W} \sin wt.$$

By solving this equation for the amplitude of the relative motion of the mass with respect to the base, the following equation is obtained:

$$(z - y_o) = Z = \frac{y_o (f/f_n)^2}{1 - (f/f_n)^2} \qquad \text{Eq. 8-9}$$

The amplitude, Z, is then the quantity measured by the gauge in Fig. 8-5. In the case of vibrometers, a dial gauge is actually used. In the case of seismographs a mirror on a lever is used. Light passing through a slit is reflected from this mirror and can be used to record the motion on photographic paper.

Equation 8-9 can be examined for the variation of Z with the ratio f/f_n. When the ratio f/f_n is large, Equation 8-9 becomes:

$$Z = -y_o \qquad \text{Eq. 8-10}$$

For this condition, the measured amplitude, Z, is the actual displacement amplitude of the ground motion. In other words, if the instrument is a seismograph it records the ground displacement; if the instrument is a vibrometer it records the displacement amplitude of the base vibrations. A little reasoning shows that this situation occurs when the natural frequency of the instrument, f_n, is small or, in other words, when the spring constant, k, is small indicating soft springs.

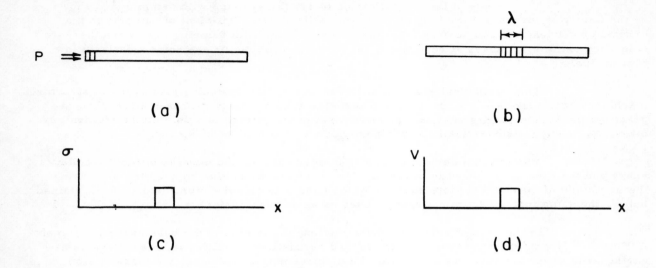

Fig. 8-6 The Initiation of an Elastic Wave

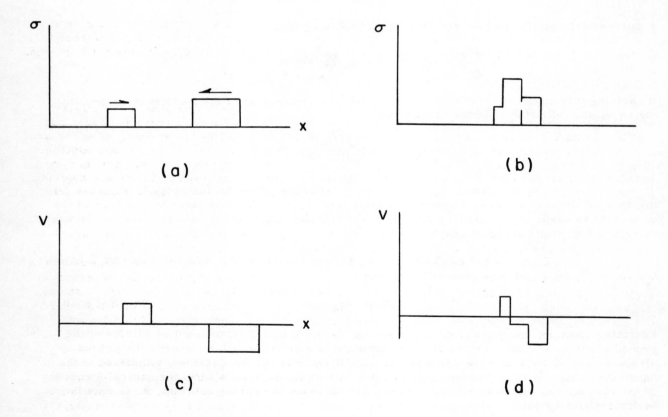

Fig. 8-7 The Interaction of Elastic Waves

As an example, consider the frequency of the ground motion to be three times the natural frequency of the instrument. The results from Equations 8-9 and 8-10 can be compared:

from Equation 8-9: $Z = y_o \dfrac{3^2}{1-3^2} = -1.12 \, y_o$

from Equation 8-10: $Z = -y_o$

A difference of this amount would not normally be too important, but, in addition, if the system included some damping the difference could be even less.

The alternate case is where f/f_n is small. In this case Equation 8-5 reduces to:

$$Z = y_o (f/f_n)^2 = y_o (w/w_n)^2 \qquad \text{Eq. 8-11}$$

When it is recognized that the acceleration of the base motion,

$$a = d^2y/dt^2 = -y_o w^2 \sin wt$$

and the amplitude of the acceleration is

$$a_o = y_o w^2$$

it can be seen that the instrument by recording Z is measuring the acceleration of the ground motion to some determinable scale. In this case, the instrument is acting as an accelerometer. Again, a little reasoning indicates that this type of instrument must have relatively hard springs.

WAVE TRANSMISSION

There are three groups of theories that can be used to determine the effects of forces on bodies. First, there is the subject of rigid body dynamics in which it is assumed that bodies cannot be deformed and that accelerations are produced instantaneously in accordance with Newton's Second Law of Motion. Second, there is the theory of elasticity, which considers the effects of force systems in static equilibrium on deformable, elastic bodies. Third, there is the field of stress wave propagation where forces are either changing rapidly or are applied for a short period of time on deformable bodies. Neither of the first two subjects covers the phenomena of interest in the third group.

As an introduction to the field of wave transmission, we can consider the mechanics of a one-dimensional problem. In this case, we examine the effects of a short duration pressure acting on the end of a rod as shown in Fig. 8-6(a) (13). The initial effect of this pressure applied for a short period of time, t, is to create a stress, σ, and to compress the end of the rod. The entire rod is not compressed, as would be the case for a static stress, as the time required for complete compression is long compared to the time of application of the stress, t; this might be considered to be a criterion for the application of this third area of theory as opposed to using the theory of elasticity.

Besides compressing a zone of the rod where the pressure is applied, the particles in this zone will be given a velocity. This zone then impinges on the zone next to it along the length of the bar causing in turn compression and velocity. In this way, a pressure wave or velocity wave is transmitted along the bar. The length of this zone or wavelength, λ, will be the product of the velocity of the wave front and the duration of the pressure on the end of the rod.

To establish the mechanics of the system, we use energy relations. The shortening of the bar due to the compression of the wave zone, assuming a unit cross-sectional area, will be $\sigma\lambda/E$. Since this shortening is the distance through which the applied stress moves, the external work done, therefore, is $\sigma^2\lambda/E$. This work done must be equal to the energy stored in the wave zone plus the kinetic energy, which produces the following equation:

$$\sigma^2\lambda/E = \sigma^2\lambda/2E + \rho\lambda v^2/2 \qquad \text{Eq. 8-12}$$

where ρ equals the mass per unit volume of the rod and v is the velocity of the particles in the wave zone. From this equation we can solve for the particle velocity:

$$v = \sigma(\rho E)^{-1/2} \qquad \text{Eq. 8-13}$$

Although a square wave has been used for this derivation, the relationship between particle velocity and stress, as shown in Equation 8-13, would still be valid for other wave shapes, provided the wave front is plane. When the wave front is sharply curved, as for a spherical wave front close to a contained explosion, Equation 8-13 would not be valid.

By using momentum relations we can deduce further information. Consider the wave of length λ, as shown in Fig. 8-6(b), during the period of time that the wave moves from its position one wavelength to the left of the position shown. During this time, t', the zone shown in Fig. 8-6(b) is acted on by the stress, σ. The result of this application of stress is to cause a change in momentum equal to $\rho\lambda v$. Hence:

$$\sigma t' = \rho\lambda v - 0 = \sigma\lambda(\rho/E)^{1/2}$$

or

$$\lambda/t' = C_p = (E/\rho)^{1/2} \qquad \text{Eq. 8-14}$$

In other words, the velocity of the wave itself, C_p, is a function of the modulus of elasticity and density of the rod and is independent of the applied stress and the duration of application. This is the velocity that is known in geophysical work as the seismic velocity or, more accurately, the compression (P-wave) or longitudinal wave velocity (seismic velocity can also refer to the shear wave velocity, which is quite different). It follows from this that the wavelength of the pulse is:

$$\lambda = C_p t = t(E/\rho)^{1/2} \qquad \text{Eq. 8-15}$$

It can be seen, by changing the sign of the stress in the above mathematics, that where the applied stress creates a tension pulse, the same particle and wave velocities will be obtained. The only difference is that, whereas the particle velocity will be in the direction of the original applied tension, the wave velocity will be in the same direction as for the compression wave.

A square wave, as indicated in the Fig. 8-6(c), is being used in this analysis. In other words, the stress, σ, is applied instantaneously, remains constant, and then is removed instantaneously. By examining the energy content in the wave, which, as shown above, is half strain energy and half kinetic energy, it follows that, with the length of the wave and the particle velocity remaining constant, the stress in the wave must also remain constant and equal to the applied pressure. This fact was assumed above, but can now be considered proven.

Equation 8-13 can be used for establishing general relations between the various aspects of the motion of the ground within the wave. For example, strain, ϵ, can be related to particle velocity, v, and the compression wave velocity:

$$\epsilon = \sigma/E = v(\rho E)^{1/2}/E = v(\rho/E)^{1/2} = v/C_p \qquad \text{Eq. 8-16(a)}$$

Also, of course, particle acceleration, a, must be similarly related:

$$a = \partial v/\partial t = C_p \partial\epsilon/\partial t = \partial\sigma/\partial t \, C_p/E \qquad \text{Eq. 8-16(b)}$$

And particle displacement must be:

$$y = \int v \, dt = C_p \int \epsilon \, dt = C_p/E \int \sigma \, dt \qquad \text{Eq. 8-16(c)}$$

As most seismic instruments measure either displacement or acceleration, it is common, using the obverse of the above equations, either to differentiate the displacement record or to integrate the acceleration record to determine the particle velocity.

By using the representation of the wave in terms of both stress and particle velocity as shown in Figs. 8-6(c) and 8-6(d), we can examine the effects of two or more waves interacting. In Figs. 8-7(a) and 8-7(c) we have the case of two waves approaching each other. In Figs. 8-7(b) and 8-7(d) the consequences of these waves overlapping is seen to be simply the algebraic sum of the individual waves. When the waves pass each other, they revert to their original form. The same type of algebra would apply to the superposition of two waves travelling in the same direction.

Still using the simple mechanics applicable to wave transmission in a one-dimensional problem, we can examine the effects of a change in properties of the bar. From the above it is known that the two material properties that are of significance in wave transmission are the modulus of deformation, or stiffness, and the density. Changes in both of these properties for wave transmission in a rod can be represented by a change in cross-sectional area, as shown in Fig. 8-8(a).

For the purpose of this analysis we return to the case described in Fig. 8-6(a) and use a rod of some other cross-sectional area than unity. The mass per unit length, instead of being the density, is m, and instead of the modulus of deformation, which is the stress required to cause a unit strain, we characterize the stiffness by k, which is the force required to produce a unit strain and is thus equal to the cross-sectional area multiplied by the modulus of deformation. With these changes, the particle velocity in the rod would be derived not as in Equation 8-13 but as:

$$v = P(mk)^{-1/2} \qquad \text{Eq. 8-17(a)}$$

where P is the force applied to the end of the rod.

If we examine the particle velocities on the element of the rod in Fig. 8-8(a) at the transition section but to the left side of it, the resultant velocity will be:

$$v = v_i - v_r = v_t \qquad \text{Eq. 8-17(b)}$$

where v_i is the particle velocity for the incident wave, v_r is for the reflected wave and v_t is for the transmitted wave. The resultant velocity of the particles to the left of the transition must be the same as the resultant velocity of the particles to the right of the transition, unless the material separates.

Using Equation 8-17(b) we can then substitute from Equation 8-17(a):

$$\frac{P_i}{(m_1 k_1)^{1/2}} - \frac{P_r}{(m_1 k_1)^{1/2}} = \frac{P_t}{(m_2 k_2)^{1/2}} \qquad \text{Eq. 8-17(c)}$$

where the subscripts 1 refer to the rod carrying the incident wave, subscripts 2 refer to the rod connected to this first rod, P_i is the force in the incident wave, P_r the force in the reflected wave and P_t the force in the transmitted wave across the interface. In addition, we can examine the forces on the particles just to the left of the transition point, as indicated by the section line in Fig. 8-8(a) and as shown by the free body diagram in Fig. 8-8(b). With our assumption of square wave shapes, the velocities within any one section of the wave will be constant; thus, no accelerations are occurring and the particles are in a state of equilibrium whereby the sum of the forces must equal 0.

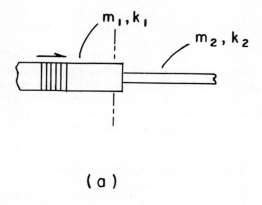

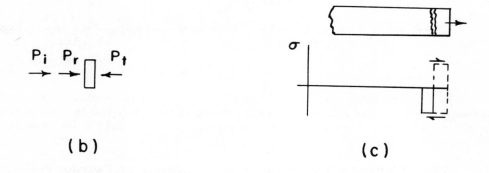

Fig. 8-8(a) to (c) The Effect of a Boundary on Wave Transmission and Reflection

The force equation is thus:

$$P_i + P_r - P_t = 0 \qquad \text{Eq. 8-17(d)}$$

It might be noted that P_r is assumed positive as its direction is not known. (If it were assumed to be acting in the negative direction and our assumption was wrong, a negative answer would be obtained. This would indicate that our assumption was wrong, not that the direction was in the negative sense. Thus, by assuming the positive direction this ambiguity is avoided.) Actually P_r may act in either direction, depending on the circumstances, as deduced below.

Combining Equations 8-17(c) and 8-17(d) and letting $n = (m_1 k_1 / m_2 k_2)^{1/2}$, we can obtain the following:

$$P_t = 2 P_i / (1 + n) \qquad \text{Eq. 8-17(e)}$$

$$P_r = P_i (1 - n)/(1 + n) \qquad \text{Eq. 8-17(f)}$$

Now, recognizing that the change in the cross-section of the rod is simply a matter of mass and stiffness, we can rewrite the equation for n as follows:

$$n = (\rho_1 E_1 / \rho_2 E_2)^{1/2} = \rho_1 C_{p1} / \rho_2 C_{p2} \qquad \text{Eq. 8-17(g)}$$

The product, ρC_p, is known as the characteristic impedance. When the materials on either side of a boundary have greatly different values for ρC_p, it is said that there is an impedance mismatch.

The equations for the transmitted and reflected stresses following Equations 8-17(e) and 8-17(f) can be written:

$$\sigma_t = 2 \sigma_i / (1 + n) \qquad \text{Eq. 8-18(a)}$$

$$\sigma_r = \sigma_i (1 - n)/(1 + n) \qquad \text{Eq. 8-18(b)}$$

Example - A P-wave travelling through Rock A with a peak radial stress of 10,000 psi impinges at right angles on a plane boundary with Rock B. The wave velocity in A is 16,000 fs and in B is 12,000 fs. The specific gravity of A is 2.9 and of B is 2.6. Calculate the peak radial stresses in the reflected and transmitted waves.

From Equation 8-17(g):

$$n = \frac{2.9 \times 16,000}{2.6 \times 12,000} = 1.485$$

From Equation 8-18(a):

$$\sigma_t = 2 \times 10,000 /(1 + 1.485) = 8,050 \text{ psi}$$

$$\sigma_r = 10,000 (1 - 1.485)/(1 + 1.485) = -1,950 \text{ psi}$$

I.e., the reflected wave will be tension.

From Equations 8-18 we can learn several things. First, the wavelength of the reflected wave, being tC_p or $t(E/\rho)^{1/2}$, must be the same for the incident and reflected waves. However, the wavelength of the transmitted wave clearly will be different.

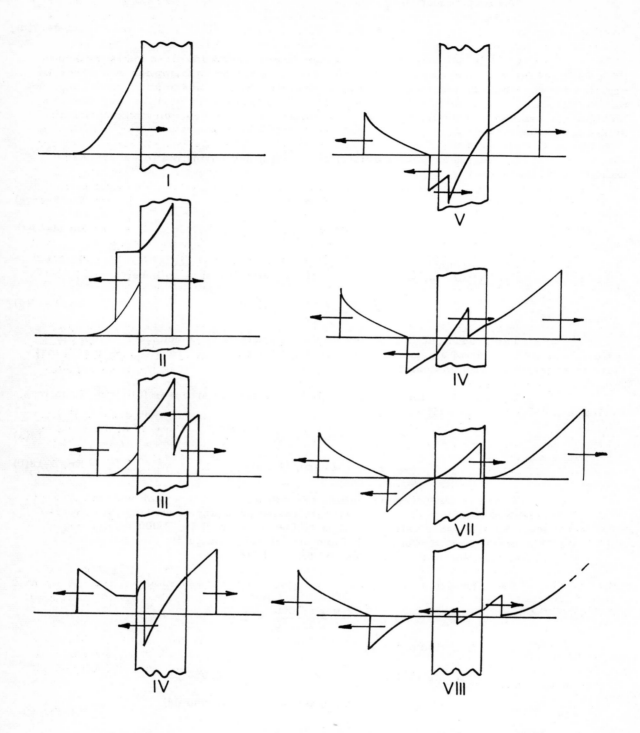

Fig. 8-8(d) Time Sequence of Wave Transmission and Reflection Through a Layer of Different Impedance

Second, when the ratio n is greater than 1, the reflected wave will be positive or, in other words, compressive. (The transmitted wave will always be positive and thus compressive.) On the other hand, when n is less than 1, the reflected wave will be negative or tension. The extreme example of this case is when n equals 0 so that the reflected wave equals the incident wave in magnitude but opposite in sign. This means that, when the wave strikes the end of the rod as a compression wave, it is reflected back from the end as a tension wave of equal magnitude.

An experiment can be set up to demonstrate this phenomenon. A rod of brittle material, such as rock, can be subjected to the impact of a hammer or explosive at one end. A compressive wave, with a stress level below the compressive strength, can be sent down the rod. At the other end, this wave will be reflected as tension, and, if the magnitude of the pulse is greater than about 10 per cent of the compressive strength, will cause a series of tension fractures as indicated in Fig. 8-8(c). Thus, when an explosion occurs close to a free face in rock, scabbing at the free face can occur for the same reason.

Fig. 8-8(d) shows a time sequence for a pulse encountering a medium of higher impedance and then of lower impedance (17). The reflections and transmissions will follow the predictions of Equation 8-18. As can be seen, the two interfaces cause the pulse to be broken up into several individual pulses that are transmitted or reflected. One source of the commonly observed random components of ground motion shown in Fig. 8-16(b), remembering the pervasiveness of joints, can now be understood.

The effect of a material having a non-linear stress-strain curve can now be examined. The simple case of a bi-linear material is shown in Fig. 8-9(a). If a stress, σ_1, is applied to a rod of this bi-linear material, the velocity of the wave will be $(E_1/\rho)^{1/2}$. If while σ_1 is acting an additional stress is applied up to the level σ_2, then the wave velocity resulting from this additional stress will be $(E_2/\rho)^{1/2}$.

Now, if the total stress, σ_2, is applied all at once, the wave front will start together and then separate into two parts as shown in Fig. 8-9(b). The part of the wave front resulting from the stress above σ_1 will be travelling at a lower velocity and consequently will lag behind the wave front due to the increment of stress σ_1. The gap between the two wave fronts will be equal to their relative velocity multiplied by the time, as shown in Fig. 8-9(b).

A similar but more complex situation will arise for a material with a curvilinear stress-strain curve, as shown in Fig. 8-9(c). Here the wave front will be smeared into a shape somewhat as shown in Fig. 8-9(d) with the average slope of the wave front decreasing with the passage of time or distance.

If, on unloading, the material provides a straight stress-strain curve, as shown in Fig. 8-9(c), then the velocity of the rear of the wave will be constant and represented by a vertical line. This also means that the total wavelength will remain constant although the shape of the wave will be continuously changing.

The above equations, derived for the one-dimensional case of the thin bar, are actually fairly similar to the equations derived for a plane wave in a three-dimensional medium and for practical purposes can normally be used. For example, the velocity of a P-wave in a solid medium is:

$$c_p = \left(\frac{E(1-\mu)}{\rho(1-\mu-2\mu^2)} \right)^{1/2} \qquad \text{Eq. 8-19}$$

where E is the modulus of deformation, μ is Poisson's Ratio, and ρ is the mass density. Using this equation would produce an answer about 5 per cent higher than using Equation 8-15 for calculating seismic velocity; this is not normally a significant difference when considering the difficulty of determining the effective E that would apply to a rock mass.

In addition to the P-waves, there are shear waves (S-waves) that can be

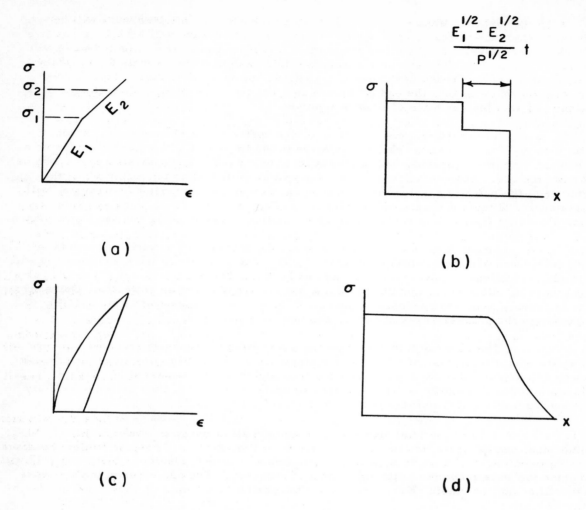

Fig. 8-9 The Effect of the Modulus of Elasticity of the Medium Carrying the Elastic Wave

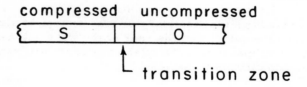

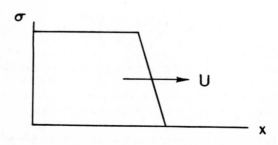

Fig. 8-10 A Shock Wave

transmitted through elastic media. The velocity of these waves in a three-dimensional solid is:

$$C_s = (G/\rho)^{1/2} \qquad \text{Eq. 8-20}$$

where G is the modulus of rigidity. These waves are often called transverse waves as the particle motion is at right angles to the direction of propagation of the wave. The velocity of the S-wave is always less than that of the P-wave and is often (depending on μ) about one-third less.

Besides the two waves described above that can be transmitted through a solid, there are several types of waves that are transmitted along the ground surface similar to waves in water. The most common of these waves is the Rayleigh wave (R-wave) where the particles vibrate in a plane parallel to the direction of the wave propagation and at right angles to the surface with an up and down and longitudinal motion. The R-waves affect the ground immediately under the surface, and the depth of influence can be calculated if the displacement amplitude and frequency of the waves are known (20).

Another type of wave that is of interest in examining the effects of explosives on rock is the shock wave associated with very high pressure. (Two differences are generally accepted between elastic waves and shock waves. All parts of an elastic wave travel at the same velocity, whereas in a shock wave the peak stress portion travels the fastest. The second difference arises from the first as the shock wave then has a very steep leading edge, whereas the elastic wave can have a finite rise time.) Hydrodynamic theory is applied to this case where it is reasonable to assume that the pressures are sufficiently high that the rock acts like a compressible fluid without shear resistance. Deformations thus occur as a result of change of density under these high pressures.

Again, in deriving the relationships for this type of wave the one-dimensional case is examined of a plane shock wave or of a shock wave in a rod as shown in Fig. 8-10. It is assumed - that the wave is travelling at a velocity U - that in the region behind the shock front (S) the particle velocity, v, pressure, σ, and mass density, ρ, are constant – that these properties are also constant in the region ahead of the shock front (O) - and that there is a transition zone at the shock front as shown in Fig. 8-10 (1).

To simplify the discussion, and considering only relative motion, the transition zone is considered to be stationary with the cylinder moving through it from right to left. The velocity of the material is thus the relative velocity (U-v).

First, the Law of Conservation of Mass is used to establish that the mass of material entering the transition zone, m, must be equal to the mass of material leaving it, i.e.:

$$\rho_o(U-v_o) = \rho_s(U-v_s) = m \qquad \text{Eq. 8-21}$$

Second, the Law of Conservation of Momentum, or the impulse-momentum equation can be written, recognizing that the period of time involved for the impulse is one second:

$$\sigma_s - \sigma_o = m(U-v_o) - m(U-v_s) \qquad \text{Eq. 8-22}$$

Finally, the Law of Conservation of Energy can be used or, in other words, the rate of work done by the cylinder is equal to the change in kinetic energy plus the change in internal energy of the material:

$$\sigma_s v_s - \sigma_o v_o = m(v_s^2 - v_o^2)/2 + m\Delta I \qquad \text{Eq. 8-23}$$

where ΔI is the change in internal energy per unit mass.

The Rankine-Hugoniot equations can be derived by solving the above three equations simultaneously. Using Equations 8-21 and 8-22, it is found that:

$$U = \frac{(\rho_s(\sigma_s - \sigma_o))^{1/2}}{(\rho_o(\rho_s - \rho_o))^{1/2}} \qquad \text{Eq. 8-24(a)}$$

where U is the velocity of the shock front, ρ_s is the mass density of material during compression due to the shock, ρ_o is the mass density before being compressed by the shock, σ_s is the stress in the shock, and σ_o the stress in the material before being compressed by the shock. Recognizing that the stress in the rock before being shocked is small with respect to the shock stress, Equation 8-24(a) can be simplified to:

$$U = \frac{(\rho_s \sigma_s)^{1/2}}{(\rho_o(\rho_s - \rho_o))^{1/2}} \qquad \text{Eq. 8-24(b)}$$

Then, again by using Equations 8-21 and 8-22, the particle velocity, v_s, behind the shock front relative to the velocity, v_o, of the undisturbed material can be obtained:

$$v_s - v_o = \frac{((\sigma_s - \sigma_o)(\rho_s - \rho_o))^{1/2}}{(\rho_s - \rho_o)^{1/2}} \qquad \text{Eq. 8-25(a)}$$

Again, recognizing that σ_o can be eliminated, as can v_o, Equation 8-25(a) can be simplified to:

$$v_s = \frac{(\sigma_s(\rho_s - \rho_o))^{1/2}}{(\rho_s \rho_o)^{1/2}} \qquad \text{Eq. 8-25(b)}$$

Then, by using Equations 8-21 and 8-23, we obtain for the change in internal energy per unit mass:

$$\Delta I = \frac{(\sigma_s + \sigma_o)(\rho_s - \rho_o)}{2 \rho_s \rho_o} \qquad \text{Eq. 8-26}$$

The curve σ_s versus ρ_s is commonly called the Hugoniot Curve, the Hugoniot Equation of State or just the Hugoniot. The term 'Equation of State' arises from the similar relation for gas from Boyles Law, PV = RT.

The Rankine-Hugoniot equations, as mentioned above, can be used in studying the effects of high magnitude shock waves. If the equation of state of the rock (i.e., the compressibility at high pressures or the relation between pressure and density) is known, then for a shock of any magnitude the velocity of the wave, the particle velocity, and the change in internal energy can be calculated. It might be noted that, except for Equation 8-26, the wave velocity, U, and the particle velocity, v_s, were derived from the conservation of mass and momentum, in which case they should apply even to explosives where chemical energy is being generated.

Example - Using the aquarium technique for studying shock waves, an explosive is detonated against a plate of rock that is in contact with a bath of water. The specific gravity of the rock is 2.70. The velocity of the shock wave transmitted to the water is measured by a photographic method and is found to be 4000 m/s. The shock velocity through the rock is measured electronically and found to be 6800 m/s. The Hugoniot for water is known in the following form (29):

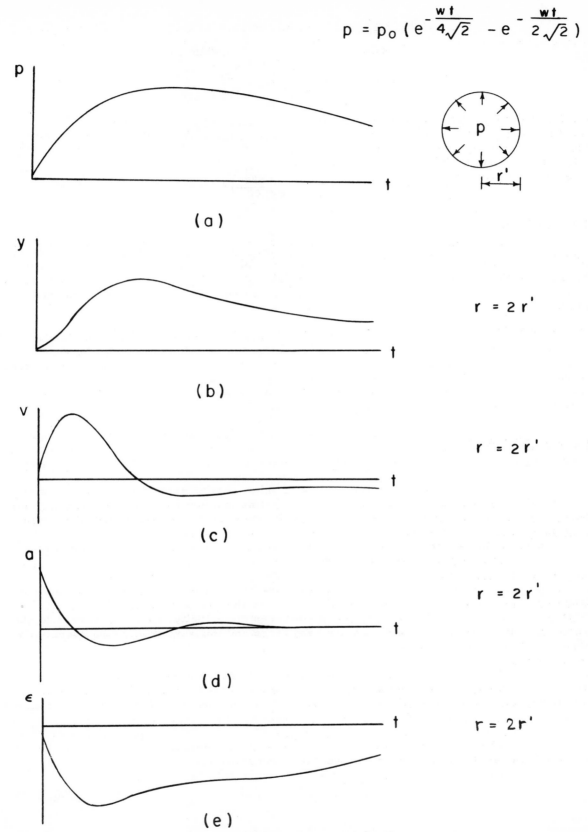

Fig. 8-11 A Pattern of Radiation from a Spherical Cavity of a Stress Wave with the Associated Particle Displacements, Velocities and Accelerations

Shock Velocity	Shock Pressure
3450 m/s	30.0 kb
3820	40.0
4120	50.0
4350	60.0

From the Hugoniot for water, the transmitted stress, σ_t, is by interpolation 46.0 kb. Then, assuming Equation 8-18(a) can be applied, the incident stress can be calculated:

$$\sigma_i = \frac{46.0}{2}\left(1 + \frac{2.70 \times 6800}{1 \times 4000}\right) = 128.5 \text{ kb}$$

EXPLOSIONS

Some theoretical work has been done on the ground motions that would occur from a fully contained explosion in a perfectly elastic medium (2). A pressure pulse of the form shown in Fig. 8-11(a) was applied inside a spherical cavity. Calculations were then made for various distances of the shape of the displacement, velocity, acceleration, and strain pulses.

In Fig. 8-11(b) the shape of the displacement pulse at a distance from the centre equal to twice the radius of the cavity is shown. From this curve it can be seen that the motion is non-oscillatory, having only a compression cycle that decreases slowly to zero. This phenomenon had been previously observed experimentally and was thought to result from the plastic nature of the ground close to the explosion. However, this theoretical solution indicates that it would occur even in a perfectly elastic medium.

In Figs. 8-11(c) and 8-11(d) the velocity and acceleration pulses are shown. These are the first and second derivatives of the displacement pulse. It might be noted that the velocity pulse has somewhat the shape of a damped oscillation vibrating through one cycle.

In Fig. 8-11(e) the strain pulse is shown for the same distance as the displacement pulse. Whereas from our previous derivation it might be expected that the strain pulse should be homologous to the velocity pulse, it should be remembered that the previous derivation was for a stress wave in a bar. For the case of waves radiating out from a spherical cavity, the fact that the wave front will have curvature, rather than being a plane front, accounts for the difference between the curves for velocity and strain.

In Fig. 8-12 the wave motions for the same pressure pulse are given for a distance from the cavity equal to infinity. Here we see that the particle displacement oscillates. Furthermore, the correspondence between the velocity and strain curves is quite close owing to the radius of curvature being very large and the wave front being essentially plane.

The time scales in Figs. 8-11 and 8-12 are the same even though the scales for the y-axis are in arbitrary units. By comparing the curves for the two distances, it is seen that the positive duration of the displacement pulse is a function of the distance from the explosion.

The following equations for the radial displacement, d, particle velocity, v, acceleration, a, and strain, ϵ, were obtained from the above theory:

$$d = \frac{k\, r'\, p_o}{\rho C_p^2}\left(\frac{r'}{r}\right) f(t, r'/r) \qquad \text{Eq. 8-25}$$

$$v = k'\frac{p_o}{\rho C_p}\left(\frac{r'}{r}\right) f'(t, r'/r) \qquad \text{Eq. 8-26}$$

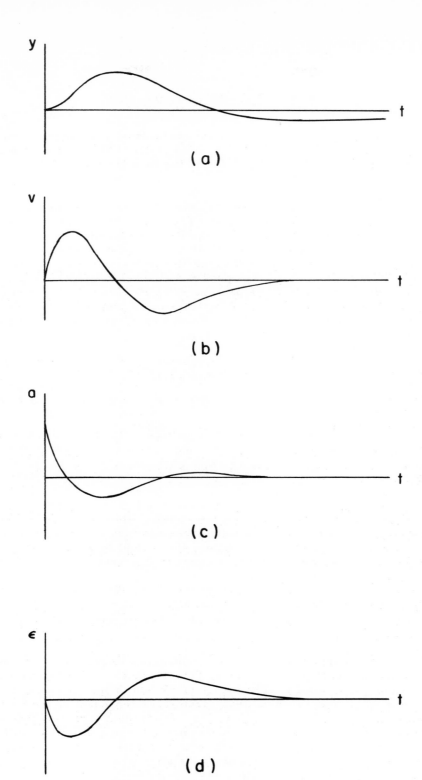

Fig. 8-12 The Wave Shapes from the Spherical Cavity at Infinity

$$a = \frac{p_o}{\rho r'}\left(\frac{r'}{r}\right) f''(t, r'/r) \qquad \text{Eq. 8-27}$$

$$= k'' \frac{p_o}{\rho C_p}\left(\frac{r'}{r}\right) f'''(t, r'/r) \qquad \text{Eq. 8-28}$$

where k, k' and k'' are numerical constants, ρ is the mass density of the medium, C_p is the P-wave velocity, r' is the radius of the cavity, r is the distance from the centre of the cavity, f, f', f'' and f''' are simply functional signs, and t is the time from the initiation of the explosion.

By examining these equations, we learn that the displacement, d, is directly proportional to the maximum pressure in the cavity (p_o is not actually the maximum pressure but is a measure of it), inversely proportional to the P-wave velocity, and inversely proportional to some power of the distance. The velocity, v, is proportional to the maximum pressure in the cavity, inversely proportional to the P-wave velocity, and inversely proportional to some power of the distance. The peak radial acceleration, a, is proportional to the maximum pressure in the cavity, independent of the P-wave velocity, and inversely proportional to some power of the distance. The strain, ϵ, is proportional to the maximum pressure in the cavity, inversely proportional to the square of the P-wave velocity, and inversely proportional to some power of the distance.

In this theoretical work the strain function was examined for distances close to the cavity and was found to be essentially as follows (2):

$$\epsilon = K/r^{1.5} \qquad \text{Eq. 8-29}$$

where K is a constant. As a check on the theory, empirical data shows that the exponent of r varies from 1.6 to 2.5 for various rock types and explosives (2). These higher values are to be expected owing to the absorption of energy by a rock mass, which is not taken into account in the above theory.

Also, whereas the theory indicates that the duration of the displacement pulse should decrease with increasing distances from the cavity, field measurements show that the pulse length decreases with distance for short distances and then at some point changes to increasing proportionately with distance (2). Again, this actual increase in pulse length with distance must result from rock properties not included in the theory.

As it is impossible at the present time, in predicting ground motions from explosions, to take into account theoretically all the significant rock properties, the alternate procedure of deriving empirical equations has been followed for certain practical requirements (3, 4).

In analysing empirical data it is advantageous to use the model laws of similitude. This means expressing all variables, both dependent and independent, as dimensionless parameters (see Appendix F). The theory of models then states that, if two experiments are conducted in which the dimensionless parameters representing the independent variables are equal, even though the magnitudes of the independent variables themselves are different, then the dimensionless parameters representing the dependent variables should also be equal. The procedure of reducing explosion effects to a common basis for comparison or, in other words, the scaling of the explosion effects, is simply to follow these model laws and express the quantities in the form of dimensionless parameters.

It is common in explosion technology to scale many of the effects by dividing by the cubic root of the yield, $W^{1/3}$. For the student newly entering this field, the rationale for this procedure is not always evident. The reason behind this practice, however, can be explained by considering the following simple case. In a cavity of volume V_a a volume of explosive, V_d, is detonated, and it is assumed that the process creates instantaneously the gas pressure p_d in the volume V_d. Then, by assuming that this detonation pressure p_d expands isothermally, the resultant pressure in the cavity will be determined by the gas laws:

$$p = p_d V_d/V_a = p_d W/(V_a m)$$

i.e., $p \propto W$

where m is the weight density of the explosive and W is the weight or yield of the explosion.

From this simple analysis, we can see that the pressure in the cavity will be proportional to the yield of the explosion. Consequently, the effects represented by Equations 8-25 to 8-28 should be proportional to the yield of the explosion, or approximately so, when the effects of the assumptions in the use of the simple gas laws are taken into account.

Now, by following the theory of models one of the first dimensionless parameters, π, representing the independent variable of distance, r, will be:

$$\pi_1' = r/V_d^{1/3}$$

In the above gas law equation the volume V_d was equated to W/m. If the weight density of explosives, m, is assumed to be constant for all cases (which is not strictly accurate but is not a bad assumption) and then if W is expressed in terms of volume, π_1, can be written:

$$\pi_1 = r/W^{1/3}$$

Thus, the common scaling of explosive effects to equivalent distances is achieved simply by dividing the distance by the cube root of the yield of the explosive, the further simplification being taken of expressing W in weight units. Although the type of explosive and the ratio of its impedance to that of the rock is important, e.g., a five-fold difference in effects can be produced by varying the explosive type (16), for a first approximation this is usually ignored.

When using this scaled distance, of course, one could expect equal values only of the dimensionless parameters that include the dependent variables. Some such dimensionless parameters for strain, ϵ, displacement, d, and acceleration, a, would be as follows:

$$\pi_2 = \epsilon$$

$$\pi_3 = d/W^{1/3}$$

$$\pi_4 = aW^{1/3}/C_p^2$$

It should be recognized that in applying the laws of similitude the determination of the dimensionless parameters is somewhat arbitrary. Very often several different dimensionless parameters could be established to include any one particular variable. For example, π_4, besides being written as above, could also be written as:

$$\pi_4' = a\rho/W$$

where W is expressed in units of weight rather than volume.

It is important to recognize this feature as it has been common to assume that, in this particular case, as the dimensionless parameter including particle acceleration normally includes in the denominator C_p^2 that the acceleration will vary with C_p^2. This deduction would not be valid for a perfectly elastic medium, as shown by Equation 8-27.

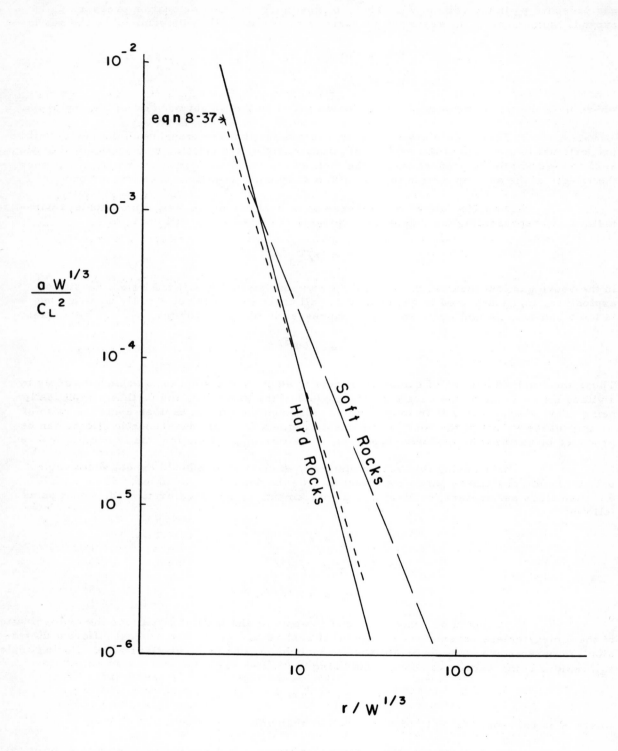

Fig. 8-13 Empirical Relation Found Between Range and Ground Acceleration from Contained Explosions

To make this point clearer, the ground motion resulting from a surface explosion can be expressed as follows (9):

$$\pi_4 = f(rW^{1/3},\ \theta,\ C_p t/W^{1/3},\ z/W^{1/3},\ p/(\rho C_p^2),\ U_d/C_p) \qquad \text{Eq. 8-30}$$

where θ is the angle to the horizontal of the path from the explosion to the point for which the calculations are being made, z is the depth below the ground surface, p is the peak pressure in the cavity, ρ is the mass density of the rock, and U_d is the detonation velocity of the charge. It should be clear from this equation that the parameters affecting such a component as acceleration cannot be determined by examining only the π-parameter that includes acceleration or any other element of motion.

With these explanations, some empirical data can be examined. Several series of TNT explosions were conducted in granite, sandstone, limestone, shale, and chalk using charges of from 10 to 320,000 lbs (3). The free field ground motions were measured at various distances with accelerometers, deformation gauges, and strain gauges.

Some of the results of this work are shown in Fig. 8-13 where the dimensionless parameter including acceleration is plotted against the dimensionless parameter representing the distance from the explosion. The curves for the soft rocks (shale and chalk) were almost identical and are well represented by the curve in Fig. 8-13. The curves for the hard rocks were close together; however, the curve in Fig. 8-13 representing them is actually that for the sandstone. The equation of this hard rock curve is as follows:

$$a W^{1/3}/C_p^2 = 0.72 (W^{1/3}/r)^{3.9} \qquad \text{Eq. 8-31}$$

where W is the yield in lb of TNT, C_p is the P-wave velocity in fs, and r is the distance in feet.

Equation 8-31 can be rewritten, taking into account the actual P-wave velocity of 7200 fs, as follows:

$$a = 3.75 \times 10^7\ W^{0.97}\ r^{-3.9}\ \text{fss} \qquad \text{Eq. 8-32}$$

It was found in this work that the duration of the initial pulse, t_d, was commonly equal to one-half the transit time (3). The deviations from this observation could be great as it was also found for some series of shots that the duration varied with distance raised to a power that varied from 1 and 2.1 (5). However, for the purpose of converting one aspect of ground motion to another, it is common to assume that duration, t_d, equals half the transit time (4):

$$t_d = r/(2C_p) \qquad \text{Eq. 8-33}$$

It is also common to assume that the rise time, t_r, is equal to one-third of the duration (3, 4). Again, the deviations from this rule can be large as it was also found in some cases that the rise time was more commonly equal to half the duration and that it could vary with the distance raised to a power that varied between 1 and 2.1 (5). With the simpler assumption of being equal to $t_d/3$ the following equation can be obtained:

$$t_r = r/(6C_p) \qquad \text{Eq. 8-34}$$

A third assumption then has to be made regarding the shape of the pulse if an empirical equation for one aspect of the ground motion is to be used to predict other aspects, e.g., v or d. A simple assumption, and one that seems to have some validity, is that the velocity pulse is parabolic in shape

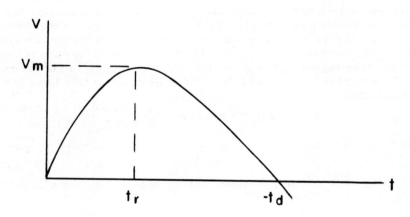

Fig. 8-14 Idealized Shape of Initial Pulse of Particle Velocity from a Contained Explosion

as shown in Fig. 8-14 (4). It can be shown that the tangent to a parabola at a point is twice the slope of the secant from the centre to the point. Consequently, the maximum acceleration for the parabola in Fig. 8-14 would be:

$$a = 2v_m/t_r = 12C_p^2 \epsilon/r \qquad \text{Eq. 8-35}$$

where v_m is the peak radial particle velocity.

This equation can be rewritten to provide an expression for peak radial strain in terms of peak radial acceleration:

$$\epsilon = ra/(12C_p^2) \qquad \text{Eq. 8-36}$$

In this way an empirical acceleration equation can be converted to the corresponding strain equation. If we substitute Equation 8-32 into Equation 8-36, we can obtain an equation which shows the peak radial strain to be proportional to the yield of the explosion and to the distance raised to the power of -2.9.

By examining independent strain records, such as are shown in Fig. 8-15, it is found that the strain varies as the scaled distance raised to the power of -1.6 to -2.8 (theoretically according to Equation 8-29 it should be -1.5) (5). Consequently, for the acceleration and strain equations to be mutually consistent and also representative of the two independent sets of measurements, it is necessary to change the empirical acceleration equation. This is accomplished by changing, as has been done by others (4), the index of the scaled distance in Equation 8-31 from 3.9 to 3.5. If we make the change at the scaled distance, $r/W^{1/3}$, of 10 as shown in Fig. 8-13, we obtain the following equation for peak radial acceleration from an underground TNT explosion:

$$a = 0.29 \, W^{0.83} \, r^{-3.5} \, C_p^2 \, fss \qquad \text{Eq. 8-37}$$

where W is in lb of TNT, r is in ft and C_p is in fs.

From Equation 8-27 it would be expected that acceleration should be independent of C_p. Hence Equation 8-37 would be multiplied by the value of C_p of the rock from which the equation was obtained. However, studies on extrapolating this equation to other rocks and to other effects (e.g., strain, displacement and velocity) show that the best results are obtained if it is assumed that acceleration varies as C_p^2.

This finding might be explained by three mechanisms. First, it is possible that soft rocks experience a lower peak pressure in the cavity of the explosion than hard rocks owing to the greater expansion of the cavity under the gas pressures. Such an expansion could be significant as it would permit greater expansion of the gas and hence lower the applied pressure. This pressure and acceleration could vary directly either with E or C_p.

Second, the fracturing of the rock around the cavity will cause the pulse to be extended (2) and thus to have a lower peak. This action would be greater for soft rocks than for hard rocks.

Third, the actual dissipative properties of rock masses are likely to be more significant in soft rocks. In other words, the pulse might be better preserved in a hard rock. Thus, out from the cavity the stress pulse and acceleration might vary for this reason with E or C_p.

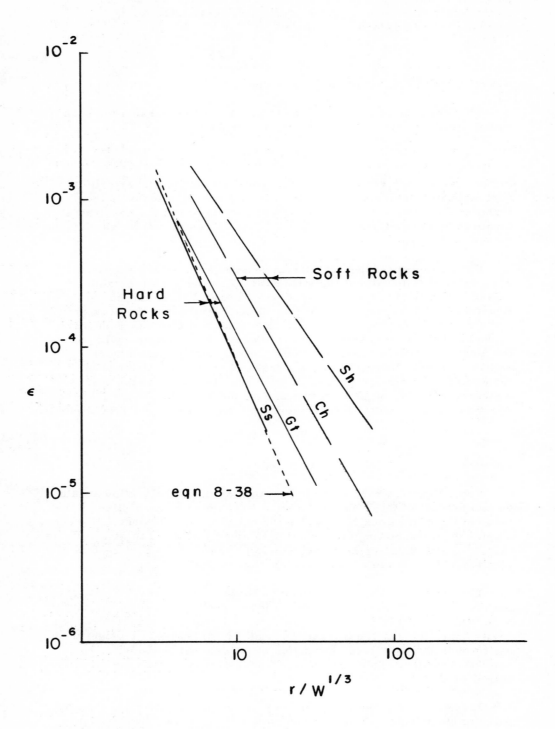

Fig. 8-15 Empirical Relations Between Ground Strain from Contained Explosions and Range

Equation 8-37 is shown in Fig. 8-13 by the dotted line. It can be seen that it represents a minor deviation from the curve fitted to the empirical data and actually provides a good representation of the data when it is considered that the individual observations of acceleration range from one-half to twice the values calculated from Equation 8-31 (3).

If we substitute Equation 8-37 into Equation 8-36, we obtain for the peak radial strain the following:

$$\epsilon = 0.024 \, W^{0.83} \, r^{-2.5} \text{ in./in.} \qquad \text{Eq. 8-38}$$

where W is in lb of TNT and r is in ft. If this equation is compared with the curve representing the field measurements, it can be seen in Fig. 8-15 that it is very close to the curve fitted to the empirical observations. In fact, this is much closer than could be reasonably expected considering the crude assumptions that were made to establish Equation 8-36.

By substituting Equation 8-38 into Equation 8-14, an expression for peak radial velocity can be obtained:

$$v = 0.024 \, W^{0.83} \, r^{-2.5} \, C_p \text{ fs} \qquad \text{Eq. 8-39}$$

where W is in lb of TNT, r is in ft and C_p is in fs. In this case, measurements which showed particle velocity varying with yield to the powers of 0.67 to 0.84 confirm the somewhat arbitrary modification of Equation 8-32 to Equation 8-37 (15).

Then, by again making use of the assumption that the velocity pulse, as shown in Fig. 8-14, is parabolic in shape, we can deduce an expression for the peak radial displacement, which should be equal to the area under the velocity curve ($2 v_m t_d/3$). Then, using Equations 8-33 and 8-39 we obtain for the peak radial displacement:

$$d = 0.008 \, W^{0.83} \, r^{-1.5} \text{ ft} \qquad \text{Eq. 8-40}$$

where W is in lb of TNT and r is in ft.

The theoretical acceleration curves of Figs. 8-11 and 8-12 can be compared with a typical actual curve shown in Fig. 8-16(c). This acceleration record is similar to that at a distance of 152 ft from a TNT explosion of 2560 lbs placed 5 ft below the surface in the same rock for which Equation 8-32 was derived (5).

Fig. 8-16(a) shows the systematic component of acceleration that is usually present in the ground motion. Fig. 8-16(b) represents the random components arising from the various deviations from the assumptions in the theory. Fig. 8-16(c) then is a typical measurement that includes both the systematic and random components of motion. It is common, particularly close to the explosion, for the systematic ground displacement to have less than a half cycle of displacement, which means one cycle of velocity and about 1-1/2 cycles of acceleration.

The curve in Fig. 8-16(c) is different from that predicted by theory and also from that which could be obtained from the assumed velocity pulse shown in Fig. 8-14. These differences simply underline that, whereas such empirical correlations are very useful, they must be considered only as first approximations indicating an order of magnitude.

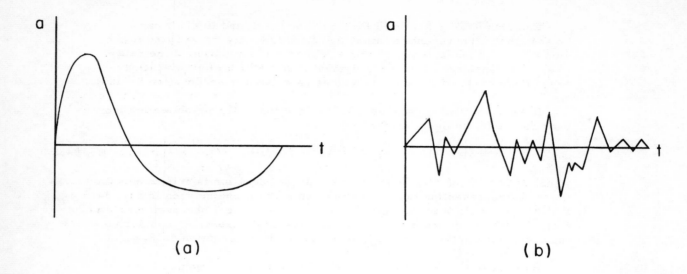

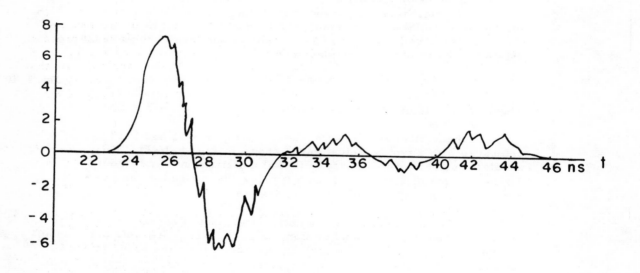

Fig. 8-16 Particle Accelerations from a Contained Explosion
  (a) Systematic Component
  (b) Random Components
  (c) Typical Measurement

The above theory and empirical data apply to fully contained explosions. It is often of interest to be able to predict the ground motions from a surface explosion. Empirical data, such as were obtained in the above mentioned program, can be used to predict these ground motions if the difference in coupling between the explosion and the rock for the surface case, as opposed to the fully contained explosion, is known. This problem can be examined theoretically using the Rankine-Hugoniot equations previously derived.

The effects of a ground surface burst compared to those from a fully contained burst can be assumed in a theoretical analysis to be proportional to the energy transmitted into the rock in the two cases. By multiplying the shock pressure, p, by the velocity of the shock front, U, the time rate of work done, $\dot{W}$, by the explosion can be obtained from Equation 8-25:

$$\dot{W} = pU = \frac{p((\sigma' - \sigma'')(\rho' - \rho''))^{1/2}}{(\rho' \rho'')^{1/2}}$$

$$= p \left(\frac{\sigma' - \sigma''}{\rho'} \frac{1-n}{n} \right)^{1/2}$$

where n is the compression under the shock pressure defined as ρ''/ρ', ρ' is the density of the medium ahead of the shock front and ρ'' is the density of the medium behind the shock front, p is the pressure in the shock pulse, U is the velocity of the shock front, σ' is the stress or pressure in the medium ahead of the shock front, and σ'' is the stress or pressure in the medium behind the shock front.

For a ground surface burst, the conservation of momentum assumed in the Rankine-Hugoniot equations still applies, i.e., the air must provide the reaction for the push that the explosion exerts on the ground. The shock front pressure is the same both in contact with the rock and with the air, and it is assumed that the pressure in front of the shock front is the same in both the rock and air. Then the ratio of the rate of energy input into the air to that put into the ground is as follows:

$$\frac{\dot{W}_a}{\dot{W}_r} = \left\{ \frac{\rho_r}{\rho_a} \left\{ \frac{n-1}{n} \right\}_a \left\{ \frac{n}{n-1} \right\}_r \right\}^{1/2} \qquad \text{Eq. 8-41}$$

where the subscripts a and r refer to air and rock respectively.

Although this ratio will be changing during the period that the explosion is occurring, owing to the change in compression at the different pressure levels, an average compression can be used over the period during which the partition of the explosive energy between air and the ground is occurring. A trial calculation has been made where the time over which the average is determined is that up to the point where the failure of the rock ceases (6). In one case that was calculated for a granite subjected to a nuclear explosion of 1 MT (based on the modulus of compressibility at low static stresses), a compression, as defined above, of 1.22 was used for the granite and 50 for air. The following ratio of energy rates was obtained:

$$\frac{\dot{W}_a}{\dot{W}_r} = \left\{ \frac{167}{0.081} \left\{ \frac{50-1}{50} \right\} \left\{ \frac{1.22}{1.22-1} \right\} \right\}^{1/2} = 106$$

Then, from this expression the actual energy input into the rock as a proportion of the total energy can be calculated as follows:

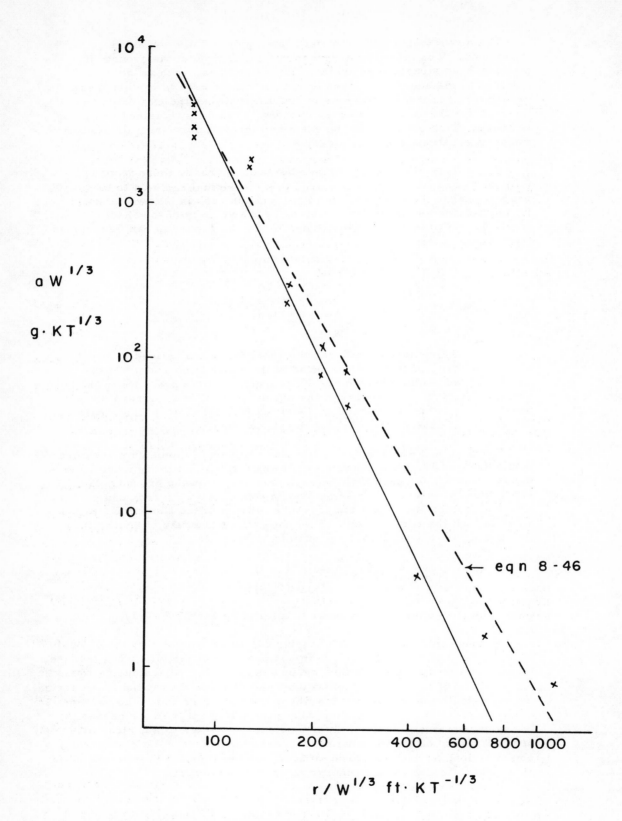

Fig. 8-17 Relation Between Ground Acceleration and Range for Ranier Nuclear Explosion (Ref. 8)

$$\dot{W}_r = \frac{\dot{W}}{\dot{W}_a/\dot{W}_r + 1}$$

hence
$$W_r = \frac{W}{\dot{W}_a/\dot{W}_r + 1} = \frac{W}{106 + 1} = 0.00935\ W.$$

From these calculations it seems that about 1 per cent of the energy of the explosion would be transmitted to the rock with 99 per cent being transmitted to the air. Measurements have shown that actually between 5 and 9 per cent of the explosive energy seems to be transmitted to the rock from a surface burst (7), and 6 per cent is a commonly used average number (4).

The equations of motion in rock resulting from a surface TNT burst can thus be determined using a 6 per cent coupling factor applied to Equations 8-37, 8-38, 8-39 and 8-40:

$$a = 0.028\ W^{0.83}\ r^{-3.5}\ C_p^{2}\ \text{fss} \qquad \text{Eq. 8-42}$$

$$\epsilon = 0.0023\ W^{0.83}\ r^{-2.5}\ \text{in./in.} \qquad \text{Eq. 8-43}$$

$$v = 0.0023\ W^{0.83}\ r^{-1.5}\ C_p\ \text{fs} \qquad \text{Eq. 8-44}$$

$$d = 0.00077\ W^{0.83}\ r^{-1.5}\ \text{ft} \qquad \text{Eq. 8-45}$$

where W is in lbs of TNT, r is in ft and C_p in fs. It has been suggested that these types of equations should only be used for calculating ground motion for points from the point of the explosion below an angle of 20 degrees to the horizon (4).

The effects of surface bursts are of more interest in military problems than in industrial blasts. Consequently, the above equations should be converted for use with nuclear explosives. In this case only a minor fraction of the total energy yield of a nuclear explosion goes into mechanical forms of energy. The estimate of what this proportion is varies from 3 to 50 per cent and depends to a large extent on the type of bomb. An average figure of 16 per cent has been recommended for predicting ground motions (4). Using this figure and converting the units of yield from pounds to kilotons (KT), Equations 8-37, 8-38, 8-39 and 8-40 can be converted to give the peak radial ground motion effects from a fully contained nuclear explosion:

$$a = 1.1 \times 10^4\ W^{0.83}\ r^{-3.5}\ C_p^{2}\ \text{fss} \qquad \text{Eq. 8-46}$$

$$\epsilon = 920\ W^{0.83}\ r^{-2.5}\ \text{in./in.} \qquad \text{Eq. 8-47}$$

$$v = 920\ W^{0.83}\ r^{-2.5}\ C_p\ \text{fs} \qquad \text{Eq. 8-48}$$

$$d = 310\ W^{0.83}\ r^{-1.5}\ \text{ft} \qquad \text{Eq. 8-49}$$

where W is in KT of nuclear explosive, r is in ft and C_p is in fs.

In this form the derived equations can be checked against the measured effects from underground nuclear trials. Fig. 8-17 shows the results of measured accelerations from the Ranier event (8). The solid line is the fitted curve through the experimental points. The dotted line is the derived Equation 8-46. Again, the agreement between the derived equation and the actual empirical data is much better than could be expected considering the various crude assumptions that have been made.

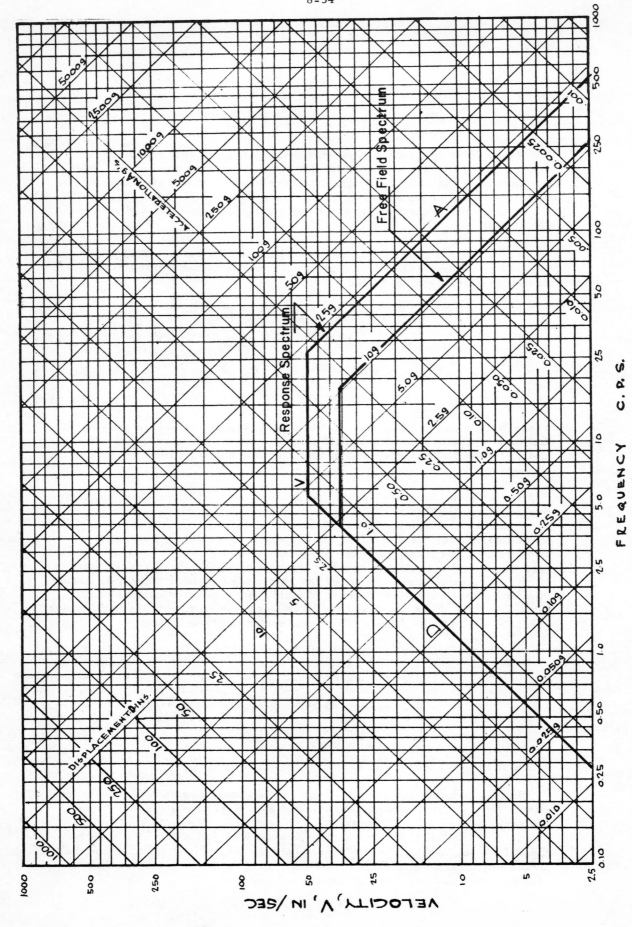

Fig. 8-18 Response Spectrum of Structural Motion Resulting from Ground Shock

However, with some substantiation for the derived equations, they can now be converted to give the peak radial ground motion effects from a ground surface nuclear explosion, again assuming that 6 per cent of the mechanical energy is transmitted to the ground.

$$a = 1060 \, W^{0.83} \, r^{-3.5} \, C_p^2 \text{ fss} \qquad \text{Eq. 8-50}$$

$$\epsilon = 89 \, W^{0.83} \, r^{-2.5} \text{ in./in.} \qquad \text{Eq. 8-51}$$

$$v = 89 \, W^{0.83} \, r^{-2.5} \, C_p \text{ fs} \qquad \text{Eq. 8-52}$$

$$d = 29 \, W^{0.83} \, r^{-1.5} \text{ ft} \qquad \text{Eq. 8-53}$$

where W is in KT of nuclear explosive, r is in ft and C_p is in fs.

The following example will indicate the use of one of the above equations and some of the problems that arise in designing underground defence installations.

<u>Example</u> - Determine the range from a 1 MT ground surface nuclear explosion at which the free field particle velocity will be 3 fs. At this range determine the particle acceleration and displacement. The ground has a C_p of 16,000 fs.

From Equation 8-52:

$$v = 89 \, W^{0.83} \, r^{-2.5} \, C_p$$

$$= 89 \times 1000^{0.83} \times r^{-2.5} \times 16,000$$

$$\therefore r = 1860 \text{ ft}$$

$$a = 1060 \times 1000^{0.83} \times 1860^{-3.5} \times 16,000^2 / 322 = 9.6 g$$

$$d = 29 \times 1000^{0.83} \times 1860^{-1.5} = 0.118 \text{ ft.}$$

As the derivations of the ground motion equations were based on empirical data, these calculated values are in the nature of envelope values representing about a 50 per cent probability of including the actual particle motion. Being based on actual measurements, they include the effects of wave forms other than that assumed in Fig. 8-14.

It has been found that a useful way to represent the ground motion from explosions as well as from earthquakes is to use the concept of a free field spectrum of motion. The free field spectrum calculated above is plotted in Fig. 8-18. This graph is simply a triaxial chart with axes for particle velocity, frequency, displacement, and acceleration. The relative positions of the various coordinates in Fig. 8-18 are governed by the assumption that they are all inter-related by the equations of simple harmonic motion.

The plot of a spectrum of ground motion in this way implies that the frequency components in the systematic and random ground motion (see Fig. 8-16) are bounded by the envelope curve.

This chart can then be compared with Fig. 8-4 where the response of an elastic system, including the magnification factor, was plotted against the frequency ratio. In Fig. 8-4 the magnification factor was applied to displacement; however, if displacement is magnified, then velocity and acceleration will be equally magnified.

A response spectrum can be plotted on Fig. 8-18 that represents the motion of a structure relative to the ground resulting from the ground shock. The relative motion is used as it is the relative displacement that produces the stresses in the structure, which is generally the main concern in these analyses. The response values following Equations 8-2 and 8-5 are related as follows:

$$V = 2\pi f D$$

$$A = 2\pi f V$$

where D is the relative displacement between a point on the structure and the ground, V is a measure of the velocity of the point and A is a measure of the acceleration.

The use of the response spectrum is based on the strong probability that any structure experiencing ground motion represented by the free field spectrum will have its greatest response to the ground motion with a frequency equal to or close to its natural frequency; in other words, resonance will provide magnification of the ground motion acting at that frequency. The assumption is also made that the peak values of motion actually occur at all the frequencies covered by the envelope.

Studies have indicated that ground wave motions are very irregular and are strongly damped so that peak values are seldom repeated. Also, as most structures include at least a moderate degree of damping, the response spectrum or envelope seems to be represented by D equal to the free field displacement, by V equal to 1.5 times the free field velocity, and by A equal to twice the free field acceleration (4). This response spectrum, which represents the maximum response values, is shown in Fig. 8-18.

It is possible that, at the particular frequency of the structure, the actual ground motion does not contain strong motion; consequently, the structure would not respond with magnification. However, as it is impossible to predict the component frequencies of any ground motion, either from explosions or from earthquakes, the spectrum values are used in design. This procedure lends an element of conservatism in the procedure.

By examining Fig. 8-18, it will be seen that structures located at a distance of 1860 ft from the 1 MT surface burst used in the above example, if they have a natural frequency less than 2.5 cycles per second, will experience the maximum displacement of 1.4 in.; however, the maximum accleration will be less than 1 g. Alternatively, for structures with a natural frequency greater than 8 cycles per second the response displacement will be less than 1 in. with the acceleration from about 7 g to 21 g, depending on the actual frequency.

These values are determined by entering the triaxial chart with the natural frequency of the structure and moving vertically until one of the boundaries of the response spectrum is obtained. For example, with a natural frequency of 8 cps the displacement will be 1 in., the velocity will be 54 is and the acceleration will be 7 g. The structure would be designed for the stresses and deformations resulting from these motions.

The range of 1860 ft, as calculated in the example, from a 1 MT burst is actually about the closest point for which an installation can be designed without requiring excessive protective measures. Installations located within this range would require for survivability very expensive provisions such as placing buildings on springs and shock mounting most equipment sitting on the ground.

The boundary line of 3 fs free field velocity also indicates the inter-relation of required depth with the accuracy of the expected attack. If it had to be assumed that the explosion would occur within a few hundred feet of the geographic location of the installation, then the full depth of 1860 ft would be required. If, on the other hand, it could be assumed that the accuracy of aiming would produce a burst on the average at a distance of about 1800 ft, then the installation could be quite near the surface. Studies can be made of this factor using the probable distribution of hits around an aiming point for any one type of bomb delivery together with an economic study of the cost of various probabilities of survival.

CRATERING

For an explosion at the ground surface the Law of Conservation of Momentum indicates that half the momentum of the explosion products should be transmitted upwards into the air and half downwards into the rock (10). Thus, the ratio of the velocity of the rock particles to the velocity of the air particles should be equal to the inverse ratio of their densities. The ratio of the kinetic energies, being proportional to the square of the ratio of velocities, will thus be proportional to the square of the inverse ratio of the densities.

At the explosion, the high pressures and temperatures in the gases cause the rock to disintegrate so that hydrodynamic theory can be used to predict pressure relations. Fig. 8-19 shows the variation of peak overpressure (in excess of atmospheric pressure) in and around the crater for a 1 MT nuclear explosion based on hydrodynamic theory (10). The two pressures shown in Fig. 8-19 are taken from these curves.

The highest pressures in the fire ball of a nuclear explosion immediately after detonation are about 7 megabars (10). The initial decrease in gas pressure is very rapid as expansion occurs and, for distances measured horizontally or diagonally, is less than for distances measured vertically owing to the relief of pressure that can occur into or towards the atmosphere. The main ground shock predicted by this theory is fairly uniform and diverges spherically downward in a vertical cone within an angle of about 90 degrees. During this initial period, the pulverized particles are being thrown upwards and outwards from the crater.

As the peak overpressure decreases, a point is reached where the phenomena cannot be considered as suitable for the application of hydrodynamics. The properties of the rock as a solid become significant. There will be, as indicated in Fig. 8-19, a zone surrounding the crater of fractured but not displaced rock. In some types of ground, particularly soils and less brittle rocks. there may be an additional zone beyond this fracture zone that has been deformed plastically, that is with permanent strain but without being fractured. This is commonly referred to as the plastic zone. These two zones can be expected to influence the subsequent strain pulse significantly, as the initial peak stress in the elastic pulse should be equal to the bearing capacity of the rock at the edge of the rupture or plastic zones.

The crushing strength of the rock will determine the extent of the failed ground; however, as we know from the theory of bearing capacity (see Chapter 1), the peak pressure sustained by a semi-infinite surface is normally much greater than the strength of a sample tested in uniaxial compression.

Thus bearing capacity is not only of importance in governing the crater size and the subsequent peak stress in the pulse, but it also influences the amount of attenuation that will have occurred at any distance beyond the crater. Fig. 8-20 shows that in the hydrodynamic state the peak overpressure varies approximately inversely with the cube of the distance. On the other hand, once the pressure is converted into an elastic pulse, then the reduction, aside from dissipation by non-elastic rock properties, would be only due to spherical expansion. As mentioned above, close to the explosion in an elastic mass the peak radial stress would vary inversely approximately as the distance raised to the power of 1.5 (2). Hence, at equal ranges out from an explosion a strong rock would have a higher stress.

As a result of examining the craters produced by nuclear explosives, it has been found in a variety of ground that the radius of the crater, R_c, for a surface burst varies approximately with the cube root of the yield (8). For hard rock the following equation has been recommended (8):

$$R_c = 50 \, W^{1/3} \qquad \text{Eq. 8-54}$$

where R_c is in ft and W is the yield in KT of nuclear explosive.

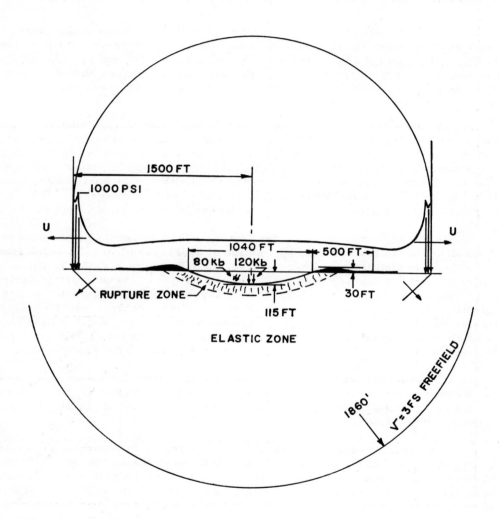

Fig. 8-19 Air Blast Ground Shock and Crater Resulting from a 1 MT Surface Nuclear Explosion

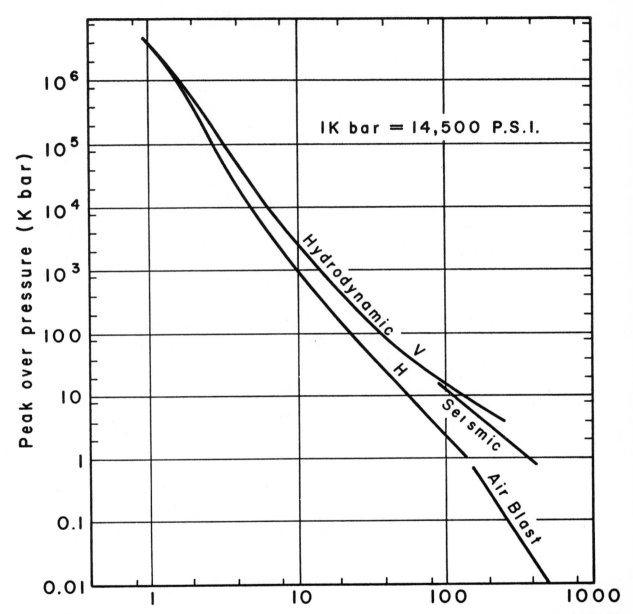

Fig. 8-20 Variation of Peak Overpressure In and Around the Crater (Ref. 10, 6)

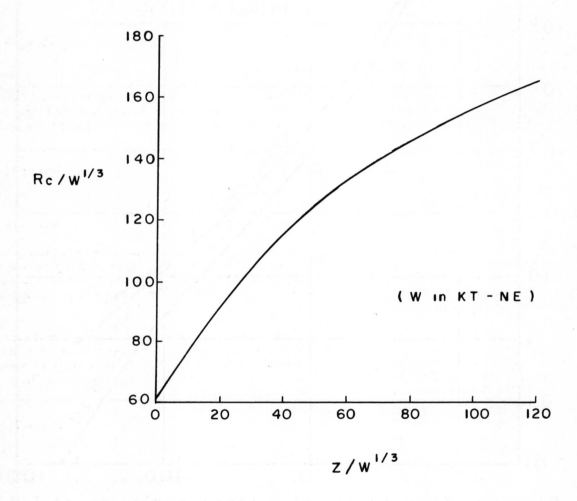

Fig. 8-21 Relation Between Crater Radius and Depth of Embedment of the Explosive (Ref. 8)

The depth of the crater, D_c, has been found to vary with the fourth root of the yield; for hard rocks the following equation has been recommended (8):

$$D_c = 20 \, W^{1/4} \qquad \text{Eq. 8-55}$$

where D_c is in ft and W is in KT of nuclear explosive.

The shape of the crater is assumed to be a paraboloid and hence the volume, V_c, is given by the equation (8):

$$V_c = \pi R_c^2 D_c / 2$$
$$= 78,500 \, W^{0.93} \qquad \text{Eq. 8-56}$$

where V_c is in cf and W is in KT of nuclear explosive.

In addition, it has been found that the average width of the lip is equal to the radius of the crater. For example, in Fig. 8-19 where the crater diameter is 1040 ft the width of the lip would be about 520 ft. Hence, the radius to the outside of the lip, R_L, would be (8):

$$R_L = 2 R_c \qquad \text{Eq. 3-57}$$

The height of the lip can be expected to be about 1/4 of the depth of the crater (8):

$$H_L = D_c / 4 \qquad \text{Eq. 8-58}$$

Fig. 8-21 shows the effect on the crater radius of the embedment of the explosion at some distance, z, below the ground surface (8). It can be seen that the radius of the crater increases with embedment but at a decreasing rate. This curve, if continued, would reach a maximum and then fall off abruptly. The appearances of typical craters as the depth z is increased are shown in Fig. 8-22 (3).

In Fig. 8-22 the distance from the centre of the charge to the edge of the crater is called the radius of rupture, R_r (3). This distance, as well as the crater radius, will increase with increasing embedment of the explosive. An extensive series of tests using TNT produced the correlation shown in Fig. 8-23 between radius of rupture and depth of embedment (3). The equation for the soft rocks in this series was determined as:

$$R_r / W^{1/3} = 3.2 \, (z/W^{1/3})^{0.29} \qquad \text{Eq. 8-59}$$

where W is in lb of TNT and the distances are in feet.

The equation for the hard rocks was established to be as follows:

$$R_r / W^{1/3} = 2.5 (z/W^{1/3})^{0.28} \qquad \text{Eq. 8-60}$$

These equations show that the radius of rupture, as was mentioned above for the depth of craters from nuclear explosions, varies approximately with the fourth root of the yield of the explosion. The curves terminate when the radius of rupture equals the depth of embedment.

Most of the rock from a crater will be deposited in the lip with the approximate dimensions as stated above. In addition, there will be some throw-out beyond the lip zone. Empirical studies have shown that the following relationship applies to one particular site where the surface material was a hard soil (11):

$$t = 2 \, W^{0.29} (R_c/r)^{2.7} \qquad \text{Eq. 8-61}$$

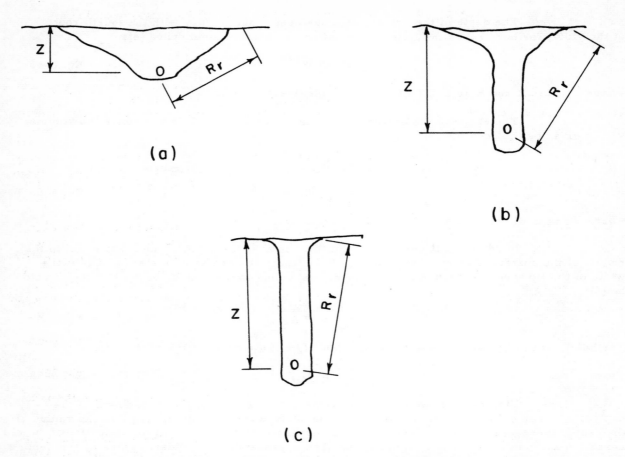

Fig. 8-22 Variation of Crater Patterns with Depth of Embedment

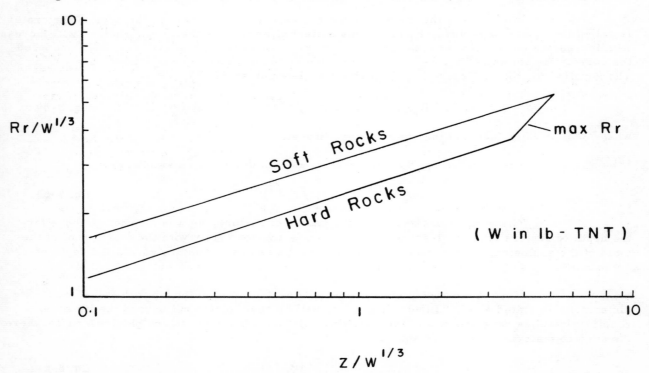

Fig. 8-23 Empirical Relations Between the Radius of Rupture and Depth of Embedment (Ref. 3)

where t is the thickness of the debris in inches, W is the yield of the explosion in lbs of TNT and r is the distance to the debris of thickness 't'.

BLASTING

The mechanics of wave propagation, explosions and cratering are of assistance in designing optimum blasting patterns for rock excavations. In addition, there is much empirical information that provides quantitative relationships between the parameters of interest for specific rock types.

Fig. 8-21 shows the effect of embedment obtained from the results of nuclear explosions (8). Fig. 8-22 also indicates the effect of increasing the embedment of the charge. In Fig. 8-22(b) the depth of embedment is greater than the depth of the resulting crater. In Fig. 8-22(c) the embedment is so great that almost no crater is created (3).

The practical implication of the ground reaction to embedment is shown in Fig. 8-24 (3). The curves in this figure show the variation of the volume of the crater with the depth of embedment. The embedment at which the volume of broken ground, which is determined mainly by the volume of the crater, is a maximum has been termed the optimum depth, z_o (12). The embedment at which the volume of the crater approaches zero has been called the critical depth, z_c (12). These depths have been shown to vary quite closely with the cube root of the yield of the explosion:

$$z_o = K_o W^{1/3} \qquad \text{Eq. 8-62}$$

$$z_c = K_c W^{1/3} \qquad \text{Eq. 8-63}$$

where K_o and K_c are parameters that vary, in different ways from each other, with the rock type.

One way that is used for determining the best combination of variables for a blasting pattern, such as is shown in Fig. 8-25, is to conduct a series of experimental blasts to determine K_o and K_c with charges weighing anywhere between 10 and 200 lbs (27).

A blasting pattern is then designed for the charge weight to be used. The burden and depth of charge are made approximately equal to z_c with the spacing up to about $1.25 \, z_c$ (27). The actual dimensions to be used in the blast pattern will be modified by such variables as the inclination (if any) of the blast hole to the vertical, the amount of sub-grade drilling that is necessary, the diameter of the blast hole, the amount of delay time (if any) between the detonation of adjacent holes and the number of rows of holes parallel to the face that are being blasted together.

<u>Example</u> - A series of crater blasts showed that the optimum depth, z_o, using 27 lb charges, was 6.5 ft. Determine the approximate burden, depth and spacing to use for a bench blast in the same rock using 200 lb of the same explosive in each hole.

From Equation 8-63 the parameter K_o can be determined:

$$K_o = 65/27^{1/3} = 2.2 \text{ ft-lb}^{-1/3}$$

For the bench blast of 200 lb the optimum depth would be:

$$z_o = 2.2 \times 200^{1/3} = 13 \text{ ft}$$

∴ Burden ≃ 13 ft

Depth of Charge ≃ 13 ft

Spacing < 1.25 × 13 < 16 ft

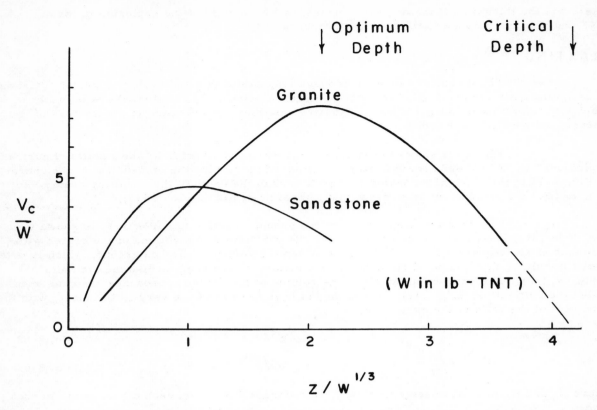

Fig. 8-24 Empirical Relations Between Volume of Crater and Depth of Embedment (Ref. 3)

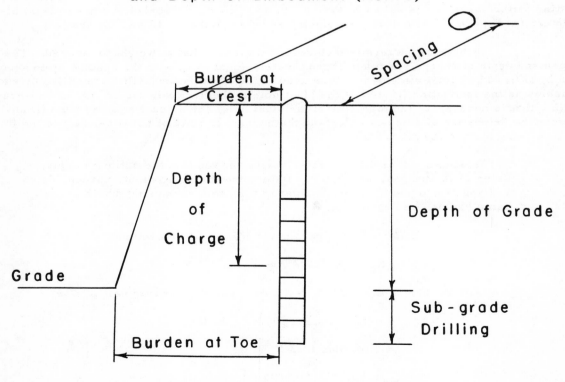

Fig. 8-25 Blast Hole in a Bench

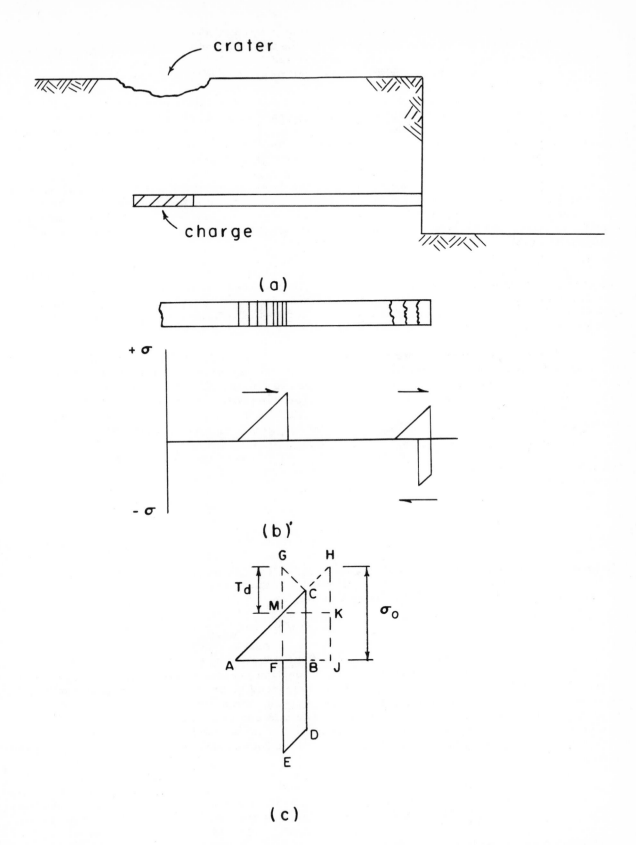

Fig. 8-26 Effects of Reflection of Shock Wave at a Free Surface

This method of designing a blasting pattern is not concerned with the detailed mechanics of the ground reaction; it is based entirely on crater experiments in the rock to be blasted using empirical Equations 8-61 and 8-62. Furthermore, the experimental blasts must be with the explosive that is to be used in production.

It is possible to obtain some idea of the comparative effects of different explosives by using Equation 8-18(a) with the impedance of the explosive being equal to its original density multiplied by the detonation velocity. This equation is not strictly applicable here as it was derived for elastic waves; however, the variation of transmitted shock with the impedance ratio is somewhat as it would be in elastic ground (16).

Another theory of blasting that has been proposed for use in designing blasting patterns is based on the assumption that all rock breakage from explosions is due to the reflection of a compressive wave at a free surface into a tensile wave, which then causes the rock to fracture. As an explanation of at least some of the fracturing, this theory has had confirmation (18).

One of the experiments in this confirmation work included the placing of the explosive, as shown in Fig. 8-26(a), under a horizontal rock surface by means of a horizontal hole from a vertical face at a much greater distance from the charge than the horizontal surface (18). The resultant crater was clearly formed by the reflected wave, and the ground between the charge and the crater remained solid.

Other experiments have demonstrated, as shown in Fig. 8-26(b), the effects of a triangular wave pulse travelling down a rod. At the end of the rod the wave is reflected back as a tensile wave and causes a series of slabs to fracture and fly off the end of the rod. In Fig. 8-26(c) the geometry of the incident and reflected waves at the point of fracture is shown. The tensile strength of the rock is T_d. Thus, if the reflected wave BDEF is superimposed on the incident wave, ACB, then the net tension, GM, will be equal to T_d.

If the reflected wave, GF, is then turned over to reform the geometry of the original wave, AJH, the thickness of the slab, FB, can be deduced. From the geometry of these triangles it follows that:

$$\overline{FB} = \overline{MK}/2$$
$$= 1/2 \, \overline{AJ} \, T_d/\sigma_o \qquad \text{Eq. 8-64}$$

where $\overline{AJ}$ is the length of the wave, T_d is the tensile strength of the rock under the dynamic loading of the wave and σ_o is the peak radial stress in the wave. It also follows that the number of slabs, n, that will be created will be as follows:

$$n = \overline{AJ}/(2\overline{FB}) = \sigma_o/T_d \qquad \text{Eq. 8-65}$$

It is thus envisaged that, after the detonation of a contained explosive, a crushed zone is produced in the rock as a result of the detonation pressure exceeding the compressive strength of the rock (14). As the gas pressure is reduced below the bearing capacity of the surrounding rock, failure by crushing no longer occurs, and a stress pulse is transmitted out from the cavity. When a free face is encountered, this compressive pulse is reflected as a tensile pulse of the same magnitude, which is likely, if the distance has not been too great, to cause fracturing owing to the tensile strength of all rocks being much less than the compressive strength.

As the stress in the compressive pulse will decrease with distance in the same way that strain was shown to decrease in Equation 8-38, a depth of embedment can be obtained at which the reflected tension created at the free face is too small to cause any fragmentation. Consequently, the determination of this critical depth would provide practical information for designing a blasting pattern.

The detonation pressure, it has been found, can be predicted approximately by the following equation (16):

$$p_d = \rho U_d^2 / 4 \qquad \text{Eq. 8-66}$$

where p_d is the detonation pressure in psi, ρ is the original mass density of the explosive in lb-sec/ft^4, and U_d is the detonation velocity in fs. Although detonation and shock phenomena are not elastic waves, it has been suggested (16, 29) that the transmitted stress can then be calculated from the impedance ratio or Equation 8-18(a). Thus, the initial stress, σ', in the rock would be:

$$\sigma' = 2 p_d / (1 + n) \qquad \text{Eq. 8-67}$$

where $n = (\rho U_d)/(\rho_r C_p)$, ρ_r is the density and C_p the P-wave velocity of the rock. The stress away from the cavity would then follow the type of relationship used in Equation 8-38 and predicted by Equation 8-29:

$$\sigma = \sigma' (r'/r)^x \qquad \text{Eq. 8-68}$$

where r' would be the radius of the explosion cavity. Whether it should be the original cavity radius or the final radius raises the question of the applicability of these elastic medium assumptions.

Theoretically, it would then be possible using Equation 8-68 and knowing the appropriate value of x to calculate the depth of embedment at which fracturing due to reflected tension would just extend back to the depth of embedment of the original charge. This depth, following Equation 8-64, should be equal to half the length of the pulse (14).

It is then suggested that, for bench blasts as shown in Fig. 8-25, the burden should be equal to the depth at which the reflected tension causes fracturing back as far as the position of the charge and that the depth of the charge should be twice this distance, which should correspond to about the optimum depth (14).

The time required for breaking by this shock wave action, t_s, is very small and can be determined by the following equation:

$$t_s = 2z/C_p \qquad \text{Eq. 8-69}$$

For example, if $z = 10$ ft and $C_p = 10,000$ fs, then t_s will be equal to 2 ms. This is to be compared with the amount of time that the expanding gases in the cavity are expending their energy, which is of the order of 10 to 100 ms (14). As opposed to the foregoing reflection theory of blasting, the push hypothesis suggests that the majority of the kinetic energy remaining in the gases from the explosion is likely to be the more important source of rock breakage. Only part of the total energy of the explosion goes into the compressive wave (10-20 per cent) with the majority of the remaining energy (about 50 per cent of the total) existing in the gas at moderately high pressures, which can apply force to lift the rock.

In Fig. 8-27(a) the push exerted by these gases is shown tending to cause a wedge of ground to break out to the free surface enclosed within a 90 degree cone. This corresponds to what can be seen to happen in many cases.

Together with the diagonal tensile stresses that will surround the boundary of the cone, there will be tangential stresses around the cavity resulting from the internal gas pressure, which will cause radial cracking. Also, it can be expected that there will be a series of circumferential cracks resulting from the change in acceleration of the particles as the shock wave moves out on a spherical front. Consequently, fragmentation can also occur from these other mechanisms.

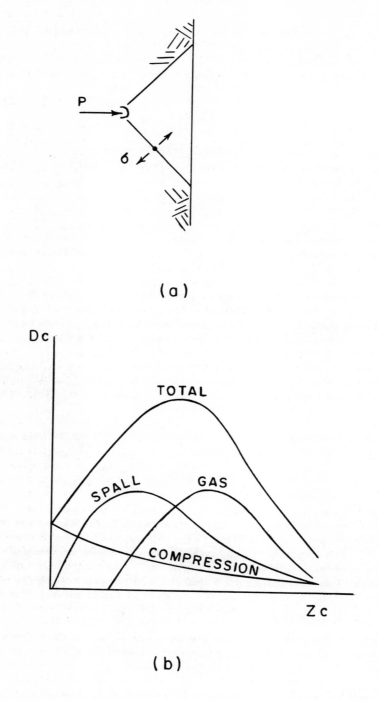

Fig. 8-27 Variation of Crater Diameter with Depth of Embedment

The total ground that is broken will have contributions from the three different mechanisms as shown in Fig. 8-27(b) (24). The crushing or compression failure around the cavity will be the principal mechanism causing broken ground for a surface burst. With embedment, the reflection of the shock wave at the ground surface can cause a considerable amount of ground to fail from the reflected tension. Then, with increased embedment, the distance over which the reflected tension is large enough to cause failure decreases; at the same time, better containment of the gases results in heave of the ground with failure by the induced radial cracking.

EXCAVATING AND BLASTING WITH NUCLEAR EXPLOSIVES

Studies have been conducted over a number of years by the U.S. Atomic Energy Commission to determine the ways in which nuclear explosions could be used for peaceful purposes. Excavating large masses of ground by the formation of craters from large explosions, has been considered one possible use for nuclear explosions. If the quantities of excavation are large enough, excavating by nuclear explosions would be cheaper than by conventional methods. In addition, there may be situations where conventional methods for large excavations would be inconvenient besides being very expensive, e.g., in such remote areas as the Arctic. Also, conventional excavation methods require a comparatively large amount of time, which on some projects may be at a premium.

The main problem in using nuclear explosives is to avoid the release of radioactive particles. In Fig. 8-28(a) a graph is shown of the variation of the amount of radioactivity released from a nuclear explosion with the depth of embedment of the charge. It can be seen that, at a scaled depth of a little more than 200 ft/$KT^{1/3}$, all the radioactivity is contained in the ground below the crater. At the same time, it can be seen that the volume of the crater, or of the excavation, varies with the depth of embedment in such a way that, when the radioactivity is fully contained, the volume of excavation is still considerable. The depth of embedment, of course, would have to be in the range where no radioactivity was released. In addition, it would be important that no air blast wave be generated by the explosion, which would be achieved by the same embedment requirements to suppress the release of radioactivity.

A 1 MT explosion would cost about $1,000,000 and would remove about 100,000,000 cubic yards of ground. The resultant cost of one cent per cubic yard provides the incentive for considering this technique. However, besides the consideration of containing the radioactivity and the air blast, the spewing out of the excavated material into an extensive lip (see Fig. 8-19) might not be acceptable in some circumstances. The analysis of any proposal could be made using the equations included in the section on Cratering.

Another possible use of nuclear explosions might be in underground mining (26). The cavity and chimney of broken ground resulting from an underground explosion would replace the conventional drilling and blasting required for stoping. With the cost of nuclear explosives varying from about $500,000 for 5 KT to $1,000,000 for an explosion anywhere between 1 MT and 20 MT, it can be seen that optimum economy is likely to occur where large blasts can be used.

It has been found that the radius of the cavity resulting from a nuclear explosion, R_s, can be related to the weight of overburden containing the explosion:

$$R_s = C W^{1/3}/(\gamma z)^{1/4} \qquad \text{Eq. 8-70}$$

where C is a constant varying with the type of rock having a value of about 760 for a rock with a P-wave velocity of 18,100 fs to 965 for a soft rock with a P-wave velocity of 7400 fs (26), W is in KT, γ is the average density of the overburden in pcf and z is the depth of embedment in ft.

The average height of the chimney above the cavity, based on the results obtained to date, is:

$$H = 4.9 R_s \qquad \text{Eq. 8-71}$$

where H is the height of the chimney above the shot point.

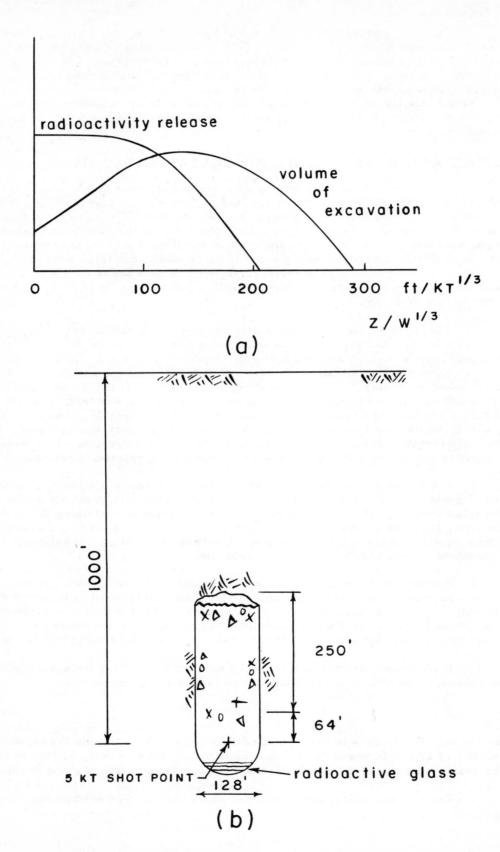

Fig. 8-28 Nuclear Explosions for Industrial Projects

In one case of a shot in hard rock, a granodiorite, the fragmentation in the chimney was such as to produce a range of particle sizes so that 10 per cent were greater than 1 ft and 50 per cent were less than 0.5 ft with an average porosity of 28 per cent (26). If this type of fragmentation can be considered typical, it would, besides eliminating conventional drilling and blasting, produce savings in the cost of crushing.

It has been found in several underground nuclear shots that the bulk of the radioactivity is concentrated in a puddle below or near the shot point as shown in Fig. 8-28(b) (26). The high temperatures and pressures of the explosion melt the immediately adjacent rock to form a glass that contains most of the fission products. Some radioactivity can penetrate along fracture and joint planes upwards and outwards from the explosion. Studies, however, indicate that very little radioactivity exists above an elevation equal to the radius of the cavity above the shot point. Furthermore, the danger of ground water contamination, even when the explosion is below the general ground water table, appears to be very remote (26).

The damaging effects from ground shock on structures or on other mining openings can be appraised by the analyses established in the other sections of this chapter.

Example - Consider as an alternative to blast hole stoping in a hard rock the use of nuclear explosions for breaking the ground. A 5 KT explosion costs $500,000. The level at which this work is to be done is at 1000 ft below the surface with the overlying ground having a density of 167 pcf. The surrounding wall rocks have a seismic velocity, C_p, of 16,000 fs and a failure strain of 0.0005. The porosity of the broken ground will be about 0.28.

From Equation 8-70:

$$R_s = 760 \times 5^{1/3}/(167 \times 1000)^{1/4} = 64 \text{ ft}$$

From Equation 8-71:

$$H = 4.9 \times 64 = 314 \text{ ft}$$

$$\text{Vol.} = \pi \times 64^2 \times (314-64) = 5 \times 10^6 \text{ cf}$$

$$\text{Wt.} = 5 \times 10^6 \times 167 \times 0.72/2000 = 302,000 \text{ T}$$

$$\text{Cost} = 500,000/302,000 = \$1.65/\text{T}$$

From Equation 8-48:

$$R(\text{to } v = 3 \text{ fs}) = \left(\frac{920}{3} \times 5^{0.83} \times 16,000\right)^{1/2.5} = 808 \text{ ft}$$

From Equation 8-47:

$$R(\text{to collapse}) = \left(\frac{920}{0.0005} \times 5^{0.83}\right)^{1/2.5} = 550 \text{ ft}$$

Fig. 8-28(b) shows the approximate geometry of the broken ground that would be obtained from the explosion in this example. By not being able to mine the bottom zone from a point 64 ft above the shot point downwards, the economy of the technique is diminished, and at $1.65 per ton for the equivalent of drilling and blasting this case would clearly not be economic. In addition, the drawing of ore broken in this way without being able to place scram drifts under the block would make it difficult if not impossible to recover all the available broken ore.

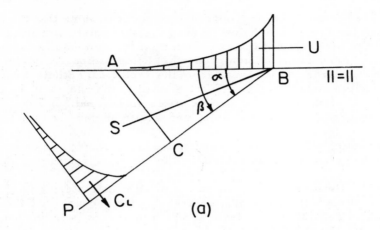

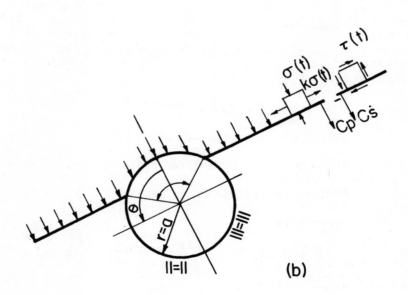

Fig. 8-29 Ground Shock Waves Generated by Air Blast Engulfing an Underground Opening

Besides these serious criticisms of the technique, the distance to which the ground shock would be quite severe with respect to any structures (e.g., electrical sub-stations) being 808 ft might be too great, as would the distance out to which all underground openings could be expected to collapse, i.e., 550 ft from the shock point. Thus, it can be seen that the use of nuclear explosives for mining would only be feasible under very unusual conditions.

UNDERGROUND OPENINGS

The explicit design of underground openings, based on a scientific analysis of loads, stresses and strength of rock, has been an objective of those working in the field of rock mechanics for some time. Under the stimulus of the military requirements for installations capable of resisting shock and blast effects, the above objective for this specialized case is close to being realized.

The mathematical analysis of the stability of underground openings under dynamic loadings is more complicated than the normal problem under static loadings. The stress distribution around the openings being engulfed by shock pulse is constantly changing with time. Also, the reaction of the rock mass, i.e., the deformation and ultimate strength, can be different under short duration loads compared to its behaviour under static loads. On the other hand, when the difficulty (at the present time) is recognized of determining the residual static stresses in which openings are placed, the dynamic problem can provide a more determinable design condition.

At the present time, only some typical cases have been examined using two-dimensional theory, in other words, assuming a plane wave front (19, 20). In Fig. 8-29(a) an air blast wave is shown passing over the surface of the ground. The velocity of the blast, U, is greater than the seismic velocities of the ground. Consequently, the ground shock fronts for both the P-wave and S-wave will be inclined back from the front of the blast. The inclination of these shock fronts can be predicted as follows:

$$\sin \beta = AC/AB = C_p t/(Ut) = C_p/U \qquad \text{Eq. 8-72(a)}$$

$$\sin \alpha = C_s/U \qquad \text{Eq. 8-72(b)}$$

In other words, in Fig. 8-29(a) when the blast front is at A, a ground shock signal will start radiating outwards. When the blast front has travelled from A to B, the P-wave front will have travelled from A to C. Thus, the sides of the triangle ABC will be governed by the air blast velocity, U, and the P-wave velocity, C_p.

In Fig. 8-29(b) the engulfment of an underground opening by a ground shock front is shown. In the P-wave there will be a normal compressive stress that varies with some function of time, $\sigma(t)$, and a tensile stress parallel to the wave front equal to some fraction of the magnitude of the compressive stress, $k\sigma(t)$.

In the case of the S-wave there will be shear stresses acting normal and parallel to the wave front, as shown in Fig. 8-29(b). These shear stresses, of course, can be resolved into equal compressive and tensile stresses at 45 degrees to the shear stresses.

Some work has been done on analysing a few cases of the diffraction problem illustrated in Fig. 8-29(b) (20). In these studies the procedure was followed of considering a succession of materials having gradually more complex properties. In this way relatively simple cases are, with a considerable amount of work, analysed first. Then, as the complicating material properties of plasticity and viscosity are introduced into the more ideal analyses, bases are obtained for judging the effects of actual rock conditions. The cases that were analysed in this work assumed a blast wave with a constant velocity greater than the seismic velocities of the ground. The ground shock pulses were assumed to be much longer than the width of the underground opening.

In Fig. 8-30(a) the tangential stresses at two points on the surface of a cylindrical opening resulting from the engulfment by the P-wave and the S-wave originating from a surface blast wave (as shown in Fig. 8-29(a)) are shown (20). In this case the radius of the opening is 17.5 ft, the rock is assumed to be perfectly elastic with C_p = 17,300 fs and C_s = 10,000 fs. The tunnel is at a depth of 500 ft below the surface. The air blast is that produced by a 20 MT surface nuclear explosion at a distance where the peak overpressure in the blast wave would be 6500 psi. The peak radial stress in the P-wave is 6300 psi, and the peak shear stress in the S-wave is 3450 psi. The air blast velocity for this overpressure is about 22,000 fs. The ratio k in these computations was assumed equal to -1/3. The air blast pressure for the 20 MT burst would have a positive duration of about 5.1 seconds, which would produce a ground shock pulse for the P-wave with a length of about 88,000 ft. This pulse length, of course, is very large with respect to the opening.

In Fig. 8-30(a) the effect at Point ② is shown. The P-wave produces a maximum compressive tangential stress of 11,930 psi at a time when the front of the pulse is about 1 1/2 diameters beyond the tunnel. The stress at this point decreases gradually until the arrival of the S-wave when the stress decreases abruptly and eventually becomes tension with a magnitude of about 2,500 psi.

Point ① in Fig. 8-30(a) is located at a position where one would expect, simply from a knowledge of static stress distribution, that the tangential stresses would be small from the P-wave. However, it happens to be the point where the combination of the compressive stresses due to both the P-wave and the S-wave produce their maximum combined effects. The graph shows that the peak compressive stress here is about 9,500 psi, which is still lower than the maximum stress produced by the P-wave at Point ②.

This analytical work, as well as other experimental work (28), has shown that, for pulses that are long with respect to the opening, these peak dynamic stress concentrations can be predicted with acceptable accuracy by using the peak radial and tangential stresses of the shock pulse in the equations established for static stress fields. Thus, for Point ② using static stress distribution theory, the peak compressive stress would be calculated as:

$$\sigma_t = 6300 (3 + (-0.33)) = 16,800 \text{ psi}$$

This figure is about 40 per cent greater than the 11,930 psi obtained from the dynamic analysis owing to the sharp decrease in the peak stress of the pulse.

The shape of the pressure pulse in the fireball of the explosion is shown in Fig. 8-30(b), where it can be seen that the peak of 6,500 psi quickly drops to about 2,500 psi (10, 20). At lower overpressures this decrease in peak pressure is not so great, and the calculations using static stress concentration factors would be in close agreement with a dynamic analysis (20).

In the case of the S-wave with a shear stress of 3,450 psi, this is equivalent to a field stress with principal stresses at 45 degrees to the wave front of 3,450 psi compression and 3,450 psi tension. Thus, at Point ① using static stress theory:

$$\sigma_t = 3450 (3 + 1) = 13,800 \text{ psi}$$

This figure is also greater than the 9,500 psi obtained from the dynamic analysis.

In addition to the compression and shear waves generated by the surface loading, there will be Rayleigh or R-waves affecting the ground close to the surface. In Fig. 8-31(a) an example is shown of the vertical stress created at depth and its variation with time. The actual magnitude of this vertical stress decreases with depth raised to the power of 1.5 (20). This figure shows that, immediately in front of the surface load, a vertical tensile stress is created that is changed into a vertical compressive stress shortly after the passage of the load over the point in question.

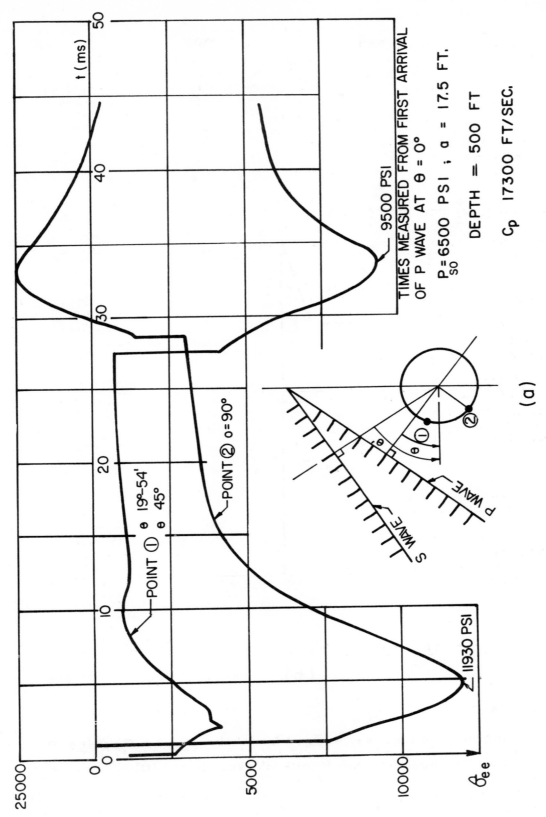

Fig. 8-30 Tangential Stresses Around an Underground Opening Resulting from Various Ground Shock Waves (Ref. 10)

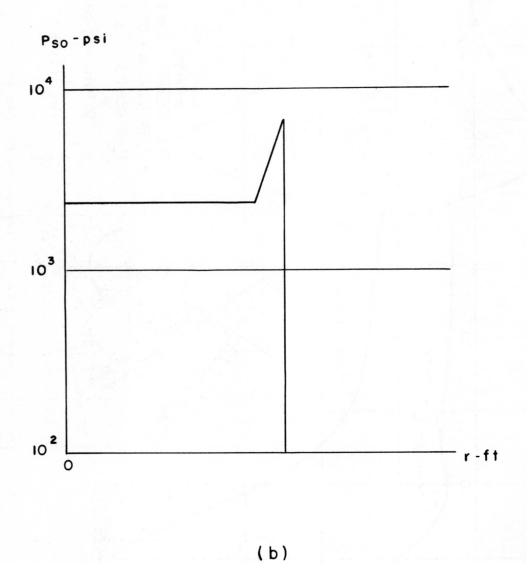

(b)

Fig. 8-30 Tangential Stresses Around an Underground Opening Resulting from Various Ground Shock Waves (Ref. 10)

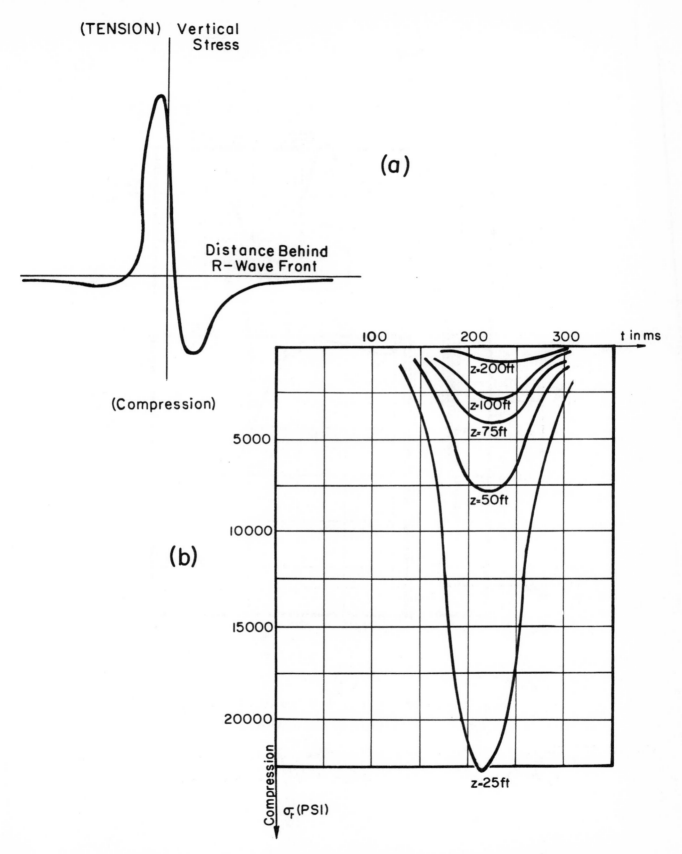

Fig. 8-31   Ground Stresses from Rayleigh Waves (Ref. 20)

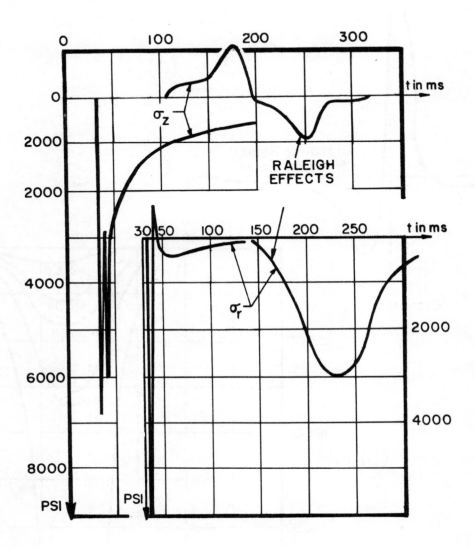

Fig. 8-32 Free Field Ground Stresses from Various Shock Waves (Ref. 20)

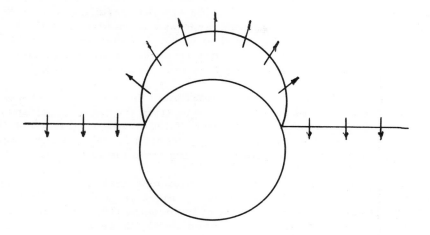

Fig. 8-33 Reflected Tension from an Underground Opening (Ref. 19)

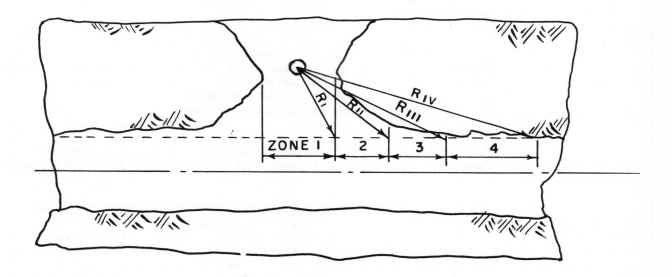

Fig. 8-34 Empirically Determined Tunnel Damage Zones (Ref. 3)

Actually, by theoretical analysis it was found that the radial stress, σ_r, is larger than other components of stress; thus, the importance of the R-wave effects is better judged by examining the magnitudes of this stress.

In Fig. 8-31(b) the variation with time of the stress σ_r at various depths is shown (20). The air blast pressure for this example was that resulting from a 20 MT surface nuclear explosion at a distance where the peak overpressure would be 10,000 psi. The rock is assumed to be perfectly elastic with C_p equal to 17,000 fs and μ equal to 0.25. This graph shows the rapid decrease in ground stress resulting from this wave as the depth below the surface is increased.

To provide a measure of the comparative effects of the three waves, Fig. 8-32 shows the vertical and radial normal stresses, arising from the same air blast as Fig. 8-31, resulting from the three waves for a point at a depth of 100 ft. From this graph it can be seen that the free field compressive stresses arising from the P-wave will be the greatest, but that the maximum free field tensile stresses arising from the R-wave might be of importance.

A subsequent study, still using a two-dimensional case or a plane wave front travelling parallel to the axis of the tunnel, included a more detailed examination of the transient effects during the early stages of the engulfment of the opening (19). The results of this study show that, besides the stresses created as described above, there would be a reflection from the top of the cavity of a tensile wave as indicated in Fig. 8-33. This reflected wave should be propagated radially with no shear component (19). It seems that the magnitude of this tension is only of significant value within a distance of about half the radius of the tunnel. The result of such a reflected wave could be to produce serious scabbing.

Besides the possibility of this type of scabbing failure, there is also the possibility that failure could occur as a result of the stress concentrations described above. In other words, if the maximum compression stress created by the ground shock at the surface of the opening was greater than the strength of the rock, taking into account the short duration of the loading, then compression failure could follow and continue for the period of time that the stress level remained above the failure level.

However, in some rocks the resistance against a short duration stress is considerably greater than that against static stress of the same magnitude. Furthermore, in actual rock, as discussed in Chapter 1, the theoretical stress concentrations may seldom occur. Consequently, the comparison of the theoretical stresses around an opening with the strength measured on a laboratory sample may be a conservative procedure. The knowledge that the strength of the rock substance, as measured on small samples, is generally much less than the rock mass provides a factor in the other direction.

Where tangential surface tensile stresses are created radial tension cracks could be induced. Some experimental data exists to indicate that all of these types of failures can actually occur.

To provide some empirical information for the design of underground defence installations, an extensive series of tests was conducted in the same program as mentioned above in Cratering (3). As a result of detonating charges adjacent to tunnels, four zones of damage were classified. Zone 1 (see Fig. 8-34) is defined as the length of tunnel where complete collapse or break-through to the crater occurs. Zone 2 is characterized by continuous rock breakage increasing in amount towards the explosion. The failed rock originates from most of the perimeter of the tunnel in this zone. Zone 3 is characterized by continuous rock breakage of a relatively uniform thickness arising from the surface of the tunnel nearest the explosion. Zone 4 is characterized by discontinuous rock failure, probably arising from the shaking down of previously loosened material.

Studies of these empirical results established the following outer limits for these zones as a function of the radius of rupture R_r (3):

$$R_1 = 0.5 R_r \qquad \text{Eq. 8-73}$$

$$R_2 = R_r \qquad \text{Eq. 8-74}$$

$$R_3 = 1.3 R_r \qquad \text{Eq. 8-75}$$

$$R_4 = 2.1 R_r \qquad \text{Eq. 8-76}$$

Example - A projected blast in an open pit is to break 250,000 tons of hard ore. The total amount of explosive used would be 120,000 lb. The blast hole pattern is designed to have a burden of 21 ft, a depth to centre of charge of 23 ft, a spacing of 28 ft, and 729 lbs of explosive in each 8 in. diameter hole. 200 ft away and parallel to the nearest row of holes (in a total of seven rows) is a 16 ft x 16 ft inclined transportation tunnel. Estimate the damage that might occur to the tunnel.

From Equation 8-60, the radius of rupture for the charge in one hole is calculated:

$$R_r/W^{1/3} = 2.5(23/729^{1/3})^{0.28} = 3.25$$

$$R_r = 3.25 \times 729^{1/3} = 29.2 \text{ ft}$$

From Equation 8-76 the maximum distance to any tunnel damage is obtained:

$$R_4 = 2.1 \times 29.2 = 61 \text{ ft}$$

If the shock waves from adjacent holes can combine within a cone of 45 degrees from each hole, the maximum that could combine would be within 200 ft along one row, i.e.:

$$N = 200/28 = 7.15 \text{ say 7 holes.}$$

Using this number of holes as the maximum possible that could combine in a horizontal direction to produce an additive effect rather than a cancelling effect, we can calculate:

$$R_4 = 2.1 \times 2.5(23/729^{1/3})^{0.28} \times (7 \times 729)^{1/3} = 117 \text{ ft}$$

Therefore, it seems unlikely that the tunnel would be damaged.

From the description of the classification of the zones, it would appear that the failure of the tunnel in Zone 1 results from compressive stresses in the walls exceeding the strength of the rock and lasting for sufficient time for failure to propagate far enough into the walls to cause the roof to collapse.

In Zone 2 it would seem that compressive failure also occurs, but that the failure zone is not sufficient to cause complete collapse of the tunnel. This zone is probably also affected by the reflected tensile stress causing scabbing.

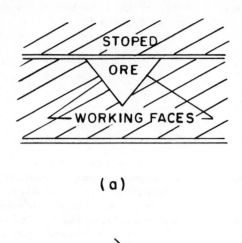

(a)

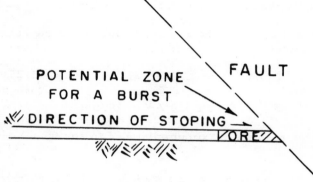

(b)

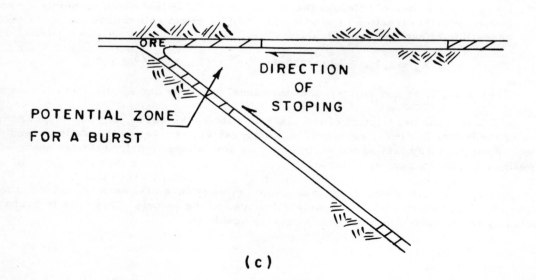

(c)

Fig. 8-35 Mining Geometries Favouring Rock Bursts

Zone 3 would seem to be the area where the compressive stresses are not high enough to cause compressive failure, but the reflected radial tension plus possibly the tangential tension is sufficient to cause failure of the roof or the surface towards the explosion.

In Zone 4 the stresses are probably inadequate to cause any failure of normally competent ground. Of course, any zones of weakness, such as dykes of altered rock or fault breccias, could at distances equivalent to Zone 4 produce Zone 3 effects.

ROCKBURSTS

Rockburst is the term applied to the cases of rock failure that combine suddenness and violence on a sufficiently large scale to endanger openings, equipment or personnel. Consequently, for a rockburst to occur it is necessary, first, for the stress in the rock to exceed the strength. Second, it is necessary for the rock to transform into strain energy the work done on it by the stress rather than to dissipate this work by some form of plastic strain. Then failure, when it occurs, is a brittle fracture accompanied by the violence associated with the sudden release of all the strain energy.

Stress in excess of the strength of the rock can arise from three different sources. First, the concentration of stress around openings and in pillars, possibly aggravated by faults (see Fig. 3-4), might be sufficient to cause rockbursts in ground around stopes. Fig. 8-35 shows three cases that have been identified as causing rockbursts in some mines.

Fig. 8-35(a) shows the vertical elevation where stoping has proceeded upwards from one level to the next with two oblique faces converging so that a pillar of ore of decreasing dimensions is created. The stress concentration in this ore pillar will become very large as the pillar becomes small if the physical properties of the ore are such as to provide a brittle, elastic reaction to this stress. A rockburst might result from the failure of this pillar.

Fig. 8-35(b) shows the not uncommon situation of a fault intersecting the vein of ore. In this case, if the direction of the stoping is towards the fault, the pillar of ore adjacent to the fault will be subjected to increasing stress as a result of the difficulty of transmitting stress across the fault, as illustrated in Fig. 3-4. Failure of this pillar of ore would then involve the rupture of a considerable amount of the wall rock.

Where dykes intersect the vein, it has been found that rockbursts occur at greater than normal frequency (23). The stress distribution in these cases may be, if the dyke is of soft, altered rock, as for the fault shown in Fig. 8-35(b). Alternatively, the dyke rock can be harder and stiffer than the country rock, in which case it would tend to attract stress rather than shed stress into the adjacent ground. Such dykes can carry a much higher stress than the surrounding rock, and they have been found to produce an explosive failure when their confinement is eliminated by mining (23).

It might be noted that both Figs. 8-35(a) and 8-35(b) could be either horizontal plans or vertical elevations, depending on the dip of the ore body, but qualitatively the same stress regimes would apply.

In Fig. 8-35(c) another situation is shown that has led to bursts in various mines (21). The simultaneous stoping of branching veins in a direction towards their intersection tends to produce an overlapping of the stress concentrations around each stope. The pillar of wall rock between the veins and near their intersection seems to be the zone that is most likely to burst under these conditions.

Aside from stress concentrations, the second source of high stress can be simply the force of gravity on the overlying rock in mines that are taken to great depths. The vertical gravitational stresses alone in some Canadian mines could be as high as 10,000 psi, while in mines in South Africa and India they could go up to 13,000 psi. In these cases,

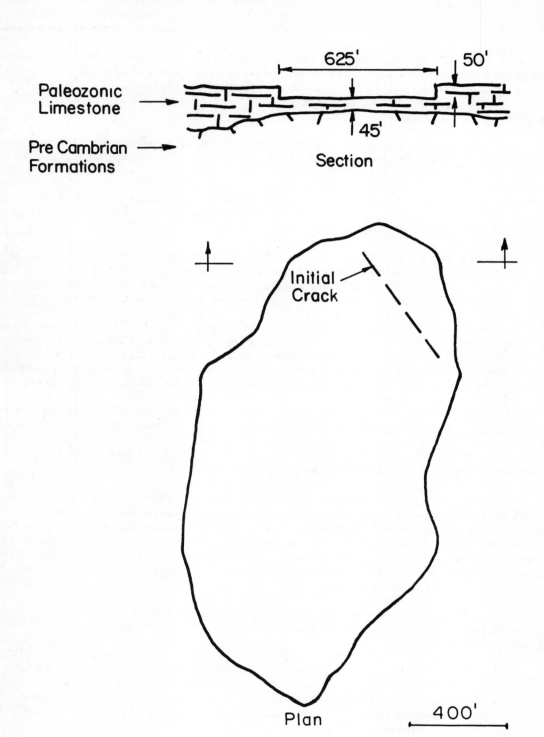

Fig. 8-36 Plan and Section of an Open Pit at the Time of Floor Upheaval (Ref. 22)

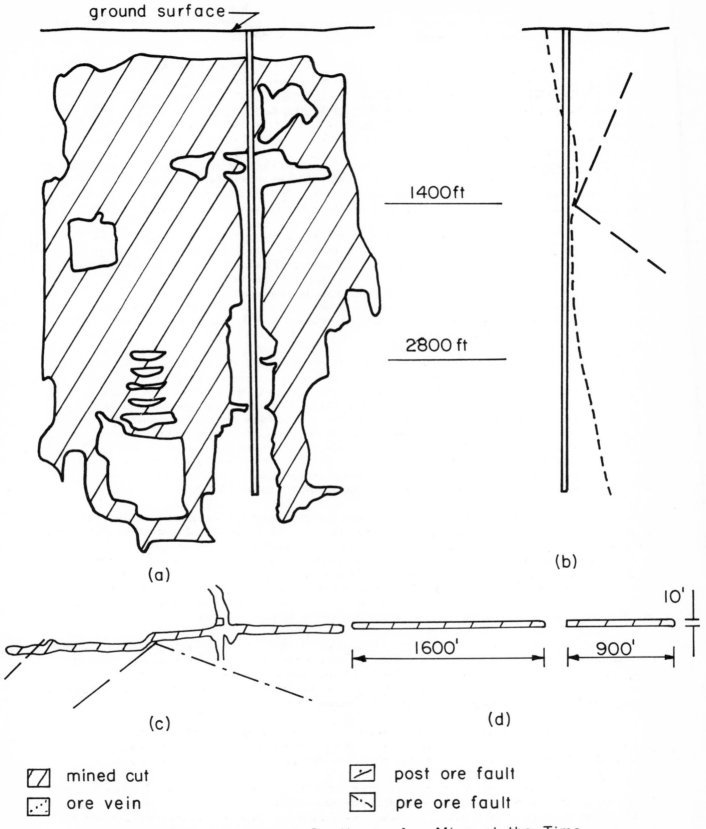

Fig. 8-37 Plans and Sections of a Mine at the Time of a Large Rockburst (Ref. 21)

with the stress concentration factor around an underground opening in elastic ground being at least two, very high surface stresses could be created in ground that did not provide any plastic reaction. In these cases, as has been experienced, rockbursts can occur in drifts as well as around extensive stopes.

The third source of high stresses is probably the legacy of tectonic or crustal adjustments that are simply classified as residual stresses. There is evidence that horizontal residual stresses in some areas can exceed the vertical gravitational stress. Under these circumstances, as for deep mines as described above, it is possible for drifts, in addition to the more critical stress conditions associated with large stopes, to burst. In fact, rockbursts in areas of high residual stresses have occurred in mines at depths of only 400 ft.

As an extreme situation, failures, caused undoubtedly by high residual stresses, have occurred in open pits. In Fig. 8-36 one case is shown where in stripping for an open pit mine no more than 50 ft of excavation had been removed before the bottom of the excavation heaved up as much as 8 ft and a large crack appeared in the floor of the excavation (22).

In Fig. 8-37 the plans and sections of a mine at the time of a particularly large rockburst are shown (21). To account for the large amount of energy released, it is thought that a residual horizontal stress field must exist in this area; however, this has not been proven. The vein in this mine dips between 75 and 85 degrees. Its width varies from small stringers to a maximum of 70 ft. A pillar had been left in the ore body opposite the shaft with the stoping having proceeded out about 1600 ft in one direction and 900 ft in the other direction.

A rockburst caused this shaft pillar to fail, and the shaft was closed completely from the 1400 ft to the 2800 ft level. Seismic records of the event were recorded as far away as 570 miles. Calculations based on the seismic records received at several stations indicated that the amount of energy that had been transmitted to the ground surrounding the mine was about 5×10^{17} ergs (22). By assuming that TNT has a total energy content of 1.7×10^6 ft-lb/lb and that of this about 5 per cent in an underground explosion would be transmitted in the form of kinetic energy to the ground (with an additional 5 per cent in strain energy), this rockburst would seem to have been equivalent to detonating about 200 tons of TNT.

Rockbursts seem to produce ground shocks with characteristics different from those produced by other causes. For example, in one mine it was found that rockburst signals had frequencies between 200 cps and 400 cps, whereas blasting induced frequencies between 40 cps and 60 cps. Other mine noises such as crushers, hoists, tramming, etc. produced signals with frequencies less than 100 cps. This would suggest that a rockburst in this mine was a rapid sequence of individual failures rather than one complete and simultaneous failure. It is quite possible, of course, that other mines could produce relatively sudden failures that would produce a shorter wave train or pulse in a lower frequency range.

Some rockbursts might be prevented or the damage minimized. Many configurations can be avoided by careful planning. For example, in Fig. 8-35(a) the working faces might be arranged so that they move from the centre of the stope towards the abutments, thus not producing a remnant that is subjected to an increasingly severe stress concentration. A longwall face would also produce this requirement.

Stoping towards a major fault, as shown in Fig. 8-35(b), might, if feasible, be avoided by arranging, after mapping all major faults and dykes, for the stoping to travel from faults, if there were not too many, outwards rather than towards these faults.

With branching veins, as shown in Fig. 8-35(c), the simple rule can be established that only one vein should be stoped at a time and, preferably, stoping should be from the intersection outwards. In general, stoping should follow a definite sequence not leaving any isolated zones of ore (21).

Another method that might be used for controlling rockbursts would be to measure the in situ field stresses before development. If this were done successfully, the areas where high residual stresses existed would be located. Then, either these areas could be avoided, or special precautions could be taken in the driving of development openings and in the mining.

A third method of control is concerned with the rock properties rather than with the stresses. A rock testing program correlated with the rock in which bursts have occurred might indicate that some ground is susceptible to bursts, whereas other formations are not. Planning of development and mining would recognize these differences.

Where it is impossible to avoid bursting rock, it might be possible to change the rock ahead of faces or around openings from a brittle elastic medium to a semi-fractured material. This has been done by drilling ahead of the face and setting off charges at the bottom of these holes. The procedure seems to produce a zone that will not sustain the high stress concentrations that otherwise occur. Consequently, the zone of high stress is shifted farther away from the face and, if failure occurs as a result of these stresses, it might result in a contained bump.

Another technique that was thought might provide some control, or at least a warning, of rockbursts was the monitoring of the microseisms that are produced by many rocks subjected to stress. The number of microseisms per unit time generally varies with the level of stress; consequently, it would be logical to assume that, as the stresses approached the failure level, the number of microseisms would pass some critical level, which would provide a warning of impending failure.

This technique has shown in some cases unusual activity preceeding a burst by periods of time from 15 minutes to a month. However, in mines where extensive monitoring has been carried out over periods of several years it was found that no reliable warning was obtained by this technique.

Furthermore, it was found that in many cases during the period immediately preceeding a severe rockburst the microseismic activity in the surrounding area had fallen to a low level. Measurements of other effects such as closure have also shown a similar pattern. It is conceivable that such bursts occur as a result of the rock not reacting to the stress even to the extent of providing some internal, microscopic cracking. Following this concept, it is possible that microseismic activity might indicate the relief of dangerous stress rather than the building up of critical stresses. Thus, the strain energy in a rock that did not produce microseisms would be a maximum, producing the explosive type of failure characteristic of severe rockbursts.

Finally, the dynamic or time-dependent effects of mining can be important. If the rock mass tends to relax, yield or flow under high stress (and mass rock formations probably include some such mechanism even if it is slight and inconspicuous), then the rate of mining can affect the concentration of strain at critical locations. At a slow rate of advance the maximum strain energy in abutment zones could be less than at high rates of advance as a result of time being provided for the decrease and spreading of stress concentrations through some yielding mechanism as shown in Fig. 3-3. On the other hand, in some locations it has been found that by increasing the rate of advance the frequency of bursts has been conspicuously decreased. Thus, few generalizations can yet be made on this subject.

REFERENCES

1. Kolsky, H., "Stress Waves in Solids", Oxford Univ. Press (1953).

2. Duvall, W.I., "Strain-Wave Shapes in Rock near Explosions", Geophysics, Vol. 18, No. 2 (1953).

3. U.S. Corps of Engineers, "Design of Underground Installations in Rock", Dept. of the Army, EM 1110-345-434 (July 31, 1961).

4. Newmark, N. and Haltiwanger, J., "Air Force Design Manual, Principles and Practices for Design of Hardened Structures", U.S. Air Force Special, Kirtland Air Force Base, TDR 62-138 (1962).

5. Engineering Research Associates, Division of Remington Rand, Inc., "Sandstone", Underground Explosion Test Program, Technical Report No. 5, Vol. II, Contract No. DA-04-167-Eng-2984, U.S. Corps of Engineers (April 15, 1958).

6. Coates, D.F., "Rock Mechanics Applied to the Design of Underground Installations to Resist Ground Shock from Nuclear Blasts", Rock Mechanics Proc. 5th Symp. Rock Mechanics, Univ. of Minnesota, Pergamon Press (1963).

7. Fogelson, D.E. et al., "Propagation of Peak Strain and Strain Energy from Explosion-generated Strain Pulses in Rock", Quart. Colorado School of Mines, Vol. 54, No. 3 (1959).

8. "The Effects of Nuclear Weapons", U.S. Department of Defence and Atomic Energy Commission, U.S. Government Printing Office, Washington, D.C. (1962).

9. Engineering Research Associates, Division of Remington Rand, Inc., "Rock", Underground Explosion Test Program, Final Report Vol. II, Contract No. DA-04-167-Eng-298, for U.S. Corps of Engineers (April 30, 1953).

10. Brode, H.L., "Weapons Effects for Protective Design, Ground Support Systems Weapons Effect", Rand Corp., p. 1951 (1960).

11. Carlson, R.H., "Local Distribution of Material Ejected by Surface Explosions", White Tribe Interim Report, Boeing Aeroplane Company Document D2-6955-2 (1961).

12. Livingstone, C., "Fundamental Concepts of Rock Failure", Quart. Colorado School of Mines, Vol. 51, No. 3 (1956).

13. Donnell, L., "Longitudinal Wave Transmission and Impact", Trans. ASME, Vol. 52 (1930).

14. Hino, K., "Fragmentation of Rock through Blasting", Quart. Colorado School of Mines, Vol. 51, No. 3 (1956).

15. Duvall, W.I. et al., "Vibrations from Instantaneous and Millisecond-delay Quarry Blasts", USBM RI 6151 (1963).

16. Atchison, T.C. and Roth, J., "Comparative Studies of Explosives in Marble", USBM RI 5797 (1961).

17. Clay, R.B. et al., "Shock Waves in Solids and Rock Mechanics", Internat. Symp. on Mining Research, Missouri School of Mines, Pergamon Press (1962).

18. Duvall, W.I. and Atchison, T.C., "Rock Breakage by Explosives", USBM RI 5356 (1957).

19. Logcher, R., "A Method for the Study of Failure Mechanisms in Cylindrical Rock Cavities due to the Diffraction of a Pressure Wave", Technical Report T 62-5 MIT, Dept. of Civil Engineering (1962).

20. Baron, M. L. et al., "Theoretical Studies on Ground Shock Phenomena", Mitre Corp., Bedford, Mass. (1960).

21. Morrison, R. G. K., "Report on the Rockburst Situation in Ontario Mines", Trans. CIMM, Vol. 45, pp. 225-272 (1942).

22. Coates, D. F., "Some Cases in Engineering Work of Residual Stress Effects", GSA Proc. Internat. Conf. State of Stress in the Earths Crust (1964).

23. Grobbelaar, C., "A Statistical Study into the Influence of Dykes, Faults and Raises on the Incidence of Rockbursts", Assoc. Mine Mgrs. South Africa, Vol. 1958/59, pp. 1033-1053 (1960).

24. Nordyke, M. D., "On Cratering - A Brief History, Analysis and Theory of Cratering", USAEC, Univ. of Calif. Radiation Laboratory 6578 (1961).

25. Harrison, M., "Excavation with Nuclear Explosives", USAEC, Univ. of Calif. Radiation Laboratory 5676 (1959).

26. Boardman, C. et al., "Characteristic Effects of Contained Nuclear Explosions for Evaluation of Mining Applications", USAEC, Univ. of Calif. Radiation Laboratory 7350 (1963).

27. Bauer, A., "Application of the Livingstone Theory", Quart. Colorado School of Mines, Vol. 56, No. 1 (1961).

28. Durelli, A. et al., "Stress Distribution in the Body of a Square Hole in a Large Plate during Passage of a Stress Pulse of Long Duration", Proc. Internat. Symp. Photo-elasticity, Chicago (1963).

29. Cook, M. et al., "Measurements of Shock and Detonation Pressures", OTS, AD 258201, Washington (1961).

APPENDIX A

SYMBOLS AND ABBREVIATIONS

(Dimensional symbols in brackets indicate the mechanic's nature of the term, i.e., L stands for any unit of length, M for mass, F for force, T for time and D is used if the term is dimensionless.)

$a(LT^{-2})$	–	acceleration
$A(LT^{-2})$	–	response acceleration obtained from a response spectrum
$b(L \text{ or } D)$	–	width of foundation, opening or pillar
$b(FL^{-2})$	–	bar or one atmosphere
$B(L)$	–	width of foundation, opening or pillar
$c(D)$	–	damping coefficient
$c_c(D)$	–	critical damping coefficient
$c\ c\ (L)$	–	centre to centre
$cc(L^3)$	–	cubic centimeter
$cf(L^3)$	–	cubic foot
$c(FL^{-2})$	–	cohesion
$c'(FL^{-2})$	–	effective cohesion intercept in terms of effective stress
$cfs(L^3T^{-1})$	–	cubic feet per second
$ci(L^3)$	–	cubic inch
$cm(L)$	–	centimeter
cpn	–	compression
cps	–	cycles per second
$cy(L^3)$	–	cubic yard
$C_b(D)$	–	coefficient of $\frac{WL^3}{EI}$ for calculating the deflection of a beam due to pure bending
$C_p(LT^{-1})$	–	velocity of P-wave
$C_s(D)$	–	coefficient of $\frac{WL^3}{EI}$ for calculating the deflection of a beam due to shear stresses
$C_s(LT^{-1})$	–	velocity of S-wave
$C_v(D)$	–	coefficient of variation
$d(D)$	–	grain size
$dia(L)$	–	diameter
$d_t(L)$	–	depth of tension crack in the crest of a slope
$d_w(L)$	–	depth of ground water below the crest of a slope
$D_c(L)$	–	crater depth
$D(L)$	–	depth of foundation or openings beneath ground surface; diameter; response relative displacement
$e(D)$	–	base of natural logarithms (2.7183...); void ratio
Eq.	–	equation
$E(FL^{-2})$	–	modulus of linear deformation (Young's modulus)

$El(L)$	-	elevation
$E_p(FL^{-2})$	-	modulus of deformation of pillar rock
$f(D)$	-	sometimes used for $\tan^2(45 + \varphi/2)$
$f(L^{-1})$	-	frequency
$f_n(T^{-1})$	-	undamped natural frequency
$f_r(T^{-1})$	-	resonant frequency
$fs(LT^{-1})$	-	feet per second
$ft(L)$	-	feet
$F_s(D)$	-	factor of safety
$g(L)$	-	the gap due to shrinkage between the concrete lining and the rock in a pressure tunnel
$g(LT^{-2})$	-	acceleration due to gravity
$gpm(L^3T^{-1})$	-	gallons per minute
$G(FL^{-2})$	-	modulus of shear deformation
$G_s(D)$	-	specific gravity of solid particles
$h(D)$	-	height of pillar
$H_c(L)$	-	critical height or maximum height of a slope for a given inclination
$H(L)$	-	a height or depth; vertical height of slope
$H_L(L)$	-	height of lip of a crater
$i_c(D)$	-	critical angle of slope to horizontal
$i(D)$	-	angle of slope to horizontal; hydraulic gradient
in.(L)	-	inch
$is(LT^{-1})$	-	inches per second
$ID(L)$	-	inside diameter
$I'(D)$	-	influence coefficient for calculating the settlement of a uniformly loaded area on the surface of semi-infinite elastic mass
$I''(D)$	-	influence coefficient for calculating the settlement of a uniformly loaded area on the surface of an elastic layer over an incompressible layer
$I(L^4 \text{ or } ML^2)$	-	moment of inertia
$j(FL^{-3})$	-	seepage force per unit volume or seepage pressure per unit length
$J(ML^3 \text{ or } L^4)$	-	polar moment of inertia
k	-	constant
$k(D)$	-	stress concentration; ratio S_t/S_o
$k(F)$	-	kip, one thousand pounds force
$k(FL^{-1})$	-	spring constant
$kb(FL^{-2})$	-	kilobar or one thousand atmospheres
$kg(F)$	-	kilogram
$k(LT^{-1})$	-	coefficient of permeability
$k_s(FL^{-3})$	-	modulus of subgrade reaction
$ksi(FL^{-2})$	-	kips per square inch
K	-	constant

$K(D)$	-	coefficient of earth pressure
$K(FL^{-2})$	-	modulus of compressibility
lb-ft(FL)	-	pound-foot
lb-in.(FL)	-	pound-inch
ln a	-	natural logarithm of a
log a	-	logarithm of a to base 10
LF	-	linear foot
$L(L)$	-	a length
max	-	maximum
$m(D)$	-	Poisson's Number
$m(FL)$	-	moment due to a unit load
$m(ML^{-1}$ or $FL^{-2}T^2)$	-	mass per unit length
min	-	minimum
$ms(T)$	-	millisecond
$m/s(LT^{-1})$	-	meters per second
$M(M$ or $FL^{-1}T^2)$	-	mass
$MF(D)$	-	magnification factor
$M(FL)$	-	moment
$n(D)$	-	impedance ratio; compression under shock pressure; number of slabs created by reflected tension; porosity; ratio of moduli of deformation of wall rocks to pillar rock.
$N(D)$	-	dimensionless coefficient used in the formula for bearing capacity (N_r, N_c, N_g); also, used for slope stability coefficient or factor
$N(F)$	-	normal force
$OD(L)$	-	outside diameter
$p(FL^{-2})$	-	contact pressure
$p_a(FL^{-2})$	-	active earth pressure; compressed air pressure above atmospheric pressure
pcf(FL^{-3})	-	pounds per cubic foot
$p_e(FL^{-2})$	-	effective pressure on a surface of failure within a slope
$p_i(FL^{-2})$	-	pressure inside a thick-walled cylinder
$p_o(FL^{-2})$	-	pressure outside a thick-walled cylinder
$p_o(FL^{-2})$	-	earth pressure at rest
$p_p(FL^{-2})$	-	passive earth pressure
psf(FL^{-2})	-	pounds per square foot
psi(FL^{-2})	-	pounds per square inch
psig(FL^{-2})	-	pounds per square inch above atmospheric pressure, i.e., gage
$P(F)$	-	a force
$P_d(FL^{-2})$	-	a detonation pressure
P-wave	-	compressional wave
$q(L^3T^{-1})$	-	a rate of discharge
$q(FL^{-2})$	-	bearing pressure; surcharge pressure on crest of slope
$q_a(FL^{-2})$	-	allowable bearing pressure

$q_f(FL^{-2})$	-	ultimate bearing capacity
$Q(F)$	-	a foundation load
$Q_B(FL^{-2})$	-	uniaxial compressive strength of a sample of width B
$Q_o(FL^{-2})$	-	uniaxial compressive strength for a sample of unit width
$Q_u(FL^{-2})$	-	uniaxial compressive strength
$r(L)$	-	radius or radial distance
$r_e(L)$	-	inside radius of the elastic zone around a shaft
$r_i(L)$	-	inside radius of a thick-walled cylinder
$r_o(L)$	-	outside radius of a thick-walled cylinder
$R(F)$	-	a reaction force
$R(L)$	-	radius or radial distance
$R_c(L)$	-	crater radius
$R_L(L)$	-	radius to outside of crater lip
$R_r(L)$	-	rupture radius
$R_s(L)$	-	radius of cavity from a fully contained explosion
R-wave	-	Rayleigh wave
$R_1(L)$	-	radial distance from an explosion within which tunnels completely collapse
$R_2(L)$	-	radial distance from an explosion within which tunnels have continuous rock fracture of increasing amount towards the explosion
$R_3(L)$	-	radial distance from an explosion within which tunnels have continuous rock fracture of equal amounts
$R_4(L)$	-	radial distance from an explosion within which tunnels have discontinuous rock fracture
$s(L)$	-	spacing
$sc(L^2)$	-	square centimeter
$sf(L^2)$	-	square foot
$si(L^2)$	-	square inch
std	-	standard
$S(L^{-3})$	-	section modulus
$S_c(FL^{-2})$	-	$c/\tan\varphi$
$S_h(FL^{-2})$	-	field stress in the horizontal direction
$S_o(FL^{-2})$	-	field stress in the horizontal direction normal to the vein
$S_r(D)$	-	degree of saturation
S-wave	-	shear wave
$S_t(FL^{-2})$	-	field stress in the direction transverse to the stoping area
$S_v(FL^{-2})$	-	field stress in the vertical direction
SD	-	standard deviation
$S_x(FL^{-2})$	-	field stress in the x-direction
$S_y(FL^{-2})$	-	field stress in the y-direction
$S_z(FL^{-2})$	-	field stress in the z-direction
$t(T)$	-	time
$t_c(L)$	-	thickness of concrete or grout
$t_d(T)$	-	duration of shock pulse
$t_r(T)$	-	shock pulse rise time
$t_s(L)$	-	thickness of steel
tsn	-	tension

$t_s(T)$	-	time for shock wave from an explosion to travel from the cavity to the free face of the ground and back to the cavity
$T(F)$	-	tangential force
$T(T)$	-	natural period of oscillation
$T_d(FL^{-2})$	-	dynamic tensile strength
$T_r(D)$	-	transmissibility
$T_s(FL^{-2})$	-	static uniaxial tensile strength
$u(FL^{-2})$	-	hydrostatic pressure; pore pressure
$U(F)$	-	buoyant force
$U(ML^2T^{-2}$ or $FL)$	-	strain energy
$U_o(ML^{-1}T^{-2}$ or $FL^{-2})$	-	strain energy per unit volume
$v(LT^{-1})$	-	linear velocity
$v_m(LT^{-1})$	-	peak radial velocity
$V_c(L^3)$	-	crater volume
$V(LT^{-1})$	-	response velocity obtained from a response spectrum
$V(LT^{-1})$	-	shock front velocity
$V(L^3)$	-	volume
$w(D)$	-	water content
$W(F$ or $MLT^{-2})$	-	load or weight
WF	-	wide-flange steel beam
$x(L)$	-	linear displacement or coordinate in direction of x-axis
$y(L)$	-	linear displacement or coordinate in direction of y-axis
$\bar{y}(L)$	-	distance to the centre of gravity of an area
$z(L)$	-	linear displacement or coordinate in direction of z-axis
$z_c(L)$	-	critical depth of explosive
$z_0(L)$	-	optimum depth of explosive
$z_o(L)$	-	depth from the crest of a slope within which horizontal tension can occur
$\alpha(D)$	-	sometimes used for $(45 + \varphi/2)$
$\beta(D)$	-	angle or inclination to the vertical of the resultant force or stress on a plane
$\gamma(D)$	-	shear strain
$\gamma(FL^{-3})$	-	unit weight (bulk density)
$\gamma_d(FL^{-3})$	-	unit dry weight (dry density)
$\gamma_s(FL^{-3})$	-	unit weight of solid particles
$\gamma_w(FL^{-3})$	-	unit weight of water
$\gamma_t(FL^{-3})$	-	total unit weight (water plus solids)
$\delta(L)$	-	deformation
$\epsilon(D)$	-	linear strain
$\epsilon_r(D)$	-	linear strain in the radial direction
$\epsilon_t(D)$	-	linear strain in the tangential direction
$\epsilon_\theta(D)$	-	linear strain in the tangential direction

$\eta\,(FL^{-2}T)$	-	coefficient of viscosity
$\lambda(L)$	-	wavelength
$\lambda(D)$	-	a dimensionless parameter influencing slope stability
μ	-	micro or one millionth
$\mu(D)$	-	Poisson's Ratio; coefficient of friction
$\mu_q(D)$	-	correction factor to slope stability factor accounting for surcharge pressure
$\mu_t(D)$	-	correction factor to slope stability factor accounting for tension cracks
$\mu_w(D)$	-	correction factor to slope stability factor accounting for water table
$\nu(D)$	-	Poisson's Ratio
$\pi(D)$	-	3.1416...
$\rho(L)$	-	radius of curvature
$\rho\,(ML^{-3} \text{ or } FL^{-1}T^2)$	-	mass density
$\sigma(FL^{-2})$	-	normal stress
$\sigma'(FL^{-2})$	-	normal effective stress
$\sigma_{cr}(FL^{-2})$	-	critical stress with respect to elastic instability
$\sigma_f, \sigma_u (FL^{-2})$	-	ultimate strength
$\sigma_i(FL^{-2})$	-	incident stress
$\sigma_r(FL^{-2})$	-	radial stress; reflected stress
$\sigma_\theta(FL^{-2})$	-	tangential stress
$\sigma_t(FL^{-2})$	-	tangential stress; transmitted stress
$\sigma_o(FL^{-2})$	-	peak radial stress in a shock
$\sigma_{yp}(FL^{-2})$	-	elastic limit or yield point
$\sigma_1(FL^{-2})$	-	major principal stress
$\sigma_2(FL^{-2})$	-	intermediate principal stress
$\sigma_3(FL^{-2})$	-	minor principal stress
$\varphi(D)$	-	angle of shearing resistance or friction
$\varphi'(D)$	-	effective angle of shearing resistance or friction in terms of effective stress
$\omega(T^{-1})$	-	angular frequency
$\tau(FL^{-2})$	-	shear stress
$\tau_f(FL^{-2})$	-	shear strength

APPENDIX B

GLOSSARY

(Dimensional symbols in brackets indicate the mechanic's nature of the term, i.e., L stands for any unit of length, M for mass, F for force, T for time, and D is used if the term is dimensionless.)

Abutment: In dam work, either the sloping sides of the valley at the dam site or the part of the dam adjacent to the sides of the valley (7)*. In a more general sense, the supporting zone of an arch, beam or bridge girder, or the area of concentration of ground stress that has been deflected by an opening or by a zone of more yielding ground.

Adit: A horizontal or near horizontal passage from the ground surface into a mine or underground installation.

Adsorb: To condense and hold a gas or liquid on the surface of a solid.

Adsorbed Water: Water, held by physicochemical forces, having physical properties substantially different from absorbed water or chemically combined water.

Aeolotropic: Varying properties in different directions.

Altered Rock: A rock that has undergone changes in its chemical and mineralogical structure since its original deposition (2).

Amplitude: The maximum value of an oscillating quantity.

Angle of External Friction: δ (degrees): The maximum angle of obliquity between the normal and the resultant stress acting between rock and surface of another material.

Angle of Internal Friction: φ (degrees): The maximum angle of obliquity between the normal and the resultant stress acting on a surface within a soil or rock.

Angle of Obliquity: (degrees): The angle between the direction of the resultant stress or force acting on a given plane and the normal to that plane (1).

Angle of Repose: The angle with a horizontal plane at which loose material will stand on a horizontal base without sliding (1),(2).

Anisotropy: Condition of having different properties in different directions; for example, the state of geologic strata of transmitting sound waves with different velocities in the vertical and in the horizontal directions (4). In mechanics it refers particularly to the situation of a material having different moduli of deformation in different directions. It is often used to mean having definite orientation of mechanical properties and fabric (7).

---

*Numbers indicate references listed at the end of this Appendix.

Arch:	1. Curved roof of underground opening (3). 2. A structure spanning an opening that is curved to reduce bending stresses.
Arching:	The transfer of stress from a yielding part of a rock mass to an adjoining less-yielding or restrained part of the mass.
Anticline:	A fold or arch of rock strata dipping in opposite directions from a ridge or axis (4).
Back:	The ore above any horizontal opening, such as a tunnel, stope or drift (2).
Barrier Pillar:	See Pillar.
Bearing Capacity:	See Ultimate Bearing Capacity.
Bed:	The smallest division of a stratified series of rock layers, marked by a divisional plane from its neighbours above and below.
Bench:	1. One of two or more divisions of a coal seam, separated by slate, etc., or simply separated by the process of cutting the coal, one bench or layer being cut before the adjacent one. 2. A ledge left in tunnel construction (2). 3. The berm or horizontal ground left between each succeeding lift or blast in open pit mining.
Biaxial Stress:	See Stress.
Block Caving:	A method of mining ore from the top down in successive layers of much greater thickness than characteristic of top slicing. Each block is undercut over the greater part of its bottom area. The ore then caves with the cover following.
Block Flow Slide:	See Slide.
Body Force: (F):	A force that arises from a field condition that results in every particle in the body experiencing an element of the total force, e.g., a gravitational force or a magnetic force.
Breasting:	Supporting the face of a working with boards across the face.
Brittleness:	A property of materials that rupture or fracture with little or no plastic flow (4).
Bulk Modulus, Modulus of Compression:	The ratio of the change in average stress to the change in unit volume.
Camouflet:	The underground cavity created by a fully contained explosive.
Caving:	The failure and sloughing in of boreholes, mine workings, or excavations (5). A method of mining in which the ore is undercut or its support is removed. It is then allowed to cave or fall.

Clay Size: That portion of a granular mass finer than 0.002 mm (0.005 mm in some cases) (1).

Closure: The relative inward movement of footwall and hangingwall, or of any two opposite surfaces in an underground opening.

Coefficient: In physics, a number commonly used in computation as a factor, expressing the amount of some change or effect under certain conditions such as to temperature, length, time, volume, etc. Multiplier of a term. It may include letters, but is usually taken as the numerical portion of the term (6).

Coefficient of Earth Pressure: K(D): The principal stress ratio at a point in a granular mass. Active: K_a: The minimum ratio of: (1) the minor principal stress, to (2) the major principal stress. This is applicable where the granular mass has yielded sufficiently to develop the lower limiting value of the minor principal stress. At Rest: K_o: The ratio of: (1) the minor principal stress, to (2) the major principal stress. This is applicable where the granular mass is in its natural state without having been permitted to yield. Passive: K_p: The maximum ratio of: (1) the major principal stress, to (2) the minor principal stress. This is applicable where the granular mass has been compressed sufficiently to develop the upper limiting value of the major principal stress (1).

Coefficient of Internal Friction: The tangent of the angle of internal friction (1).

Coefficient of Permeability: $k(LT^{-1})$: The rate of discharge of water under laminar flow conditions through a unit cross-sectional area of a porous medium under a unit hydraulic gradient and standard temperature conditions (usually 20°C)(1).

Coefficient of Subgrade Reaction: $k_s(FL^{-3})$: Ratio of: (1) load per unit area of surface to (2) corresponding settlement of the surface (1).

Coefficient of Variation: The ratio, expressed as a percentage, of the standard deviation to the arithmetic mean.

Coefficient of Viscosity: $n(FTL^{-2})$: The shearing force per unit area required to maintain a unit difference in velocity between two parallel layers of a fluid a unit distance apart (1).

Cohesion: (FL^{-2}): The portion of the shear strength, S, indicated by the term c, in Coulomb's Equation, $S = c + \sigma \tan\varphi$, where σ is the normal stress on the incipient surface of failure and φ is the angle of internal friction. It has the nature of an intergranular binding force.

Competent: See Incompetent for Obverse.

Compressional Wave, Longitudinal Wave, Dilatational Wave, P-Wave, Pressure Wave, Irrotational Wave: A travelling disturbance in an elastic medium characterized by volume changes (and hence density changes) and by particle motion in line with the direction of travel of the wave.

Consolidated: In geology, having been pressed into a hard rock. In soil mechanics, having simply been brought into equilibrium with the applied forces causing a decrease in volume (7).

Controlled-strain Test:	A test in which the load is so applied that a controlled rate of strain results (1).
Controlled-stress Test:	A test in which the stress to which a specimen is subjected is applied at a controlled rate (1).
Country Rock:	Rock adjacent to or surrounding a mineral deposit or dyke in which no minerals of economic interest occur (5).
Creep: (L):	The term, like 'plastic', has many meanings and is often used ambiguously. 1. Slow deformation that results from long application of a stress. By many it is limited to stresses below the elastic limit. Part of the creep is usually a permanent deformation. 2. An imperceptibly slow, more or less continuous downward and outward movement of slope-forming soil or rock. The movement is essentially viscous under shear stresses sufficient to produce permanent deformation, but too small to produce shear failure (2).
Crest:	1. The top of an excavated slope or the summit land of any eminence. 2. The highest point on an anticline. 3. The line connecting the highest points on the same bed in an infinite number of cross-sections across a fold (4).
Critical Circle (Critical Surface):	The sliding surface for which the factor of safety is a minimum assumed in an analysis of a slope in weak ground where average stresses can be used (1).
Critical Slope:	The maximum angle with the horizontal at which a sloped bank of given height will stand unsupported (1).
Crosscut:	A level driven across a vein or, in general, across the direction of the main workings.
Deviator of Stress:	The difference between a normal stress and the average normal stress at that point.
Deviator Stress:	A corruption of Deviator of Stress used in soil mechanics to mean the difference between the major and minor principal stresses.
Crown:	Another term for the roof of a tunnel (7).
Crown Pillar:	See Pillar.
Cut and Fill Stoping:	The cutting of a series of slices from the bottom of a block to the top with the filling of the excavated zones with waste.
Cycle of a Vibration: (D):	One repetition of oscillatory motion.
Deflection: (L):	The movement of a point on a body.
Deformation: (L):	The change in a linear dimension of a body, or the absolute movement of a point on a body.
Degrees of Freedom: (D):	Equals the number of variables (coordinates) needed to describe the motion of a system.

Density (Mass Density): (ρ) (ML^{-3}):   Mass per unit volume.

Dilatational Wave: See Compressional Wave (4).

Dip: The angle of a slope, vein, rock stratum, or borehole as measured from the horizontal plane downward (5). Where applicable, the dip is measured normal to the strike.

Displacement: The straight line distance between two points or positions.

Drawdown: (L):   Vertical distance the free water elevation is lowered or the reduction of the pressure head due to the removal of free water (1).

Ductile: 1. Capable of being drawn through the opening of a die without breaking and with a reduction of the cross-sectional area. 2. In mineralogy, capable of considerable deformation, especially stretching, without breaking, said of several native metals and occasionally said of some tellurides and sulphides. 3. Pertaining to a substance that readily deforms plastically (4).

Duration of Shock Pulse: t_d(T):   The time required for the pulse (generally the particle acceleration) to rise from some stated fraction of the maximum amplitude and to decay to this value.

Dynamometer: An instrument for measuring force.

Earth Pressure: The pressure or force exerted by a granular mass on any boundary (see Coefficient of Earth Pressure).

	Symbol	Unit
Pressure	p	FL^{-2}
Force	P	F or FL^{-1}

Elasticity: The property or quality of being elastic. An elastic body returns to its original form or condition after a displacing force is removed (4).

Elastic Limit: σ_y(FL^{-2}):   The maximum stress that a specimen can withstand without undergoing permanent deformation either by solid flow or by rupture. Also called Yield Point (4).

Elastic State of Equilibrium: State of stress within a mass when the internal resistance is not fully mobilized and the deformation is fully recoverable.

Ellipse of Stress: See Stress Ellipse.

Ellipsoid of Stress: See Stress Ellipsoid.

Energy: (ML^2T^{-2} or FL):   The ability to do work; when due to motion, it is called kinetic energy; when due to position or state, it is called potential energy.

Entry: An underground passage used for haulage or ventilation, or as a manway (2).

Equipotential Line: Line along which water will rise to the same elevation in piezometric tubes (1).

Extensometer: Instrument used for measuring small deformations, deflection, or displacements (3).

Face: 1. In any adit, tunnel, or stope, the end at which work is progressing or was last done. 2. The face of coal is the principal cleavage-plant at right angles to the stratification. Driving on the face is driving against or at right angles with the face. 3. The surface exposed by excavation. 4. The working face, front, or forehead, is the face at the end of the tunnel heading; or at the end of the full-size excavation (2).

Fault: A fracture or fracture zone along which there has been displacement of the two sides relative to one another parallel to the fracture. The displacement may be a few inches or many miles (2). All faults are the same mechanically, i.e., the ratio of the major to the minor principal stress has been great enough to cause shear failure. However, the resultant geometries have been differentiated by the geologist. Closed: A fault in which the two walls are in contact. Dip: A fault whose strike is approximately at right angles to the strike of the strata. Dip Slip: A fault in which the net slip is practically in the line of the fault dip. Hinge: A faulting about an axis normal to the plane of faulting, which may produce a fault that on one side of the pivotal axis would be called normal and on the other side reverse, yet there may not be any differential movement in the centre of the mass of the two parts of the faulted body. Strike-Slip: A fault with no vertical displacement. Longitudinal: A fault whose strike is parallel with the general structure. Normal: A fault in which the hanging wall has been depressed relatively to the foot wall. Oblique: A fault whose strike is oblique to the strike of the strata. Oblique Slip: A fault in which the net slip is between the direction of dip and the direction of strike. Open: A fault in which the two walls are separated. Parallel Displacement: A fault in which all straight lines on opposite sides of a fault and outside of the dislocated zone, that were parallel before the displacement, are parallel afterward. Pivotal: see Hinge fault. Reverse: A steeply dipping fault in which the hanging wall has been raised relatively to the foot wall. Rotary: A fault in which some straight lines on opposide sides of the fault and outside of the dislocated zone, parallel before the displacement, are no longer parallel, that is, where one side has suffered a rotation relative to the other. Step: A series of closely associated parallel faults. Strike: A fault whose strike is parallel to the strike of the strata. Strike Slip: A fault in which the net slip is practically in the direction of the strike. Tangential: Fault with dominantly horizontal movement; contrasts with radial fault. Tension: A fault produced by tension, sometimes used incorrectly as synonymous with gravity fault or normal fault. Thrust: A shallow dipping fault in which the hanging wall has been raised relatively to the foot wall. Transcurrent: See Strike-Slip fault. Translatory: See Rotary fault. Transverse: A fault whose strike is transverse to the general structure. Vertical: A fault in which the dip is 90 degrees.

Fault Breccia: The breccia which is frequently found in a shear zone, more especially in the case of thrust faults (2).

Field Stress: See Stress.

Flow: 1. That which flows or results from flowing; a mass of matter moving or that has moved in a stream, as a lava-flow (2). 2. Continuous straining at constant stress with stress relations governed by the strength or plastic parameters of the rock; internal deformation is a necessary part of flow.

Flow Line: The path that a particle follows under laminar flow conditions (1).

Force: (F): The action of one body upon another body, which changes or tends to change their relative motions, positions, sizes or shapes. A force has four characteristics: 1) magnitude, 2) direction, 3) sense, and 4) point of application. Therefore, it is a vector quantity.

Forced Vibration: The oscillation of a system whereby the response is governed by the external disturbing force.

Forepoling: A method of securing drifts in progress through loose ground by driving ahead poles, laths, boards, slabs, etc. to prevent the inflow of the ground on the side and top, the face being protected by breastboards (2).

Formation: An assemblage of rock masses grouped together into a unit that is convenient for description and mapping (3).

Foundation: Lower part of a structure that transmits the load to the soil or rock (1). The supporting ground is also often called the foundation; however, the term subgrade is less ambiguous for this zone.

Fracture: 1. The general sense of the term is failure by parting of the material. 2. In geology, the manner of breaking and the appearance of a mineral when broken, which is distinctive for certain minerals, as conchoidal fracture. Also, breaks in rocks due to intense folding or faulting. 3. In petroleum work, the process of breaking oil, gas, or water-bearing strata by injecting a fluid under such pressure as to cause partings in the rock (4). 4. In mechanics, sometimes restricted to tension failures, i.e., a break in the continuity of a body of rock not attended by a movement on one side or the other and not oriented in a regular system (3).

Fractured: Rock cracked or broken into fragments along planes other than joints or bedding (5).

Free-body Diagram: A diagram showing all forces acting on a body or portion of a body that has been isolated by replacing all contacting or attached objects with forces representing their effects on the isolated body.

Frequency: $f(F^{-1})$: The number of cycles in a unit time.

Free Vibration: Vibration that occurs in an elastic system in the absence of forced vibration or external disturbing forces.

Free Water Elevation: See Ground Water Level.

Friction, Coefficient of: $\mu(D)$: The frictional resistance divided by the normal force between the contacting surfaces.

Friction: The resistance to motion tangential to and between two surfaces arising from the nature of the surface and the normal force between the surfaces.

Full Face Method: A method of tunnelling or drifting whereby the whole area of the face is blasted out each round (7).

Function: If two variables are so related that there is a value of one variable for each value of the second variable in a given range, then the first variable is called a function of the second. The first variable is called the dependent variable and the second, the independent variable.

Gallery: In mining, a level or drift (2). In construction, a tunnel or a formed opening in a concrete dam.

Geophone: A device to detect sound waves transmitted through ground.

Gouge: A layer of soft material along the wall of a vein or fault.

Ground: Any rock or soil material.

Ground Shock: Suddenly initiated ground motion. It can be caused by an earthquake, explosion or rockburst.

Ground Water Level: The level below which the pores and fissures of the rock and subsoil, down to unknown depths, are full of water (5). Or elevation at which the pressure in the water is zero with respect to the atmospheric pressure (1) or see Line of Seepage.

Grout: Usually a thin mixture of Portland cement and water with or without sand, but more generally a fluid for filling cracks with or without some cementing action.

Grouting: The process of filling in or finishing with grout (2).

Gunite: A mortar applied with compressed air to the rock surfaces of an underground opening.

Hanging Wall: The mass of rock above a fault plane, vein, lode or bed of ore (4).

Hardness: As used for rocks in drilling and bit-setting, it is the relative ability of a mineral to scratch another mineral (5).

Hardness Scale: See Mohs' Scale (2).

Hard Rock: Rock that requires drilling and blasting for its economical removal (4).

Head: (L): Elevation: Head due to elevation above any datum. Pressure: Head due to pressure. Velocity: Head due to velocity, equal to $v^2/2g$.

Homogeneous: Having the same properties at all points (3).

Hydraulic Gradient: i(D): The loss per unit distance of elevation head plus pressure head. Critical Hydraulic Gradient: Hydraulic gradient at which the intergranular pressure in a mass of cohesionless soil is reduced to zero by the upward flow of water (1).

Hooke's Law: Stress is proportional to strain.

Hysteresis: 1. In physics, a lagging or retardation of the effect when the forces acting upon a body are changed as if from viscosity or internal friction. 2. In a magnetic material, as iron, a lagging in the values of resulting magnetization due to a changing magnetizing force (2). In rocks, the recovery of strain on reduction of stress often lags until the stress has been reduced to zero.

Impact: A collision of two masses.

Impulse: (FT): The product of force and the time during which the force acts.

Incompetent: A term often used without intending a precise meaning. 1. In geology, strata and rock structure not combining sufficient firmness and flexibility to transmit a thrust; consequently, admitting only the deformation of flowage (2). Such layers suffer plastic flow on folding (8). 2. Soft or fragmented rocks in which an opening, such as a borehole or an underground working place, cannot be maintained unless artifically supported by casing, cementing or timbering (5).

Inertia: The fundamental property of matter, which resists a change in the state of rest or motion of a body (6).

In situ: In its natural position or place (3).

Isotropic: Having the same properties in all directions. Said of a medium with respect to elasticity, conduction of heat or electricity, or radiation of heat or light (2).

Joint: 1. In geology, a plane, or gently curved crack or fissure, which is one of an approximately parallel set of fissures ranging from a few inches to many feet apart (2). Joints occur in rocks of nearly all kinds and generally in two or more sets that divide the rocks into polyhedral blocks (3).

Joint System: Consists of two or more joint sets or any group of joints with a characteristic pattern, such as a radiating pattern, a concentric pattern, etc. (4).

Kiloton: KT: One thousand tons.

Kinematics: Physics of motion (4).

Kip: k(F): 1. A force equal to 1000 lb. 2. A level or gently sloping roadway, at the extremity of an engine plane, upon which the full cars stand ready to be sent up the shaft (2).

Layer: Any bed or stratum of rock separated from the adjacent rock by a plane of weakness (3).

Level: A horizontal passage or drift into or in a mine. It is customary to work mines by levels at regular intervals in depth.

Line of Seepage (Seepage Line) (Phreatic Line): The upper free water surface of the zone of seepage (1).

Linear Material: A material that will either react to stress so that strain varies linearly (i.e., the stress-strain curve is a straight line) with stress or the time rate of strain varies linearly with stress.

Lithology: The study of rocks as such; a branch of geology much developed in recent years. By making thin sections and examining them under the microscope the nature of a rock substance may be determined (2).

Longwall: A system of working a seam or vein in which the whole seam is taken out and no pillars left, excepting the shaft pillars, and sometimes the main-road pillars. Longwall advancing is mining outward from the shaft pillar and maintaining roadways through the worked-out portion of the mine. Longwall retreating is first driving haulage road and airways to the boundary of a tract of coal or vein of ore and then mining it in a single face without pillars back toward the shaft (2).

Mass: (M): The quantity of matter containing within a body. In dynamic problems the acceleration caused by a force varies inversely with mass of the body. The units of mass are equivalent to the weight of the body divided by the acceleration due to gravity.

Massive: 1. In petrology, of homogeneous structure, without stratification, flow-banding, foliation, schistocity, and the like; said of the structure of some rocks; often, but incorrectly used as synonymous with igneous and eruptive. 2. Occurring in thick beds, free from minor joints and lamination; said of some stratified rocks (2).

Mean: Generally refers, unless otherwise specified, to the arithmetic mean or average, which is sum of a number of values for a quantity divided by the number.

Mean Deviation: $\overline{MD}$: The sum of the absolute differences between the arithmetic mean, A, and a number of values of a quantity, x, divided by the number n. It is a measure of the dispersion of the number of values about the mean, e.g.,

$$\overline{MD} = \frac{\Sigma(A - x)}{n}.$$

Mechanics:	The branch of physics that deals with the action of forces on bodies. It is subdivided into statics and dynamics (2).
Mechanical Vibration:	Motion of a body, which repeats itself in a definite time interval.
Megaton: MT:	One million tons.
Micro:	1. One millionth. 2. In lithology, indicating that the structure designated is so minutely developed as not to be recognized without the help of the microscope (2).
Microseism:	A slight tremor or vibration of the earth's crust (2).
Mineralogy:	The science that deals with minerals, which together in rock masses or in isolated form make up the material of the crust of the earth (2).
Modulus:	A number or quantity that measures a force or function (4).
Modulus of Compression:	See Bulk Modulus.
Modulus of Deformation:	See Modulus of Elasticity.
Modulus of Elasticity (Modulus of Deformation): E (FL^{-2}):	The ratio of normal stress to normal strain for a material under given loading conditions; numerically equal to the slope of the tangent (hence 'tangent modulus') or the secant (hence 'secant modulus') of a stress-strain curve. The use of the term Modulus of Elasticity is recommended for materials that deform in accordance with Hooke's Law, the term Modulus of Deformation for materials that deform otherwise (1).
Modulus of Rigidity (Shear Modulus): $G(FL^{-2})$:	The ratio of shear stress to shear strain for a material determined either from the slope of the tangent or of the secant of a stress-strain curve (4).
Modulus of Rupture: (FL^{-2}):	The maximum extreme fibre stress at failure in bending of a test beam. As such it is a measure of the tensile strength of the material.
Mohr Circle:	A graphical representation of the stresses acting on the various planes at a given point (1).
Mohr Envelope (Rupture Envelope) (Rupture Line):	The envelope of a series of Mohr Circles representing stress conditions at failure for a given material. According to Mohr's Rupture Hypothesis, a rupture envelope is the locus of points the coordinates of which represent the combinations of normal and shearing stresses that will cause a given material to fail (1).
Mohs' Scale:	Arbitrary quantitative units by means of which the scratch hardness of a mineral is determined. The units of hardness are expressed in numbers ranging from 1 through 10, each of which is represented by a mineral that can be made to scratch any other mineral having a lower-ranking number; hence, the minerals are ranked from the softest, as follows: Talc (1) ranging upward in hardness through gypsum (2), calcite (3), fluorite (4), apatite (5), orthoclase (6), quartz (7), topaz (8), corundum (9), to the hardest, diamond, with the highest ranking number (10) (5).

Moisture Content (Water Content): The ratio, expressed as a percentage, of: (1) the weight in a given granular mass, to (2) the weight of solid particles (1). In assay work, the ratio is the weight of water to the weight of solid particles plus water.
w (D):

Moment: (FL): The product of the force and the perpendicular distance from the axis to the line of action of the force.

Moment of Inertia: $I(L^4$ or $ML^2)$: The second moment of an area about any axis, equal to the summation of the products of each elemental area multiplied by the square of its moment arm (6). Or the second moment of a mass about any axis, equal to the summation of the products of each elemental mass multiplied by the square of its moment arm.

Muck: Broken ground from an underground operation.

Multiple Openings: Any series of underground openings separated by rib pillars or connected at frequent intervals to form a system of rooms and pillars (3).

Natural Frequency: $f_n(T^{-1})$: The frequency of a system undergoing free vibrations.

Normal Distribution, Standard Normal, Normal Curve, Gaussian: The distribution of numerical data, x, about an average value, A, that follows the Gaussian Equation:

$$y = \frac{1}{\overline{SD}\sqrt{2\pi}} e^{-1/2(x-A)^2/\overline{SD}^2}$$

where y is the frequency of occurrence and $\overline{SD}$ is the standard deviation of the data.

Normal Line, or Normal: The line drawn to a curve that is perpendicular to the tangent to the curve at this point.

Octahedral Shear Stress: The stress that can be used for the strength criterion according to the maximum energy of distortion theory. The octahedral plane is at equal angles to the three principal planes, and the normal component of stress is equal to the average of the three principal stresses.

Ore: A mineral of sufficient value as to quality and quantity that may be mined with profit (2).

Orogenic: Pertaining to the processes by which ranges of mountains are formed (3).

Orthogonal: At right angles.

Oscillation: A variation, usually with time, of the magnitude of a quantity with respect to a specified reference when the magnitude is alternately greater and smaller than the reference.

Ovaloid: A plane figure resembling a rectangle with semi-circular ends (3).

Overbreak: Rock excavated in tunnel work beyond the payline or beyond the perimeter required by the contract or by the use of the tunnel.

Overcoring: The act of drilling around a feature with a coring bit; generally used to describe the operation of cutting away the constraint on a small hole in rock by drilling over the small hole with a larger coring bit. By measuring the change in reaction of the walls of the small hole, a measure is obtained of the original constraint.

Overpressure: The pressure in an air blast wave in excess of the atmospheric pressure.

Parameter: A quantity constant in a special case but variable in different cases (4). Alternatively, it can be considered as an arbitrary constant, i.e., it is arbitrary before being specified; e.g., angle of internal friction, modulus of deformation, etc.

Parting: A small joint in coal or rock or a thin layer of different rock in a rock bed (2).

Payline: The perimeter used in tunnel contracting with unit prices beyond which payment is not made for any volume of rock excavated or concrete placed.

Penstock: A closed conduit, tube, pipe or tunnel for conducting water to a power house (2).

Perched Water Table: A water table, usually of limited area, maintained above the normal free water elevation by the presence of an intervening relatively impervious confining stratum (1).

Percolation: The movement of gravitational water through soil.

Period of a Vibration: $T(T)$: The time for the vibration to repeat itself.

Permeability Coefficient: See Coefficient of Permeability.

Petrofabrics: The study of spatial relations, especially on a microscale, of the units that comprise a rock, including a study of the movements that produced these elements. The "units" may be rock fragments, mineral grains, or atomic lattices (4).

Petrology: A general term for the study by all available methods of the natural history of rocks, including their origins, present conditions, alterations and decay.

Photoelasticity: Pertaining to a method of determining stress distributions by optical means (3).

Piezometer: An instrument for measuring pressure head (1).

Pillar: In situ rock between two or more underground openings. The terms Height, Thickness and Width should be restricted to the dimension normal to the plane of the workings or openings. The Length is the greater dimension in this plane, and the Breadth can be used for the lesser dimension in this plane. Barrier Pillar: Large pillars used to isolate the effects of mining zones for purposes of ground control, ventilation, drainage or explosion control. Crown Pillars: Pillars between the back of the completed stope in a steeply dipping seam and either the level or stope floor above. Rib Pillar: Long pillars with their length in

Pillar: (concluded)   the direction of the dip of the ore body. **Sill Pillars**: Pillars below a stope floor or sill in steeply dipping seams. In seams with low dip angles these pillars are sometimes called <u>Chain Pillars</u>.

Plane Strain: The case of a body that can be considered as long with the ends fixed between smooth, rigid abutments so that the deformation and strain in the long direction is zero and the forces applied to the body are all perpendicular to the long direction and do not vary along the body.

Plane Stress: The case of a body where the forces that are applied are all within one plane, as would be the case with forces applied to the edges of a plate and in the plane of the plate.

Plastic: The term has so many meanings that it is a great source of confusion. The simplest meaning, incorporating the common elements of all other definitions, is that the material or the state of the material is such that on application of stress some of the strain will be irrecoverable. Hence, <u>Plastic Strain</u> is irrecoverable strain; <u>Plastic Deformation</u> is irrecoverable deformation. To those working with metals the term has come to mean a material that will fail by yielding rather than fracturing; furthermore, the theoreticians idealize these materials so that beyond the yield point the reaction to stress is constant, i.e., there is no strain hardening. To geologists the term can imply gliding, rotation of grains, recrystallization or just permanent deformation with the retention of cohesion (4). Some geologists now recommend the use of the term 'ductile' for a material that has a large amount of strain at failure and the avoidance of the term 'plastic'. To those working with ceramics and soils the term can mean a material that can be deformed beyond the point of recovery without cracking or appreciable volume change. In soil mechanics, <u>Plastic Equilibrium</u> means the state of stress where sufficient deformation has occurred so that the ultimate shearing resistance of the soil mass has been mobilized (1).

Plane Shear Slide: See Slide.

Plutonic: Of igneous origin. A general name for those rocks that have crystallized in the depths of the earth, and have therefore assumed, as a rule, the granitoid texture (2).

Poise: (FTL^{-2}):   The unit of absolute viscosity, equal to one dynesecond per square centimeter. Named from the physicist Poiseuille. The centipoise, one one-hundredth of a poise, is a more convenient unit and the one customarily used (4).

Poisson's Effect: A term sometimes used to describe the lateral deformation resulting from longitudinal stress.

Poisson's Number: m(D):   The reciprocal of Poisson's Ratio.

Poisson's Ratio: μ, v(D):   The ratio of the transverse normal strain to the longitudinal normal strain of a body under uniaxial stress (4).

Pore:
Interstice or void. A space in rock or soil not occupied by solid mineral matter (4). A pore would not be of a larger order of magnitude than the grain size and thus would not refer to solution cavities or other large natural openings.

Porosity: $n(D)$:
The ratio, usually expressed as a percentage, of: (1) the volume of voids of a given mass to (2) the total volume of the mass.

Preconsolidation Pressure:
The maximum pressure that produced consolidation exerted on unconsolidated sediment by overlying material; the overburden may have been removed later by erosion, but the consolidation remained.

Pressure: $p(FL^{-2})$:
Force per unit area applied to outside surface of a body (4).

Pressure Grouting:
The act or process of injecting, at high pressures, a thin cement slurry through a pipeline or borehole to seal the pores or voids in the rock or to cement fragmented rocks together (5).

Pressure Head:
Equivalent to the height of a column of water than can be supported by the pressure.

Principal Plane:
Each of three mutually perpendicular planes through a point in a body on which the shearing stress is zero. Intermediate Principal Plane: The plane normal to the direction of the intermediate principal stress. Major Principal Plane: The plane normal to the direction of the major principal stress. Minor Principal Plane: The plane normal to the direction of the minor principal stress.

Principal Stresses:
See Stress.

Progressive Failure:
Failure in which the ultimate resistance is progressively, rather than simultaneously, mobilized along the ultimate failure surface.

Proportional Limit: (FL^{-2}):
The greatest stress that a material is capable of developing without any deviation from proportionality of stress to strain (Hooke's Law) (4).

P-Wave:
See Compressional Wave.

Quicksand:
Sand that is (or becomes, upon the access of water) "quick", i.e., shifting, easily movable or semi-liquid (4). Actually, it is a condition of water seepage whereby the upward seepage pressure is equal to or greater than the buoyant density of the sand.

Radial Stress: $\sigma_r(D)$:
Stress normal to the tangent to the boundary of any opening (3).

Radius of Gyration: (L):
Of an area or mass is equal to the square root of the moment of inertia divided by the area or mass.

Radius of Influence of a Well: Distance from the centre of the well to the closest point at which the ground water is not lowered when pumping has produced the maximum steady rate of flow.

Raise: An opening, like a shaft, made in the back of a level to reach a level above (2).

Ravelly Ground: Rock that breaks into small pieces when drilled and tends to cave or slough into the hole when the drill string is pulled. Similarly, ground that produces falling chunks or flakes on exposure in an underground opening (7).

Rayleigh Wave (R-wave): A surface seismic wave propagated along the plane surface of a homogeneous elastic solid (4).

Relaxation Time: (T): The time taken by an exponentially decaying quantity to decrease the amplitude by a factor of $1/e = 0.3679$.

Response: The motion of an elastic system resulting from an excitation.

Reflection: 1. In seismic prospecting, the returned energy (in wave form) from a shot that has been reflected from a velocity discontinuity back to a detector. 2. In seismic prospecting, the indication on a record of reflected energy (4).

Refraction: The deflection from a straight path suffered by a ray of light, heat or sound, in passing obliquely from one medium into another in which its velocity is different (4).

Repose, Angle of: See Angle of Repose.

Resistance Gages: See Electrical Resistance Gages.

Residual Stress: See Stress.

Residual Strain Energy: The energy in a rock arising from the strain and stress presently existing due to previously applied forces or deformations.

Resonance: This occurs when the frequency of the forced vibrations coincides or at least approaches the natural frequency of the system.

Resultant of Two or More Forces: The single force that will produce the same external effect on a body as the given forces.

Rheology: The study of flow in materials. It usually and preferably implies the study of time-dependent strain in both solids and liquids. However, it is sometimes used to describe the study of all types of deformation - elastic, plastic, and viscous.

Rib: 1. In coal mining, the solid coal on the side of a gallery or long wall face; a pillar or barrier of coal left for support. 2. The solid ore of a vein; an elongated pillar left to support the hanging-wall, in working out a vein (2).

Rib Pillar: See Pillar.

Rigidity: The property of stiffness or of not yielding.

Rise (of an arch): The maximum height that the undersurface of an arch rises.

Rise Time of Shock Pulse: $t_r(T)$: The interval of time required for the leading edge of the pulse (generally the particle acceleration) to rise from some specified small fraction to some specified larger fraction of the maximum value.

Rock: Geologically, any naturally formed aggregate of mineral matter constituting an appreciable part of the earth's crust (2).

Rock Mass: The in situ rock made up of the rock substance plus the structural discontinuities.

Rock Substance: The solid part of the rock mass typically obtained as drill core.

Rock Physics: The study of the physical characteristics of rock and the methods that can be used for their determination.

Rockburst: The violent failure of rock, when on a scale sufficient to endanger material or personnel.

Rockfall: The relatively free falling of a newly detached segment of rock of any size from a cliff, steep slope, or an underground opening (4).

Room: A wide working place in a flat mine corresponding to stope in a steep vein (2).

Rupture: Fracture.

Running Ground: Rock that breaks off readily and falls into underground openings (2).

Safety Factor: The ratio of the ultimate (fracture or yield) stress to the working stress.

Scalar Quantity: One which has only magnitude.

Scale:
1. To remove surface, loose rock from excavation faces.
2. Loose, thin fragments of rock on either roofs or walls (2).
3. To reduce or increase according to similitude requirements.

Section Modulus: S, $Z(L^3)$: A measure of the resistance offered by a beam to a bending moment. It is equal to the area moment of inertia divided by the distance from the neutral axis to the extreme fibre.

Seepage (Percolation): The slow movement of gravitational water through ground (1).

Seepage Force: $J(F)$: The force transmitted to ground by the seepage of water through the natural openings, voids or pores.

Seismic: Pertaining to, characteristic of, or produced by earthquakes or earth vibration (4).

Seismic Velocity: (FT^{-1}): The velocity through ground of either the P-wave or the S-wave.

Seismograph: An apparatus for indicating the sense, duration and intensity of an earthquake or similar ground shock.

Separation: The distance between any two parts of a bed, vein, etc. disrupted by a fault (4).

Set: A frame for supporting the ground around a shaft, tunnel or other excavation (2).

Shear Failure: Failure resulting from shear stresses. To distinguish it from micro-fracturing or local, insignificant fracturing, it is usefully defined as being of sufficient magnitude to destroy or seriously endanger the surrounding area (1).

Shear Stress: See Stress.

Shear Modulus: See Modulus of Rigidity.

Shear Strength: (FL^{-2}): The internal resistance offered to shear stress. It is measured by the maximum shear stress, based on original area of cross-section, that can be sustained without failure (4).

Shear Wave, Distortional Wave, Equivolumnar Wave, Secondary Wave, Transverse Wave, S-Wave: 1. In seismology, a wave motion in which the motion of the particles, or the entity that vibrates, is perpendicular to the direction of progression of the wave train. 2. In geophysics, a body seismic wave advancing by shearing displacements (4).

Sheeting: 1. To the geologist in a restricted sense, the gently dipping joints that are essentially parallel to the ground surface; they are more closely spaced near the surface and become progressively farther apart with depth. Especially well developed in granitic rocks. 2. To the geologist in a general sense, a set of closely spaced joints (4). 3. To the engineer, a series of boards placed behind supports, such as sets, to retain fragments of ground.

Shield: In mining or tunnelling, a framework or hood generally of steel protecting the workers, pushed forward as the work advances (2).

Sill Pillar: See Pillar.

Simple Harmonic Motion: Oscillatory motion of a particle or mass with constant amplitude, which is sinusoidal with time. Such motion results when the restoring force is proportional to the displacement (as in a simple pendulum of small amplitude).

Shock Wave: See Ground Shock.

Skin Friction: (FL^{-2}): The frictional resistance developed between a granular mass and a structure (1).

Slabbing: The development and falling of thin fragments of rock from the surfaces of an underground opening.

Slickenside: Polished and striated surface that results from friction along a fault plane (4).

Slide: A relatively deep-seated failure of a slope. Three main types can be identified: **Block Flow Slide**: A slide resulting from internal deformation leading to failure of a hard rock, jointed, blocky mass. It is believed that failure is initiated by the concentration of stress on corners of the individual blocks bounded by joint planes and that, when a general breakdown has occurred, a flow of blocks and pulverized material to the bottom of the slope occurs. **Plane Shear Slide**: A slide resulting from the presence of a plane of weakness, e.g., fault, dyke or soft layer, in a critical orientation within the slope. Large segments of the slope move down along this plane. **Rotational Shear Slide**: A slide resulting from the yielding and redistribution of shear stresses in a soft rock or soil so that a more or less circular surface of failure develops before the cohesion of the rock breaks down and permits a comprehensive, circular sector of the slope to fail by rotating.

Soft Ground: Heavy ground. Rock above underground openings that does not stand well and requires heavy timbering (2).

Specific Gravity: G(D): Ratio of the mass of a body to the mass of an equal volume of water at a specified temperature (4).

Specific Volume: $V(F^{-1}L^3)$: Volume per unit weight of a material.

Square-Set: A set of timbers composed of a cap, girt and post. These members meet so as to form a solid 90° angle. They are so framed at the intersection as to form a compression joint and join with three other similar sets (2).

Squeezing Ground: Ground that slowly moves into a tunnel without fracturing.

Stability Factor (Stability Number): N (D): A pure number used in the analysis of the stability of slopes defined by the following equations: $N = H_c \gamma_e / c$, where H_c = critical height of the sloped bank, γ_e = effective unit weight of the ground and c = cohesion of the ground (1). It varies with the angle of internal friction, the ground water level, the slope angle, the surcharge on the crest and the depth of tension crack.

Standard Normal Distribution: See Normal Distribution.

Statics: The branch of mechanics that deals with systems of forces acting on bodies in a state of equilibrium.

Steady-State Vibrations: Vibrations that continue to repeat themselves with time. Forced vibrations are steady-state vibrations.

Stope: A cavern, chamber, or room from which ore has been extracted (3).

Stoping: The loosening and removal of ore in a mine (4).

Standard Deviation: $\overline{SD}$: The square root of the quotient of the sum of the squares of the difference between the arithmetic mean, A, and a number of values of a quantity, x, divided by the number, n. It is a measure of the dispersion of the number of values about the mean, i.e.,

$$\overline{SD} = \sqrt{\frac{(A-x)^2}{n}}.$$

Stiffness: (FL^{-1}): The ratio of change of force to the corresponding change in deflection of an elastic element.

Strain: E(D): Deformation per unit of length. <u>Normal Strain</u> is the deformation per unit of length in the direction of the deformation. <u>Shear Strain</u> is the deformation per unit of length at right angles to the deformation or more commonly the relative change in the angle defining the sides of an infinitesimal element.

Strain Energy: The energy stored in a body arising from the strain and stress existing in the body. This is a form of potential energy that is theoretically available for doing work on release of the stress.

Strain Gage: An instrument for measuring the expansion or compression occurring parallel to a surface of a body and over a known gage length, which can then be converted into strain.

Strain Gage Rosettes: A surface strain gage that provides sufficient information to determine the direction of the two principal strains tangential to the surface and, in some types, to determine their magnitude.

Strength: (FL^{-2}): The maximum stress that a body can withstand without failing by rupture or continuous deformation. Rupture strength or breaking strength refers to the stress at the time of rupture. If a body deforms, after a certain stress has been reached, continuously without any increase in stress, this is also called strength (4). By common usage, it can be described as the greatest stress that a substance can withstand under normal short-time experiments, or the highest point on a stress-strain curve.

Stress: (FL^{-2}): The force per unit area, when the area approaches zero, acting within a body. <u>Biaxial Stress</u>: The state of stress where either the intermediate or the minor principal stress equals zero. <u>Effective Stress</u> (Effective Pressure)(Intergranular Pressure), $\sigma(FL^{-2})$: The average normal force per unit area transmitted from grain to grain in a granular mass. It is the stress that is effective in mobilizing internal friction. <u>Field Stress</u>: The stress existing in a rock mass independent of any man-made works. <u>Residual Stress</u>: Stress that exists in a formation owing to previously applied forces or deformations. <u>Triaxial Stress</u>: A state of stress where the three principal stresses have finite magnitudes, or simply a three-dimensional state of stress. <u>Neutral Stress</u> (Pore Pressure)(Pore Water Pressure) u, (FL^{-2}): Stress transmitted through the pore water (water filling the voids of the mass). <u>Normal Stress</u> (FL^{-2}): The stress component normal to a given plane. <u>Principal Stress</u>, σ_1, σ_2, $\sigma_3 (FL^{-2})$: Stresses acting normal to three mutually perpendicular planes intersecting at a point in a body, on which the shear stresses are zero. <u>Major Principal Stress</u> $\sigma_1(FL^{-2})$: The largest (with regard to sign) principal stress.

Stress: (FL^{-2}): (concluded)	Minor Principal Stress, $\sigma_3(FL^{-2})$: The smallest (with regard to sign) principal stress. Intermediate Principal Stress, σ_2 (FL^{-2}): The principal stress whose value is neither the largest nor the smallest (with regard to sign) of the three. Shear Stress (Shearing Stress)(Tangential Stress), $\tau(FL^{-2})$: The stress component tangential to a given plane. Total Stress, $\sigma(FL^{-2})$: The total force per unit area acting within a granular mass. It is the sum of the neutral and effective stresses (1).
Stress Concentration: k(D):	Ratio of the stress at any point to the applied or principal field stress (3).
Stress Difference:	The algebraic difference between the maximum and minimum principal stresses (4).
Stress Ellipse:	In two dimensional analysis, the ellipse whose half-axes are the major and minor principal stresses.
Stress Ellipsoid:	The ellipsoid whose half-axes are the principal stresses (4).
Stressmeter:	An instrument for measuring stress within a body.
Strike:	The course or bearing on a level surface of the outcrop of an inclined bed or a structure; the direction or bearing of a horizontal line in the plane of an inclined stratum, joint, fault, cleavage plane, or other structural plane; it is perpendicular to the direction of the dip (2).
Subgrade:	1. The ground that supports, and is significantly stressed by, a structure. 2. It is also used in blasting work to mean the ground below the toe of a bench.
Sublevel:	An intermediate level opened below the main level, or, in the caving system of mining, below the top of the ore body, preliminary to caving the ore between it and the level above (2).
Subseismic:	Travelling at less than the velocity of the S-wave in the adjacent or in the undisturbed ground.
Subsidence:	Failure of ground above the workings resulting from the removal of ore (or any other ground) by underground operations.
Superseismic:	Travelling at greater than the velocity of the P-wave in the adjacent or in the undisturbed ground.
S-Wave:	See Transverse Wave.
Swelling Ground:	A soil or rock that expands when wetted (5).
Tangential Stress: $\sigma_t, \sigma_\theta (FL^{-2})$:	Stress parallel to the tangent to the boundary of any opening (3).
Tectonic:	Pertaining to the rock structures resulting from the deformation of the earth's crust (3).
Thickness:	The distance at right angles between the hanging and the footwall of a lode or lens (2).
Toe:	1. The downstream zone at the base of a dam. 2. The bottom of a slope.

Ton: T(F): An avoirdupois unit weight. A short or net ton equals 2,000 pounds (907.20 Kg). A long or gross ton equals 2,240 pounds (1,016.6 Kg). A metric ton equals 2,204.6 pounds (1,000 Kg) (2). A kiloton equals one thousand short tons. A megaton equals one million short tons.

Top Heading Method: A method of tunnelling or drifting whereby the top part of the face is excavated ahead of the lower part and in a separate operation (7).

Transient Vibrations: Vibrations that disappear with time. Free vibrations are transient in character.

Transeismic: Travelling at greater than the velocity of the S-wave and less than that of the P-wave.

Transverse Wave: See Shear Wave.

Trepanning: In rock mechanics, to cut out a block of rock. As a special case, see Overcoring.

Triaxial Stress: See Stress.

Tubbing: A lining of timber or metal for a shaft; especially a water tight shaft lining consisting of a series of cast-iron cylinders bolted together; used in sinking through water-bearing strata (2).

Tunnel: A single underground opening approximately horizontal, with few or no intersecting openings (3).

Ultimate Bearing Capacity: q_u, q_f (FL^{-2}): The average load per unit area required to produce failure of a supporting rock mass.

Ultimate Strength: σ_f, σ_u (FL^{-2}): See Strength.

Unconfined Compressive Strength: See Compressive Strength.

Unconsolidated: An accumulation of sediment that has not been cemented and/or compacted (5).

Undercut: 1. To undermine, to hole, or to mine. 2. To cut below or in the lower part of a coal bed by chipping away the coal with a pick or mining machine. It is usually done on the level of the floor of the mine, extending laterally over the entire face and 5 or 6 feet into the material (2). 3. To mine under a block of ore.

Underground Openings: Natural or manmade excavations under the surface of the earth (3).

Uniaxial Stress: A state of stress where the minor and intermediate principal stresses are zero.

Unit Weight: (FL^{-3}): Weight per unit volume. <u>Dry Unit Weight (Unit Dry Weight)</u>, γ_d, (FL^{-3}): The weight of solids per unit of total volume of mass. <u>Effective Unit Weight</u>, γ_e(FL^{-3}): The unit weight that, when multiplied by the height of the overlying column of ground, yields the effective pressure due to the weight of the overburden (1). <u>Saturated Unit Weight</u>, γ_{sat} (FL^{-3}): The wet unit weight of a granular mass when saturated (1). <u>Submerged Unit Weight (Buoyant Unit Weight</u>, γ_b(FL^{-3}): The weight of the solids in air minus the weight of water displaced by the solids per unit of volume of soil mass; the saturated unit weight minus the unit weight of water. <u>Wet Unit Weight (Mass Unit Weight)</u>, γ_m, γ_t (FL^{-3}): The weight (solids plus water) per unit of total volume of mass, irrespective of the degree of saturation.

Vector Quantity: One that has magnitude, direction, and sense, and can, therefore, be expressed graphically.

Velocity: The time rate of change of displacement.

Vibrating-Wire Gage: A strain gage that detects change in strain by measuring the change in frequency of a wire under tension whose length is changed by the strain of the surface on which the gage is attached. The frequency is measured by setting the wire vibrating with an electro-magnet and then measuring the frequency of vibration through an electro-magnetic pickup.

Viscosity: The property of liquids and solids that causes them to resist instantaneous change of shape but to produce strain that is dependent on time and the magnitude of stress.

Viscosity Coefficient: See Coefficient of Viscosity.

Viscous: See Viscosity.

Viscoelastic: Describes a material or a condition where strain resulting from stress is partly elastic and partly viscous.

Viscous Flow: See Viscosity and Creep.

Void: See Pore.

Void Ratio: e(D): Ratio of volume of intergranular voids to volume of solid material in a rock (4).

Voussoir: Any of the tapering or wedge-shaped pieces of which a masonry arch or vault is composed. The middle one is usually specifically called the Keystone (2).

Wall: 1. The side of a level or drift. 2. The country rock bounding a vein laterally. The side of a hole. The overhanging side is known as the hanging wall, and the lower lying one as the footwall. 3. The face of a long-wall working or stall, commonly called Coal Wall. 4. A rib of solid coal between two boards (2).

Wall Rock: See Wall.

Waste: That which has no real value, as barren rock in a mine, or the refuse from ore dressing and smelting plants (2).

Water Table:	See Free Water Elevation.
Wavelength: $\lambda(L)$:	The perpendicular distance between two wave fronts of a periodic wave in an isotropic medium in which the displacements have a difference in phase of one complete period.
Winze:	A vertical or inclined opening, or excavation, connecting two levels in a mine, differing from a raise only in construction. A winze is sunk underhand and a raise is put up overhand. When the connection is completed, and one is standing at the top, the opening is referred to as a winze, and when at the bottom, as a raise. Also called an underground shaft (2).
Work: (FL):	The product of force and the distance through which it moves (4).
Working:	1. A working may be a shaft, quarry, level, opencut, or stope, etc. Usually used in the plural. 2. A name given to the whole strata excavated in working a seam. 3. Making a noise before falling such as unsupported roof strata (2).
Yield Point:	See Elastic Limit.
Young's Modulus:	See Modulus of Elasticity.

REFERENCES

(1) "Glossary of Terms and Definitions in Soil Mechanics", Proc. ASCE SM 4 (1958).

(2) Fay, A.H., "A Glossary of the Mining and Mineral Industry", USBM Bull. 95 (1948).

(3) Obert, L. et al., "Design of Underground Openings in Competent Rock", USBM Bull. 587 (1960).

(4) "Glossary of Geology and Related Sciences", 2nd Ed., American Geological Institute, Washington (1960).

(5) Long, A.E., "A Glossary of the Diamond-Drilling Industry", USBM Bull. 583 (1960).

(6) Krynine, D. and Judd, W., "Principles of Engineering Geology and Geotechnics", McGraw-Hill (1957).

(7) Turner, F. and Weiss, L., "Structural Analysis of Metamorphic Tectonics", McGraw-Hill (1963).

APPENDIX C
FLOW NETS

From fluid mechanics we know that flowing water contains energy that arises from three different sources. It has kinetic energy owing to its velocity. It has potential energy owing to its state or position, in other words, its height above an arbitrary datum. It also has pressure energy owing to the pressure in the water. This pressure can be measured by the height to which the water will rise in a standpipe measured above some arbitrary datum. This height is normally called the piezometric head.

It is the difference in piezometric head between two points within a body of water that causes flow to occur from the point of high head to the point of low head. From Darcy's Law the rate of flow, q, can be calculated from the equation:

$$q = kiA \qquad \text{Eq. C-1}$$

where k is the coefficient of permeability expressed in units of velocity, i is the hydraulic gradient and A is the cross-sectional area through which the flow is occurring. If the rate of flow is divided by the cross-sectional area, the velocity of the flow, v, is obtained:

$$v = ki \qquad \text{Eq. C-2}$$

Also from fluid mechanics we obtain the concept that the rate of flow into any volume of ground must equal the rate of flow out of the volume. From this concept the various equations of continuity have been obtained. One deduction from this principle is of interest for the construction of flow nets.

In Fig. C-1 an element of ground is shown through which water is flowing. The element is not oriented with respect to the direction of flow. Consequently, flow occurs into the element through faces AB and AD and out of the element through faces BC and CD. The total inflow thus is as follows:

$$Q = v\,dy \times 1 + u\,dx \times 1$$

where v is the component of velocity flowing into the element through face AD, dy is the width of the face AD with breadth being equal to a unit distance; u and dx apply to face AB.

The outflow from the element is equal to the following:

$$Q = v\,dy + (\partial v/\partial x)\,dx\,dy + u\,dx + (\partial v/\partial y)\,dy\,dx$$

From these two equations for Q the following equation is obtained:

$$\partial v/\partial x + \partial u/\partial y = 0$$

From Darcy's Law it is known that:

$$v = k\,\partial h/\partial x \qquad u = k\,\partial h/\partial y$$

By combining these equations the following expression is obtained:

$$\partial^2 h/\partial x^2 + \partial^2 h/\partial y^2 = 0 \qquad \text{Eq. C-3}$$

Equation C-3 is a Laplacian equation that is identical in form to the condition of compatibility of principal strains in a stressed body. In both cases the equation establishes a condition of orthogonal trajectories. This means, in the case of fluid flow, that there will be two sets of

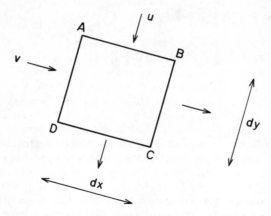

Fig. C-1 Flow Velocities Through an Element of Ground

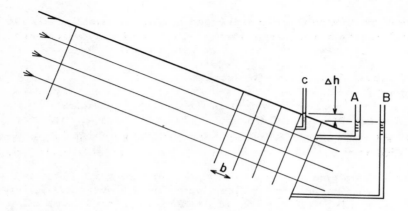

Fig. C-2 Flow Parallel to a Long Slope

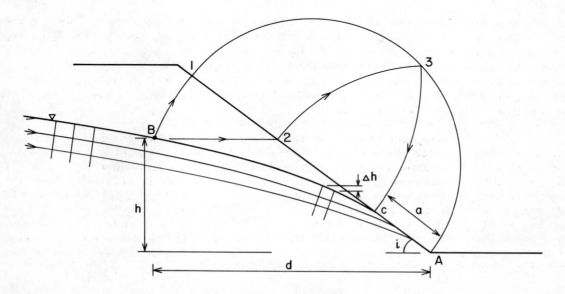

Fig. C-3 Flow Under Seepage Line in a Slope

curves describing the fluid flow; one set is known as the flow lines and the other set, at right angles, is known as the equipotential lines.

In Fig. C-2 a simple case of fluid flow parallel to a slope is shown. The flow lines are spaced at equal distances so that the quantity of flow between any two lines is equal. At right angles to the flow lines the set of equipotential lines is shown. These are lines along which the piezometric head is constant and consequently flow should occur at right angles to them.

In Fig. C-2 two piezometer tubes are connected to the equipotential line on the right hand side. It can be seen from these tubes that the height to which the water rises is equal in both cases in spite of Piezometer A originating at a higher level than Piezometer B.

It is convenient to draw the equipotential lines at equal distances apart and to have this distance equal to the spacing of the flow lines. In this case squares are formed to produce the conventional flow net. If between two points in the rock mass a difference in head is h, then it follows:

$$h = \Delta h \, N_e$$

where Δh is the loss in head between any two equipotential lines (see Fig. C-2 for the difference in head between Piezometers A and C) and N_e is the number of equipotential lines between the two points.

The total rate of flow, q, can be expressed as:

$$q = \Delta q \, N_f$$

where Δq is the element of flow between any two flow lines and N_f is the number of flow lines. If the sides of the elements in the flow net are equal to b, then from Darcy's Equation it follows that:

$$\Delta q = k((h/N_e)/b) \, b \times 1$$

It then follows that:

$$q = k \, h \, N_f / N_e \qquad \text{Eq. C-4}$$

Equation C-4 then makes it possible, in spite of complex geometry in some cases, to calculate the flow rate through a rock simply by sketching in a compatible flow net that fulfils the requirements of having orthogonal trajectories enclose areas whose sides are equal. Then, by counting the number of flow lines and the number of equipotential lines between the two points where a total head loss of h occurs and knowing the coefficient of permeability, k, the flow rate can be calculated.

In Fig. C-3 the case is shown of flow occurring under the seepage line in a slope with the exit being at the toe of the slope. In drawing a flow net, it should be appreciated that, although an infinite number of flow lines can be drawn, the number that is selected is purely a matter of convenience. However, once the number of flow lines is selected at one point with the spacing of the equipotential lines being the same distance apart as the flow lines at that point, then the requirement that these trajectories throughout the flow path must remain at right angles will require that the flow lines either converge or diverge with the spacing of the equipotential lines changing correspondingly. If the flow lines converge with the rate of flow remaining constant between any two flow lines, it follows that the velocity must increase. The reverse, of course, occurs when the flow lines diverge.

When the flow lines curve, it will be impossible to form squares with the flow lines and equipotential lines; however, the main requirements are that the trajectories be at right angles and that the sides of the elements be equal so far as possible. In sketching a flow net, it is necessary to go through several trials before a compatible net is obtained for the particular geometry of the problem. However, once such a compatible net is obtained, even approximately, the resulting computation using Equation C-4 is amazingly accurate considering the rather kindergarten procedure that has been followed.

As the most common case in rock mechanics for which flow nets are of interest is that of slope stability, the problem of the exit point from a slope can be considered in some detail. As a result of a considerable amount of theoretical and experimental work, the following construction has been found to provide a good approximation for the exit point on a slope (1). In Fig. C-3 a point B of known head, h, provides the starting point. With A, the toe of the slope, as the centre an arc of diameter AB is struck off from B to Point 1. A semi-circle is then drawn with the distance A-1 as diameter. The point B is projected horizontally to intersect the slope face at Point 2. With A as centre the arc 2-3 is then struck off. Then with Point 1 as the centre the arc 3-C is struck off. C is then the point at which the seepage or phreatic line intersects the face of the slope.

When drawing the flow net in the area from C to A in Fig. C-3, it must be recognized that a discharge face under atmospheric pressure is neither a flow line nor an equipotential line. The squares of the flow net are thus incomplete at this boundary and need not intersect it at right angles. This is not the case for other boundaries; e.g., if there were a horizontal impervious layer under the slope at the elevation of A, then this surface would of necessity be a flow line. The knowledge of such a surface would be of importance and of assistance in drawing the flow net. Also, the upstream sloping face of a dam under a head of water must be an equipotential line as the head is equal at every point down the slope. Consequently, with this slope surface being an equipotential line, all flow lines must be at right angles to this surface.

Returning to the construction in Fig. C-3, it has been established that the rate of discharge through the toe of the slope can be calculated using the following equation (1):

$$q = k\, a\, \sin i\, \tan i \qquad \text{Eq. C-5}$$

where a is equal to the distance from the toe, A, to the point C where the seepage line intersects the slope surface and i is the inclination of the slope.

REFERENCE

1. Casagrande, A., "Seepage through Dams", J. New England Water Works Assoc. (June, 1937).

APPENDIX D

Stress Concentration Factors

Symbols a is the radius of a circular opening or the major semi-axis of an ellipse; B_o is the breadth of an opening; B_p is the breadth of a pillar; H is the height of an opening or pillar; K equals the ratio S_h/S_v; r is the radial distance from the centre of an opening; S_h is the horizontal field stress; S_v is the vertical field stress; S_x is the horizontal field stress in the x-direction; S_y is the horizontal field stress in the y-direction; S_z is the vertical field stress; δ is the angle to the horizontal of the major axis of an opening; θ is the angle from the horizontal to a point; σ_r is the radial stress; σ_t and σ_θ are tangential stresses; $\tau_{r\theta}$ is the shear stress on radial-tangential planes.

STRESS CONCENTRATION FACTORS IN PLANE STRESS

1. Single Circular Opening[1]

$$\sigma_r = \left[\frac{S_h + S_v}{2}\right]\left[1 - \frac{a^2}{r^2}\right] + \left[\frac{S_h - S_v}{2}\right]\left[1 - \frac{4a^2}{r^2} + \frac{3a^4}{r^4}\right]\cos 2\theta$$

$$\sigma_\theta = \left[\frac{S_h + S_v}{2}\right]\left[1 + \frac{a^2}{r^2}\right] - \left[\frac{S_h - S_v}{2}\right]\left[1 + \frac{3a^4}{r^4}\right]\cos 2\theta$$

and $\tau_{r\theta} = \left[\dfrac{S_v - S_h}{2}\right]\left[1 + \dfrac{2a^2}{r^2} - \dfrac{3a^4}{r^4}\right]\sin 2\theta$

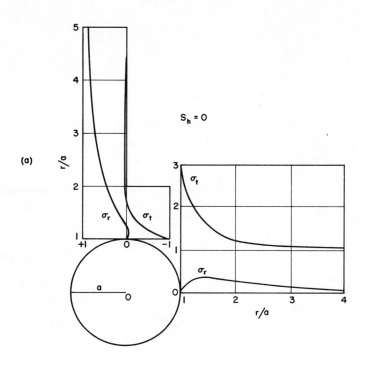

(a)

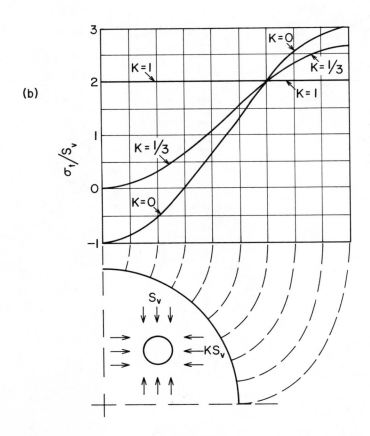

(b)

2. Single Elliptical Opening (2)

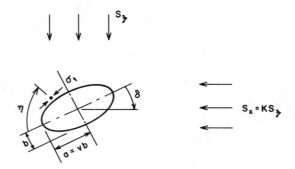

$$\frac{\sigma_t}{S_y} = \frac{2\nu(1-K)+(1+K)(1-\nu^2)\cos 2\delta +(1-K)(1+\nu)^2\cos 2(\delta-\eta)}{(1+\nu^2)+(1-\nu^2)\cos 2\eta}$$

if $\delta = 0$

$$\frac{\sigma_t}{S_y} = \frac{2\nu(1+K)+(1-K)(1-\nu^2)+(1-K)(1+\nu)^2\cos 2\eta}{(1+\nu^2)+(1-\nu^2)\cos 2\eta}$$

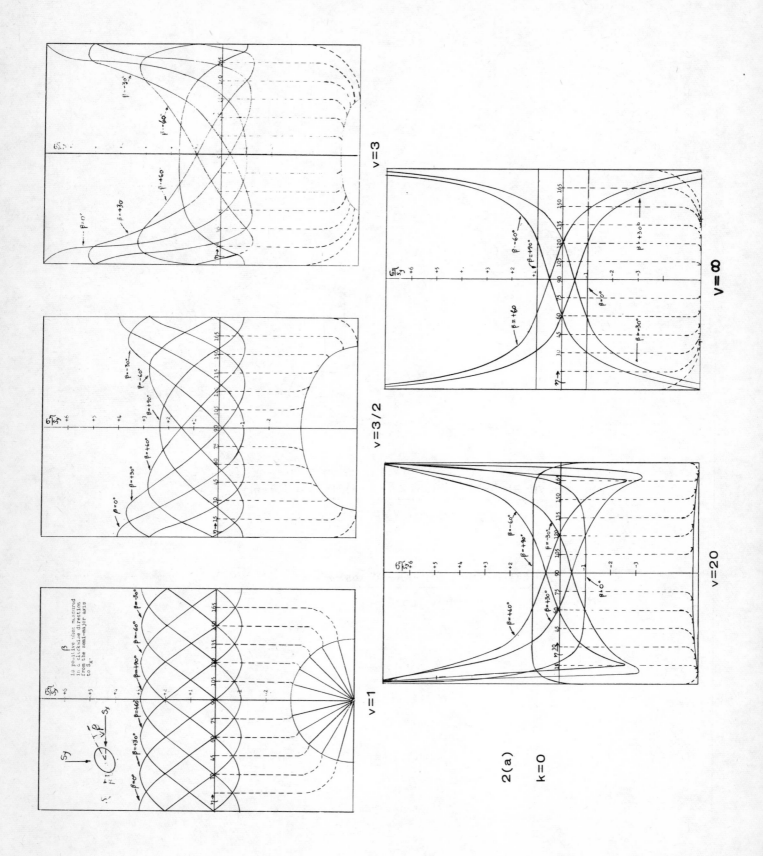

2(a) k=0

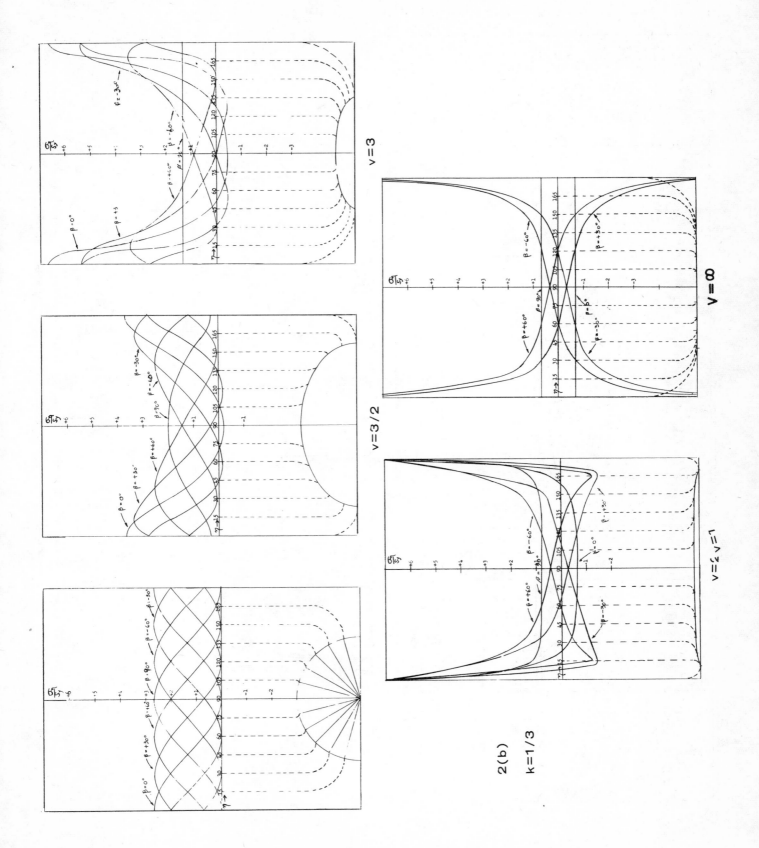

2(b)
k=1/3

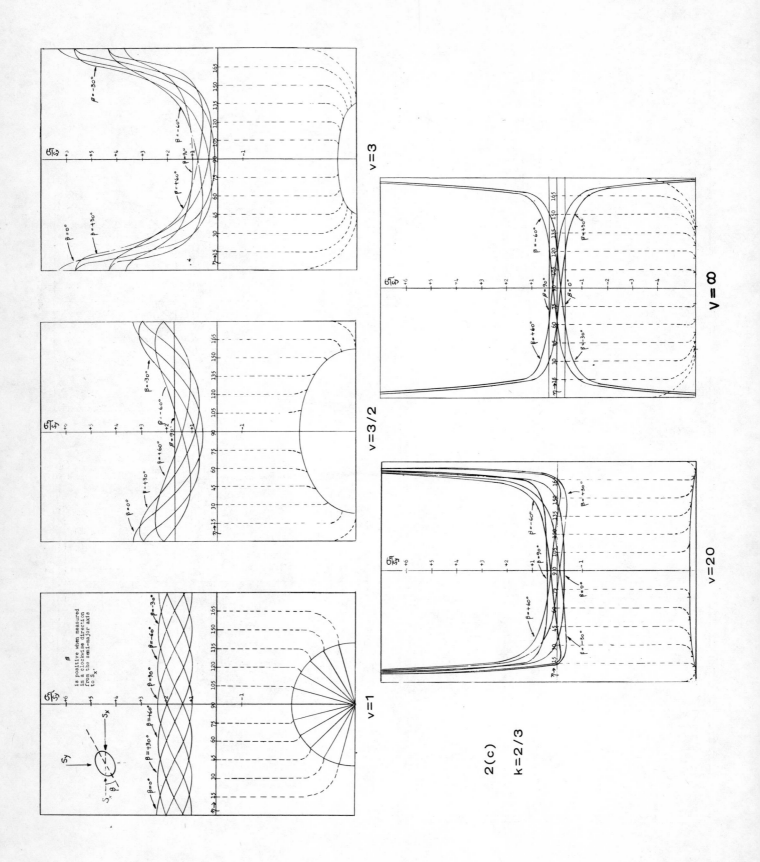

2(c) k=2/3

D-7

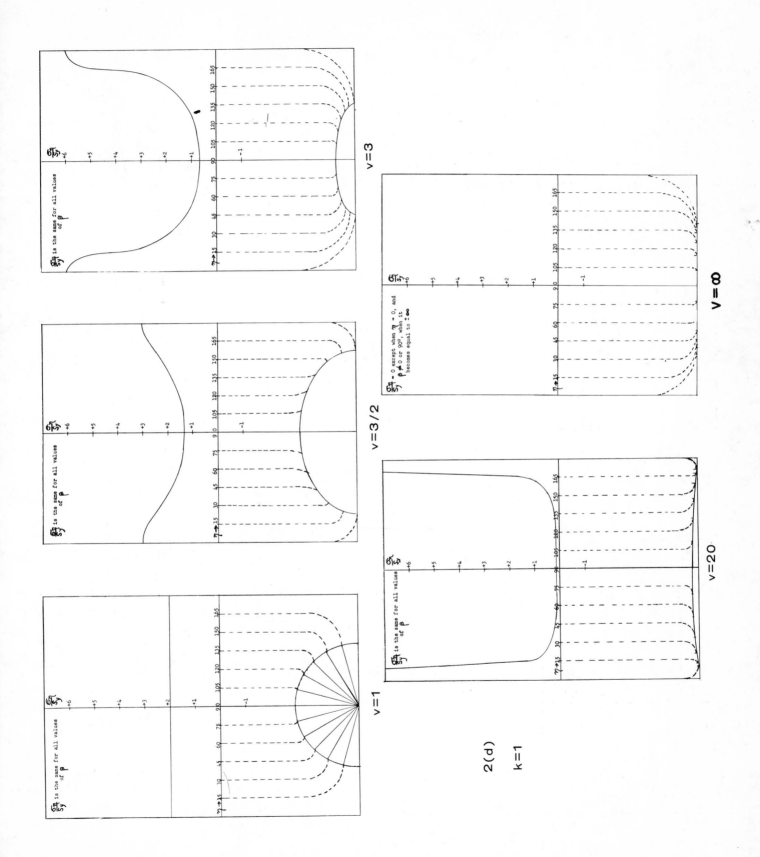

2(d)
k=1

3. Single Rectangular Opening with Fillet Radii = 1/6 (minimum of B_o or H) (1)

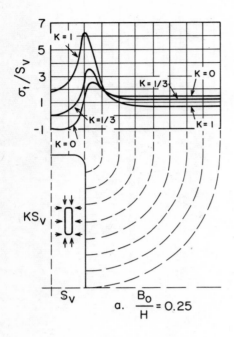

a. $\dfrac{B_o}{H} = 0.25$

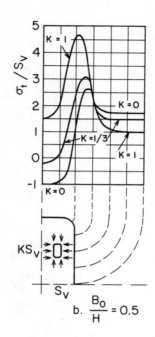

b. $\dfrac{B_o}{H} = 0.5$

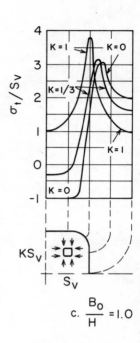

c. $\dfrac{B_o}{H} = 1.0$

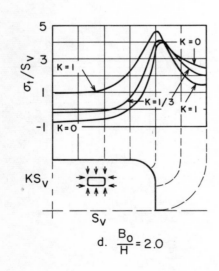

d. $\dfrac{B_o}{H} = 2.0$

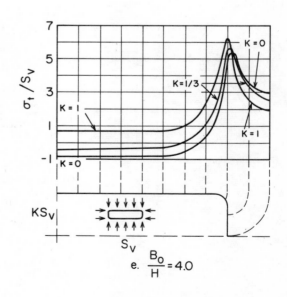

e. $\dfrac{B_o}{H} = 4.0$

f.

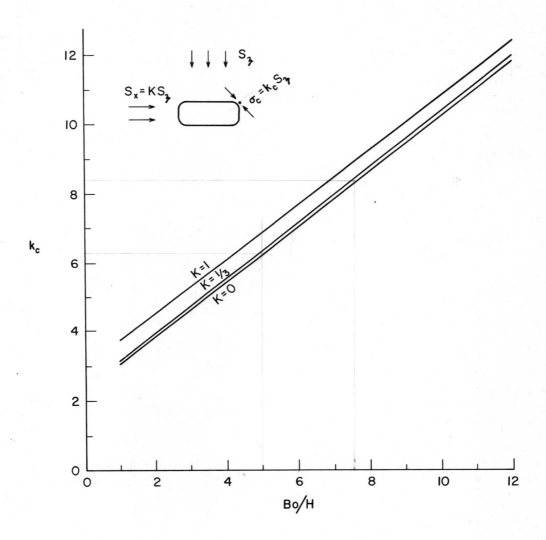

D-10

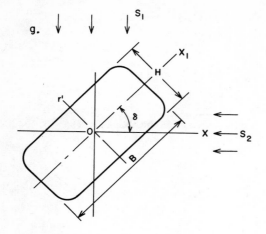
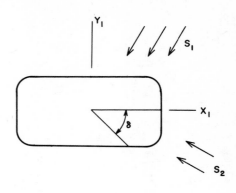

B_o/H	$\delta°$	K = 0		K = 1/3		K = 1
		Tension	Comp.	Tension	Comp.	Comp.
1 – – – –	0	−1.0	3.1	−0.3	3.1	3.8
	22.5	−0.5	3.9	0	3.8	3.7
	45	−1.1	4.7	0	4.3	3.6
	67.5	−1.3	4.0	−0.3	3.9	3.7
	90	−1.0	3.1	−0.3	3.1	3.8
2 – – – –	0	−0.8	4.0	−0.1	4.1	4.7
	22.5	−0.7	5.0	0	4.7	4.6
	45	−1.6	5.7	0	5.2	4.5
	67.5	−1.4	4.5	−0.1	4.5	4.6
	90	−1.0	2.7	−0.2	3.1	4.7
3 – – – –	0	−0.8	4.6	−0.4	4.7	5.2
	22.5	−0.7	5.9	−0.1	5.5	5.2
	45	−1.8	6.5	−0.4	6.0	5.3
	67.5	−1.6	5.0	0	5.0	5.2
	90	−1.0	2.6	−0.1	3.3	5.2
4 – – – –	0	−0.9	5.4	−0.4	5.6	6.2
	22.5	−1.0	6.5	−0.1	6.0	5.9
	45	−2.0	7.1	−0.5	6.6	5.9
	67.5	−1.9	5.5	0	5.5	5.9
	90	−1.0	2.5	0	3.5	6.2

4. Multiple Circular Openings — Infinite Number $B_o/B_p = 1$ (1)

(a)
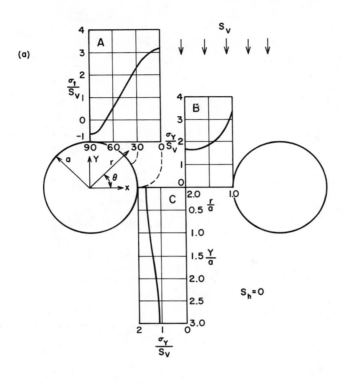

(b)
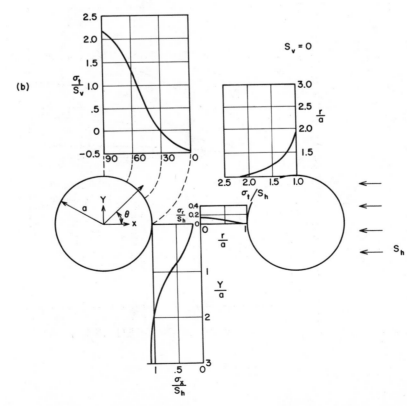

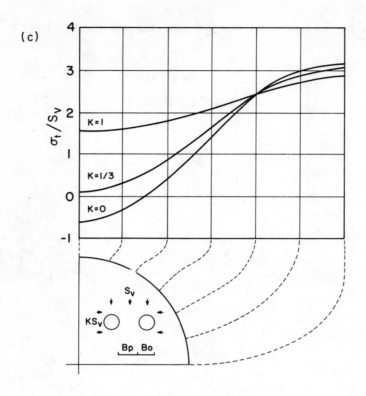

(c)

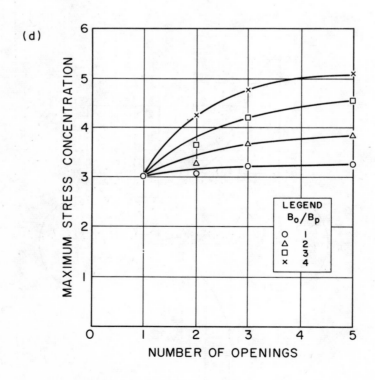

(d)

5. Multiple Ovaloid Openings (1)

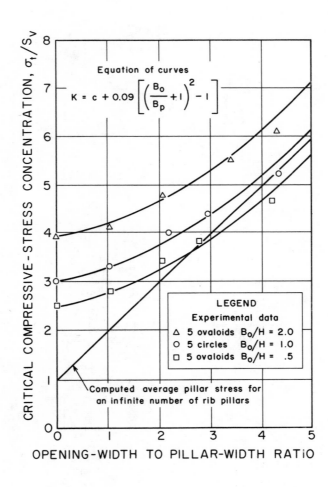

STRESS CONCENTRATION FACTORS IN TRIAXIAL STRESS

Symbols: S_h' and S_h'' are horizontal, orthogonal field stresses

I. Single Spherical Opening (3)

(a)
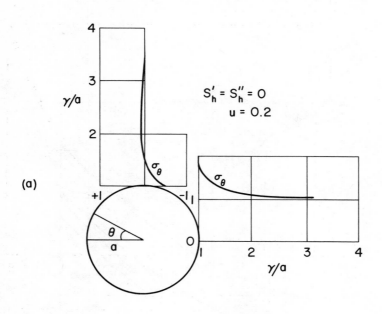

$$\sigma_\theta \text{ (at } \theta = 0 \text{ and } r = a) = \frac{(27-15\mu)}{2(7-15\mu)} S_v$$

$$\sigma_\theta \text{ (at } \theta = \pi/2 \text{ and } r = a) = -\frac{(3+15\mu)}{2(7-5\mu)} S_v$$

(b)
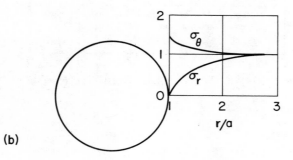

$$\sigma_\theta = \frac{2r^3 + a^3}{2r^3} S$$

$$\sigma_\theta = \frac{r^3 - a^3}{r^3} S$$

2. Single Spheroidal Opening (circular in plan, elliptical in any vertical section) (4)

$\mu = 0.2$

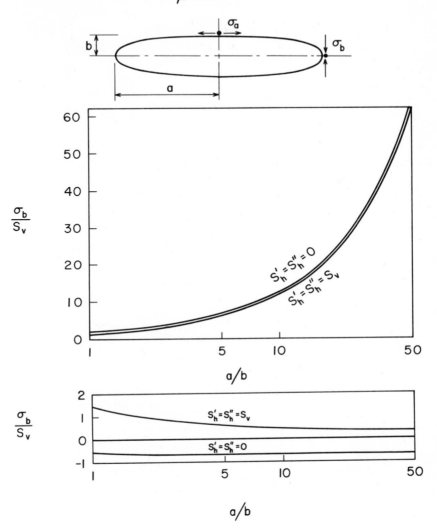

REFERENCES

1. Obert, L. et al "Design of Underground Openings in Competent Rock" USBM Bull. 587, 1960

2. Geldart, L. and Udd, J. "Boundary Stresses around an Elliptical Opening in an Infinite solid" Proc. Rock. Mech. Symp. McGill Univ., Mines Branch, 1962

3. Timoshenko, S. and Goodier, J. "Theory of Elasticity" McGraw Hill 1951

4. Terzaghi, K. and Richart, F.E. "Stresses in Rock About Cavities" Geotechnique vol. 3 no. 2 p. 57 1952-53

APPENDIX E
BEAM FORMULAE

Symbols: A is the transverse, cross-sectional area of the beam; δ_b is the deflection due to pure bending; δ_s is the deflection for a rectangular cross-section due to shear stresses; E is the modulus of deformation; G is the modulus of rigidity; I is the area moment of inertia; L is the span; M is the bending moment; p is the distributed pressure per unit length; R is the reaction at the support; V is the shear force; P is the concentrated load.

1. Simply Supported Beam —
uniformly distributed pressure:

$$R = V = \frac{pL}{2}$$

$$M \text{ (at center)} = \frac{pL^2}{8}$$

$$\delta_b \text{ (at center)} = \frac{5pL^4}{384EI}$$

$$\delta_b \text{ (at x)} = \frac{px(L^3 - 2Lx^2 + x^3)}{24EI}$$

$$\delta_s \text{ (at center)} = \frac{0.15pL^2}{AG}$$

$$\delta_s \text{ (at x)} = \frac{0.6px(L-x)}{AG}$$

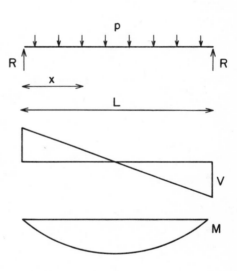

E-2

2. Simply Supported Beam –
   concentrated load at center:

$R = V \quad = \dfrac{P}{2}$

$M \text{ (at center)} \quad = \dfrac{PL}{4}$

$\delta_b \text{ (at center)} \quad = \dfrac{PL^3}{48EI}$

$\delta_b \text{ (at } x < L/2 \text{)} \quad = \dfrac{Px(3L^2 - 4x^2)}{48EI}$

$\delta_s \text{ (at center)} \quad = \dfrac{0.3\,PL}{AG}$

$\delta_s \text{ (at } x < L/2 \text{)} \quad = \dfrac{0.6\,Px}{AG}$

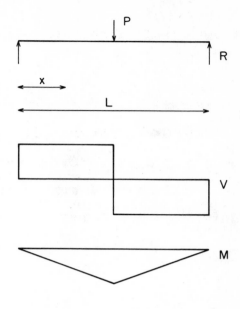

3. Simply Supported Beam –
   concentrated load

$R_1 = V_1 \quad = \dfrac{Pa}{L}$

$M \text{ (at } P\text{)} \quad = \dfrac{Pab}{L}$

$\delta_b \text{ (at } P\text{)} \quad = \dfrac{Pa^2 b^2}{3EIL}$

$\delta_b \text{ (at } x < a\text{)} \quad = \dfrac{Pbx(L^2 - b^2 - x^2)}{6EIL}$

$\delta_s \text{ (at } P\text{)} \quad = \dfrac{1.2\,Pa(1 - a/L)}{AG}$

$\delta_s \text{ (at } x < a\text{)} \quad = \dfrac{1.2\,Px(1 - a/L)}{AG}$

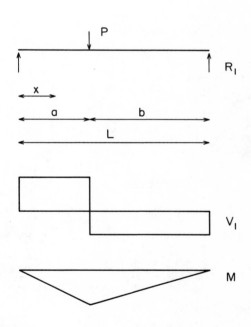

4. Fixed Beam — uniformly distributed pressure:

$R = V = \dfrac{pL}{2}$

M (at ends) $= \dfrac{pL^2}{12}$

M (at center) $= \dfrac{pL^2}{24}$

δ_b (at center) $= \dfrac{pL^4}{384 EI}$

δ_b (at x) $= \dfrac{p x^2 (L-x)^2}{24 EI}$

δ_s (at center) $= \dfrac{0.15 pL^2}{AG}$

δ_s (at x) $= \dfrac{0.6 px(L-x)}{AG}$

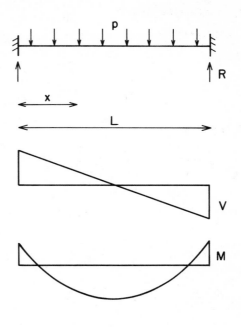

5. Fixed Beam — concentrated load at center:

$R = V = \dfrac{P}{2}$

M (at center and ends) $= \dfrac{PL}{8}$

δ_b (at center) $= \dfrac{PL^3}{192 EI}$

δ_b (at x) $= \dfrac{Px^2}{48 EI}(3L-4x)$

δ_s (at center) $= \dfrac{0.3 PL}{AG}$

δ_s (at $x < L/2$) $= \dfrac{0.6 Px}{AG}$

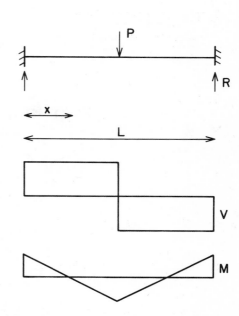

6. Fixed Beam — concentrated load:

$R_1 = V_1 = \dfrac{Pa^2}{L^3}(3b+a)$

$M \text{ (at P)} = \dfrac{2Pa^2b^2}{L^3}$

$M_1 = \dfrac{Pa^2b}{L^2}$

$\delta_b \text{ (at P)} = \dfrac{Pa^3b^3}{3EIL^3}$

$\delta_b \text{ (at } x < a) = \dfrac{Pb^2x^2}{6EIL^3}(3aL - 3ax - bx)$

$\delta_s \text{ (at P)} = \dfrac{1.2\,Pa}{AG}(1 - a/L)$

$\delta_s \text{ (at } x < a) = \dfrac{1.2\,Px}{AG}(1 - a/L)$

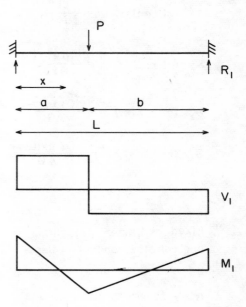

APPENDIX F

DIMENSIONAL ANALYSIS AND SIMILITUDE

Dimensional analysis is based on the requirement for homogeneous equations to have consistent units. For example:

$$5 \text{ apples} + 2 \text{ cows} = ?$$

is not a homogeneous equation as the units of each term are different. The units in mechanics are normally those of force (F), mass (M), length (L) and time (T). For example, the units of displacement can be represented by L and the units of velocity by LT^{-1}.

It is common practice to represent the units of any quantity in either terms of force or mass but not both. Force and mass are related by Newton's Second Law, and thus one can be expressed in terms of the other. For example:

$$F = Ma$$

where F is the force on a mass, M, causing an acceleration, a. In the English system of units:

$$1 \text{ poundal} = 1 \text{ lb-mass} \times 1 \text{ fss}$$

In the American system:

$$1 \text{ lb-force} = 1 \text{ slug} \times 1 \text{ fss}.$$

Thus, through Newton's Second Equation the units of force and mass are defined in terms of each other. Consequently, if the units of force are established with respect to some standard, the units of mass must be related as follows:

$$M = F/a$$

It should be noted that the acceleration due to gravity, g, only becomes involved in these units when weight is considered. Weight, or the force of gravity, gives every body on the surface of the earth an acceleration of approximately 32.2 fss. Hence, the mass of a body can be determined by measuring the force of gravity on it and dividing this force by the acceleration due to gravity. In the American system, if lbs-force are divided by fss then the units of mass are slugs.

Through Newton's Second Law of Motion, as mentioned above, the units of force can be expressed in units of mass or vice versa. For example, from:

$$F = Ma$$
$$F = MLT^{-2}$$
$$\text{or} \quad M = FL^{-1}T^{2}$$

Therefore, a quantity such as stress can be expressed in the units FL^{-2} or $ML^{-1}T^{-2}$.

The requirement for an equation to have consistent units means that in the following equation:

$$x = 5y + z^2$$

the terms $5y$ and z^2 must have the same units as x. Similarly, in the equation

$$x = y + \log z$$

the terms y and $\log z$ must have the same units as x. It can be seen that the requirement of homogeneity is most easily fulfilled if x, y and z are dimensionless.

In searching for functional relations in a process represented by the general type of equation $x = f(y, z)$, or in processing experimental data, it has been found useful not only to fulfil the requirements of homogeneity but also to provide guidance in determining the functional relations, to combine the variables into dimensionless groups. For this purpose, Buckingham's Pi Theorem is one method of obtaining the minimum number of significant dimensionless groups of variables involved in any process.

In applying Buckingham's Theorem, we let:

n = the number of variables,
m = the number of primary dimensions,
    i.e., F, L or T,
π = a dimensionless group or parameter.

From the theorem it has been established that the number of dimensionless parameters, or π's, is equal to $(n-m)$. The procedure is then to select m variables, X, Y, Z, which together include the m primary dimensions, but preferably cannot be formed into dimensionless parameters themselves. It is useful at this stage also to arrange for this group to include the principal independent variable and the principal dependent variable. This selected group is then combined with each of the remaining variables, A, B ... N, into $(n-m)$ π equations as follows:

$$\pi_1 = X^{x1} Y^{y1} Z^{z1} A$$

$$\pi_2 = X^{x2} Y^{y2} Z^{z2} B$$

$$\cdot\cdot\quad\cdot\cdot\cdot\cdot\cdot\cdot\cdot\cdot\cdot$$

$$\pi_{n-m} = X^{x_{n-m}} Y^{y_{n-m}} Z^{z_{n-m}} N$$

Then, into each of the above π-equations the dimensions for each quantity are inserted and the indices are solved for by recognizing that the π's are dimensionless and hence the sum of the indices of F, L and T must equal zero.

The use of this theorem is most easily explained through an example. Suppose a research project is established to explore the possibility of relating the compression strength of a rock to the rebound of a drop hammer. By using one's imagination, the following list of possibly significant variables can be written down:

Variables	Symbols	Units
1. Uniaxial-compression strength of rock	Q	FL^{-2}
2. Modulus of deformation of rock	E	FL^{-2}
3. Viscosity of rock	n	$FL^{-2}T$
4. Mass density of rock	ρ	$FL^{-4}T^2$
5. Mass of hammer	M	$FL^{-1}T^2$
6. Height of drop of hammer	H_o	L

Variables	Symbols	Units
7. Rebound height of hammer	H_r	L
8. Acceleration due to gravity	g	LT^{-2}

From this list it is determined that the number of dimensionless or π-parameters being (n-m) is equal to (8-3) or 5.

The five π equations can then be written as follows:

$$\pi_1 = Q^{x_1} H_o^{y_1} g^{z_1} M = (FL^{-2})^{x_1} (L)^{y_1} (LT^{-2})^{z_1} FL^{-1}T^2.$$

$$\pi_2 = Q^{x_2} H_o^{y_2} g^{z_1} E = (FL^{-2})^{x_2} (L)^{y_2} (LT^{-2})^{z_2} FL^{-2}.$$

$$\pi_3 = Q^{x_3} H_o^{y_3} g^{z_3} \rho = (FL^{-2})^{x_3} (L)^{y_3} (LT^{-2})^{z_3} FL^{-4}T^2.$$

$$\pi_4 = Q^{x_4} H_o^{y_4} g^{z_4} H_r = (FL^{-2})^{x_4} (L)^{y_4} (LT^{-2})^{z_4} L.$$

$$\pi_5 = Q^{x_5} H_o^{y_5} g^{z_5} n = (FL^{-2})^{x_5} (L)^{y_5} (LT^{-2})^{z_5} FL^{-2}T.$$

Now, as mentioned above, as each group must be dimensionless, the sum of the indices for each dimension must be equal to zero. Therefore, for each π-factor there will be three equations from which to solve for the magnitude of the indices. Thus, starting with the first equation:

(1) $F: x_1 + 1 = 0 \qquad x_1 = -1$

$L: -2x_1 + y_1 + z_1 - 1 = 0$

$\qquad y_1 + z_1 = -1$

$T: -2z_1 + 2 = 0 \qquad z_1 = 1$

$\qquad\qquad\qquad\qquad y_1 = -2$

hence

$$\pi_1 = Q^{-1} H_o^{-2} g^1 M = \frac{Mg}{QH_o^2}$$

(2) $F: x_2 + 1 = 0 \qquad x_2 = -1$

$L: -2x_2 + y_2 + z_2 - 2 = 0$

$\qquad y_2 + z_1 = 0$

$T: -2z_2 = 0 \qquad z_2 = 0$

$\qquad\qquad\qquad\qquad y_2 = 0$

hence

$$\pi_2 = Q^{-1} E = E/Q$$

(3) F: $x_3 + 1 = 0$ $\qquad x_3 = -1$

L: $-2x_3 + y_3 + z_3 - 4 = 0$

$\qquad y_3 + z_3 = 2$

T: $-2z_3 + 2 = 0$ $\qquad z_3 = 1$

$\qquad y_3 = 1$

hence $\pi_3 = Q^{-1} H_o^1 g^1 \rho = \dfrac{H_o g \rho}{Q}$

(4) F: $x_4 = 0$ $\qquad x_4 = 0$

L: $-2x_4 + y_4 + z_4 + 1 = 0$

$\qquad y_4 + z_4 = 1$

T: $-2z_4 = 0$ $\qquad z_4 = 0$

$\qquad y_4 = 1$

hence $\pi_4 = Q^o H_o^1 g^o H_r^{-1} = H_o/H_r$

(5) F: $x_5 + 1 = 0$ $\qquad x_5 = -1$

L: $-2x_5 + y_5 + z_5 - 2 = 0$

$\qquad y_5 + z_5 = 0$

T: $-2z_5 + 1 = 0$ $\qquad z_5 = 1/2$

$\qquad y_5 = -1/2$

hence $\pi_5 = Q^{-1} H_o^{-1/2} g^{-1/2} n$

$\qquad = \dfrac{n}{Q} \sqrt{g/H_o}$

As we have assumed that these variables are related, we can write the general equation:

$$\pi_1 = f(\pi_2, \pi_3, \pi_4, \pi_5)$$

$$\dfrac{Mg}{QH_o^2} = f\left(\dfrac{E}{Q}, \dfrac{H_o g \rho}{Q}, H_o/H_r, \dfrac{n}{Q}\sqrt{g/H_o}\right)$$

or

$$Q = \dfrac{Mg}{H_o^2} f'\left(\dfrac{E}{Q}, \dfrac{H_o g \rho}{Q}, H_o/H_r, \dfrac{n}{Q}\sqrt{g/H_o}\right)$$

The experimental data from such a project would then be grouped into the various π-factors, and the curves obtained by plotting one π-factor against another analysed for their functional relationships.

In the theory of models, ideally it is necessary for the model and prototype to be geometrically, kinematically, and dynamically similar. In other words, the model should be scaled down geometrically so that the ratios of all corresponding dimensions are equal. Kinematic similarity is obtained when the ratios of the velocities of all points in similar positions are equal and if the paths of motion of points in similar positions are geometrically similar. Dynamic similarity requires that the ratio of all forces be equal. If all these requirements can be fulfilled, it is said that similitude has been established. However, it is sometimes difficult if not impossible to fulfil all of the requirements for similitude.

Another way of insuring that the similitude is fulfilled is to maintain the significant π-factors equal between model and prototype. For example, if a model was to be set up to examine the deflection, d, of a simple beam under a concentrated load, two of the dimensionless parameters would be:

$$\pi_1 = d/B \text{ and } \pi_2 = E D^2/P$$

where B is the span of the beam, E is its modulus of elasticity, D is its depth, and P the load. If π_1 is the same for the model and prototype, geometrical similarity is being maintained. If π_2 is constant, the dynamic similarity is obtained.

Then, if the model beams were to be made of the same material as the prototypes, the modulus of deformation, E, would not be scaled down. Thus, if the geometrical scale was to be 1/50, the load P would have to be scaled up 50^2 to keep π_2 constant. Incidentally, this example indicates the reason why structural model work often requires the use of a centrifuge to achieve the large increase in loading that becomes necessary to fulfil similitude requirements.

DFC/DV

ON HER MAJESTY'S SERVICE

SERVICE DE SA MAJESTÉ

CANADA
POSTAGE PAID
PORT PAYÉ

Mines Branch

555 Booth Street

Ottawa, Canada

Attn.: Dr. D.F. Coates

Corrections Required to "Rock Mechanics Principles",
Mines Branch Monograph No. 874, 1965

Page	Para.	Line	Comments